Tourism Case Studies
Latin America Region

Tourism Case Studies Latin America Region

Edited by

Thiago Allis and Sandro Carnicelli

CABI

CABI is a trading name of CAB International

CABI
Nosworthy Way
Wallingford
Oxfordshire OX10 8DE
UK

Tel: +44 (0)1491 832111
E-mail: info@cabi.org
Website: www.cabi.org

CABI
200 Portland Street
Boston
MA 02114
USA

T: +1 (617)682-9015
E-mail: cabi-nao@cabi.org

A catalogue record for this book is available from the British Library, London, UK.

ISBN-13: 9781800628533 (hardback)
9781800628540 (ePDF)
9781800628557 (ePub)

DOI: 10.1079/9781800628557.0000

Commissioning Editor: Victoria Taylor
Editorial Assistant: Helen Elliott
Production Editor: Helen Elliott

Typeset by Exeter Premedia Services Pvt Ltd, Chennai, India
Printed and bound in the UK by CPI Group (UK) Ltd, Croydon, CR0 4YY

Contents

Contributors

Bárbara Catalano – National Scientific and Technical Research Council, Universidad Provincial de Córdoba, Córdoba, Argentina

Beatriz-Helena Díaz Solano – Universidad Autónoma del Caribe, Barranquilla, Colombia

Brendali Carrillo – School of Hospitality and Tourism Management, Universidad Peruana de Ciencias Aplicadas, Lima, Peru

Bruna C. Mendes – Federal Institute of Education, Science and Technology of São Paulo, Campos do Jordão (SP), Brazil

Camila Cardoso – Comunidade Cultural Quilombaque, São Paulo, Brazil

Carla Guerrón Montero – Department of Anthropology, University of Delaware, Newark, Delaware, USA

Carolina I. Garcia – School of Management and Marketing (Tourism Research Cluster), Faculty of Business and Law, Curtin University, Western Australia, Australia

Cassiana Gabrielli – Departamento de Turismo, Universidade Federal do Paraná, Brazil

Cecilia Pérez Winter – Institute of Geography, Faculty of Philosophy and Literature, University of Buenos Aires/National Scientific and Technical, Research Council (CONICET), Buenos Aires, Argentina

Cecilia Rivas Medina – School of Hospitality and Tourism Management, Universidad Peruana de Ciencias Aplicadas, Lima, Peru

Christof Pforr – School of Management and Marketing (Tourism Research Cluster), Faculty of Business and Law, Curtin University, Western Australia, Australia

Claudia A. Troncoso – Consejo Nacional de Investigaciones Científicas y Técnicas (CONICET), Universidad de Buenos Aires, Buenos Aires, Argentina

Debora Cordeiro Braga – Public Relations, Propaganda and Tourism Department, University of Sao Paulo, Sao Paulo, Brazil

Domingo Martínez – LIIGH, UNAM Juriquilla, Querétaro, Mexico

Driselda P. Sánchez-Aguirre – Escuela Nacional de Ciencias de la Tierra UNAM, Ciudad de México, Mexico

Edilaine Albertino de Moraes – Department of Tourism at the Institute of Humanities, Federal University of Juiz de Fora, Minas Gerais, Brazil; TBC-REDE: Laboratory of Community-Based Tourism, Sustainability, and Networks, Brazil

Eduardo S. Sant'Anna – Tourism and Hospitality School, Universidade Federal Fluminense (UFF), Niterói, Rio de Janeiro, Brazil

Elias José Mediotte – Universidade Federal de Viçosa, Viçosa, Minas Gerais, Brazil

Erin Sharpe – Department of Recreation and Leisure Studies, Brock University, St. Catharines, Ontario, Canada

Fiona Eva Bakas – Tourism Department, ECEO, Lusófona University, Lisbon, Portugal

Freddy Giovanny Quiroga Nieto – Faculty of Social and Human Sciences, Universidad Externado de Colombia, Bogota, Colombia

Gabriela Landini – Instituto de Geografía, Facultad de Filosofía y Letras, Universidad de Buenos Aires/CONICET

Gabriela Santos – Departamento de Línguas e Cultura, Universidade de Aveiro, Portugal

João A. de Freitas – Social Memory Graduate Program, Federal University of the State of Rio de Janeiro, Rio de Janeiro, Brasil

Kenia Ramos – La Quinta Gran Bahía, Cuastecomates, Mexico

Lady Soto Navarrete – Faculty of Tourism and Geography, Universitat Rovira i Virgili, Vila-seca, Spain; Faculty of Social Sciences and Humanities ESPOL Polytechnic University, Campus Gustavo Galindo, Guayaquil, Ecuador

Laiara Amorim – Quilombo Aéreo, Brazil

Larissa Almeida – RECRIA – Brazilian Creative Tourism and Experiences Network, Recife, Brazil

Luis Leonardo Garavito González – Faculty of Social and Human Sciences, Universidad Externado de Colombia, Bogota, Colombia

Luis Rubén Pérez – Universidad Autónoma de Bucaramanga, Colombia

Magnus Luiz Emmendoerfer – Federal University of Viçosa, Viçosa, Minas Gerais, Brazil; Federal University of Ouro Preto, Ouro Preto, Minas Gerais, Brazil

Marco Ocke – Business Management Department, Federal University of Santa Catarina, Florianopolis, Brazil

Marcos Eduardo Carvalho Gonçalves Knupp – Federal University of Ouro Preto, Ouro Preto, Minas Gerais, Brazil

Maria Laura Gasparini – Centre for Advanced Studies in Tourism, Department for Life Quality Studies, University of Bologna, Italy

Maria Xaus – Universidad de Buenos Aires, Buenos Aires, Argentina

Matías O. Ciaschi – Instituto de Investigaciones Económicas, Universidad Nacional de La Plata, Provincia de Buenos Aires, Argentina

Maurício Dias Duarte – Ururay Group, São Paulo, Brasil

Michael Volgger – School of Management and Marketing (Tourism Research Cluster), Faculty of Business and Law, Curtin University, Western Australia, Australia

Milagros Granda – School of Hospitality and Tourism Management, Universidad Peruana de Ciencias Aplicadas, Lima, Peru

Narcís Bassolsi Gardella – Division of Quality Science, Uppsala University, Campus Gotland, Visby, Sweden

Natália Oliveira – Centro de Ciências Sócio-Organizacionais, Universidade Federal do Rio Grande do Sul, Brazil

Natalia Porto – Instituto de Investigaciones Económicas, Universidad Nacional de La Plata, Provincia de Buenos Aires, Argentina

Òscar Saladié – Faculty of Social Sciences and Humanities, ESPOL Polytechnic University, ESPOL, Campus Gustavo Galindo, Guayaquil, Ecuador

Osiris Marques – Tourism and Hospitality School, Universidade Federal Fluminense (UFF), Niterói, Rio de Janeiro, Brazil

Pedro Vianna Godihho Peria – Escola de Administração de Empresas de São Paulo da Fundação Getulio Vargas (FGV EAESP), São Paulo, Brazil

Renato de Oliveira dos Santos – School of Arts, Sciences and Humanities, University of São Paulo, São Paulo, Brazil; TBC-REDE: Laboratory of Community-Based Tourism, Sustainability, and Networks, Brazil

Sandra Cherro Osorio – Business and Construction, Melbourne Polytechnic, Melbourne, Australia

Sandra Tolosa – Institute of Anthropological Sciences, Faculty of Philosophy and Literature, University of Buenos Aires, Buenos Aires, Argentina

Sidnei Raimundo – School of Arts, Sciences and Humanities, University of São Paulo, São Paulo, Brazil

Silvia A. Tapia – National Scientific and Technical Research Council, Universidad de Buenos Aires, Buenos Aires, Argentina

Stefan Lazic – Department of Economics and Management, University of Trento, Italy

Tatiana Marchetti Panza – Federal Institute of Education, Science and Technology of São Paulo, Campos do Jordão (SP), Brazil

Teresa Cristina de Miranda Mendonça – Department of Administration and Tourism, Federal Rural University of Rio de Janeiro, Rio de Janeiro, Brazil; TBC-REDE: Laboratory of Community-Based Tourism, Sustainability, and Networks, Brazil

Thiago Allis – University of Sao Paulo, Brazil

Thiago Chagas de Almeida – Federal University of Viçosa, Viçosa, Minas Gerais, Brazil

Thiago Duarte Pimentel – Federal University of Juiz de Fora, Juiz de Fora, Minas Gerais, Brazil

Sandro Carnicelli – University of the West of Scotland, United Kingdom

Verônica Feder Mayer – Tourism and Hospitality School, Universidade Federal Fluminense (UFF), Niterói, Rio de Janeiro, Brazil

Foreword

The nascent idea of *Tourism Cases* began to take shape in 2017, when Claire Parfitt and I talked at the annual CAUTHE conference about various ideas. Fast forward to 2020 and transformation of the idea into reality began and by 2021 completed cases were being uploaded for the world to read. The year 2022 saw *Tourism Cases* take the decision to run a series of special issues, each focusing on different parts of the world, deliberately eschewing the dominant narrative in the international and English literature to help provide a venue for other stories to be presented and heard. One of the results of this was a special issue dedicated to tourism in Latin America. The success of the call for this special issue was simply mind-blowing, with almost fifty cases submitted. This book stems from the special issue and is a fitting ending (or new beginning) to the work.

One of the best privileges of being an editor of an academic journal or a repository of case studies is that you get to read a wide range of material that you simply never would in your own little research bubble. Doing so exposes you to all sorts of new ideas, and ways of being, thinking and seeing. This has certainly been true in relation to the case studies in this book. In this way, the published cases are about so much more than just giving a venue for material that may otherwise be shunted to the side by the dominance of the Global North. Rather, it is also about challenging dominant narratives from the Global North in tourism studies and the tourism industry, showing that there are alternatives that are at least as good and may even be better.

So, my first thanks go to all the authors of the cases, both in this book and the special issue more broadly. You have enriched this reader's appreciation of Latin America and all its diversity. You have also challenged me to think about how I see the world and everything in it, forcing me to see new ways of being and thinking. There simply is no space to say how each case study has done this but I do want to especially pick out the one by Laiara Amorim *et al.* Yes, the story of the case study is great and absolutely inspiring, but for me personally the biggest eye-opening moment is associated with what I have learned about skin colour and how it is seen and presented in the Latin American context. As an editor from a Western/Global North context, I am worried about the use of the term 'Black', associating it with all the worst racial bigotry and hatred. Yet through the case study and the responses of the authors to my concerns I have learned to appreciate a new way of seeing Black, of appropriation rather than misappropriation of the term, of empowerment rather than disempowerment.

It is the story of power that comes through so strongly in the case studies, in novel and innovative ways that speak of a breaking of dominant and disempowering narratives. That this is set within a Latin American context certainly does not mean that those living elsewhere in the world, this individual included, cannot learn from it and help to apply it elsewhere. Yet, in recognising this, we must also appreciate the inherently complex dynamics at play in the publishing of these cases. Do those researching tourism or working in the industry within Latin America *need* their work to be published in English? To say yes is to align with forces of globalisation that are intimately linked to neo-colonialism. Arguably it is more beneficial to be writing in the languages of the region, not just Spanish and Portuguese, but the myriad indigenous languages as well. So the publishing of these case studies is not, I hope, about seeing in binaries but of enabling another venue for information dissemination/story sharing, that allows us all, however imperfectly, to bring a little more cross-cultural understanding into the world at local and global levels.

The work that the guest editors of the special issue and the editors of the book (Thiago Allis and Sandro Carnicelli) have put in is simply awesome and to have done it all on a voluntary basis is just a testament to how wonderful they both are, as academics and people. In bringing this book and special issue together they have helped to show to the world the richness of the community working in various ways with tourism in Latin America.

The journey of this special issue may now be over in one sense but in another it is only just beginning. When I was a younger, more ignorant academic I wondered at the value of case studies, indoctrinated as I was into the belief that *real* academic research was published in double blind, peer-reviewed journals or books. Now I am hopefully a little less ignorant. Case studies are such an excellent way to find out about the richness and diversity of the world and to be able to tell meaningful stories to the world. To be able to strip back the academic banal and focus on telling the story instead in a way that engages with the audience and enriches their lives can be liberating and empowering, for both the writer and the reader. I am finding many journal articles increasingly stuck in rigid academic paradigms, conforming to writing styles apparently designed to bar access to everyone but the initiated few. This, I feel, stands in opposition to the idea that as researchers we have a duty to help disseminate ideas and help make a positive difference beyond our academic ivory towers. Case studies, I suggest, offer just such an alternative. So please, engage with the book as a rich source of learning, one driven by stories of the real as opposed to being weighed down by structures and theories imposed on us from above (such theories and structures may have value but are valueless without an appreciation of the world that lies beyond them). Hopefully, the stories inspire you, as students, researchers, those with links to the industry, and general members of the public. They have the power to help teach us about tourism in Latin America and also new ways of seeing and interacting with tourism wherever we live.

Professor Neil Carr, University of Otago
Editor-in-Chief for CABI's *Tourism Cases*

Tourism Cases in Latin America: An Introduction

Thiago Allis[1] and Sandro Carnicelli[2]
[1]University of Sao Paulo, Brazil, [2]University of the West of Scotland, United Kingdom

Introduction

With almost one billion inhabitants, Latin America forms a very complex mosaic of nations, people and landscapes, with unique cultural, scenic and economic resources. According to the World Bank Open Data, in 2019 the region represented no more than 6% of global international tourist arrivals. Latin America hosts several top international destinations – such as the Caribbean islands and the sacred city of the Incas, Machu Picchu (Peru), not to mention the MICE (meetings, incentives, conferences and exhibitions) industry in large urban areas such as São Paulo and Mexico City, and areas of significant ecological importance like Galapagos (Ecuador) and Fernando de Noronha (Brazil). Its internal tourism dynamics – both between and within countries – is not negligible. Yet all forms of tourism in the region are influenced by social, economic and environmental issues. Even though the very notion of Latin America is somewhat arbitrary and contested, the region is seen as a source of many possibilities for tourism at global and regional scales. Despite this, it is still under-represented in international and English-written tourism literature. This gap needs to be filled. To share practices from this region it can help to inform tourism developments but also support tourism resistance and changes in the wider region but also in other parts of the world. This book aims to contribute to a shift away from an overdependence on models and ideas developed and tested in different social and cultural settings that may not be appropriate for a Latin American context. This book brings together researchers, students, practitioners and other actors involved with research, teaching and the operation of tourism. They share their research cases and work experiences in Latin America qualifying the debate for audiences inside and outside the region. This book offers a window through which to spread to the rest of the world the research and work produced in Latin America.

Tourism Research and Global South: From Epistemologies to Cases

In the last 20 years, English-written tourism literature is giving more attention to the epistemologies of the South (Chambers and Buzinde, 2015). Indeed, it is possible to observe a growing number of special issues focusing on the Global South and an increasing number of papers and books published by academics from regions such as Africa, Latin America, and South-east Asia. In 2024, for example, the journal *Tourism Planning & Development* launched a special issue on tourism policy and planning in Brazil; the *World Leisure Journal* introduced a section called 'Global South Leisure Perspectives'; while Saarinen and Rogerson (2021) published the edited book *Tourism, Change and the Global South* with many of the chapters authored by researchers based in the Global South.

However, the development of tourism research in Latin America is not that recent. Geographers in Latin America from the early twentieth century, such as Romualdo Ardissone (in Argentina) and Milton Santos (in Brazil), were extremely important for the development of tourism research in the region. Their work on human (Ardissone) and critical (Santos) geography became the foundation of many tourism studies. Indeed, the Geography Institute of the University of Buenos Aires in Argentina is named after Romualdo Ardissone and it is the home of one of the most important tourism research groups in the continent. Milton Santos's work contributed significantly to the understanding of the epistemologies of the North in relation to the epistemologies of the South when analysing urban spaces, dependency, shared-space, politics and inequalities.

On Milton Santos, Melgaço (2017, p. 946) wrote 'Despite often being labelled a specialist in globalisation, urban studies, or economic development, Santos was much more than that. He was one of those few thinkers who created a school of thought, a set of concepts that together form a robust theory'. The work of Santos helped geographers but also tourism researchers to understand the process of globalization from the perspective of the Global South and its impacts on places, spaces and society. So, in the current hyper-globalized world, it is still very important to understand the socio-political developments, issues regarding mobilities and migration, and the economic transformations happening in the Global South. But it is also relevant to understand how tourism is developing in this part of the world. The management of visitors and visitor attractions, the impacts on local communities, on fragile environments, and the action to increase (or regulate) domestic and international visitors in the countries of the Global South are worth understanding and communicating to the Global North because they are also in relation to the Global North.

On a more systematic review on literature, one can realize that even if in the 1960s some tourism research was already focusing on Latin America (Nuñez, 1963), it was only in the late 1970s and early 1980s that the academic production about Latin America started to appear more frequently in English-written academic journals. However, most of the research was carried out by academics based in universities in the United States of America and in Germany. In one of the first publications, Hudman (1978) invited readers to learn about the regional tourism development in the small community of Chichicastenango, in Guatemala, 'one of the outstanding market towns in all Latin America' (p. 115). During the 1980s, there were very few publications focusing on this region and mostly focusing on aspects such as economic development, impacts and tourist behaviour, as well as measures to control and stimulate tourism in this part of the world (Mings, 1981; Francisco, 1983; Gibbons and Fish, 1984; Gormsen, 1988).

In the 1990s two women were key in presenting tourism research in Latin America to international audiences: Regina Schlüter and Doris v.d.M. Ruschmann. In 1991, Schlüter, an Argentinian scholar, published a paper about the challenges of countries in the region to attract and maintain the flow of tourists who are not from the region – issues that, apart from certain contexts such as the Caribbean islands – are still very significant in the tourism agenda in Latin America. In 1992, possibly aligned to the United Nations Conference on Environment and Development (UNCED), also known as the Earth Summit and as Rio-92, Ruschmann (1992) – a Brazilian scholar based at the University of São Paulo – published a short paper in *Tourism Management* on ecological tourism in Brazil. In the paper Ruschmann emphasizes the 'necessity of monitoring the installation of ecological tourism infrastructure so as to provide adequate profitability and support socioeconomic development without harming the natural environment' (Ruschmann, 1992, p. 125). In 1993, Schlüter wrote a research note presenting some initial considerations about the historical development of tourism in some of the countries in Latin America. She concludes by affirming that 'even though Latin American countries have attempted to model their tourism development after that of the European countries, they have depended on the American market to implement the development' (Schlüter, 1993, p. 366).

Both Ruschmann and Schlüter would become beacons for tourism researchers in the South Cone (Argentina, Brazil, Uruguay, Paraguay and Chile).

Moreover, the research published in the 1990s helped to raise awareness about Latin America, its resources, characteristics and issues. At that time most tourism publications used social sciences and geographical perspectives to understand the region (Meyer-Arendt, 1991). Issues like drug use and trafficking as well as sexual exploitation have been presented in the literature around the same years (Herold, 1992) as well as debates about perceptions of Latin America as a destination and market for tourism expansion (Brown, 1998; Krohn and O'Donnell, 1999; Costa and Bauer, 2001). Towards the end of the 1990s, Pizam (1999) published a paper about the status of human resources in the Latin America tourism industry, discussing a shortage of skilled human resources, regarding mainly deficiencies in foreign languages, computers and marketing skills.

The reality is that until the 2000s publications on tourism in Latin America in English and in international outlets were very rare. Nevertheless, the volume of international tourism literature focusing on the region published by international journals and books gained traction with the dawn of the 21st century. If in the previous years, publications depicted fragmented approaches on disparate topics – either because local scholars were looking for some recognition in the international journals, or, inversely, international scholars were, by any reason, interested in Latin America – now the topics, approaches, areas of interest and journals became more diverse. A search on the Scopus database for research on tourism in Latin America showed 318 papers published between January 2000 and August 2024. As per Fig. 1, the number of papers on the topic increased significantly in the last decade, going from 50 in the period 2011–2015 to 113 so far between 2021 and 2024.

Critical and more situated approaches focusing on tourism in Latin America emerged, replacing more descriptive work from initial years – see, for instance, the volume edited by Mowforth *et al.* (2008) on tourism and responsibility; the research from Florence Babb and other scholars that discussed gender and identities in several countries in the region (Babb, 2010, 2012; Vizcaino-Suárez and Díaz-Carrión, 2019; Días-Carrión and Vizcaino, 2022; Khoo *et al.*, 2024); the debates on ascending tourism gentrification and dispossession on Latin American cities (Piñeros, 2017;

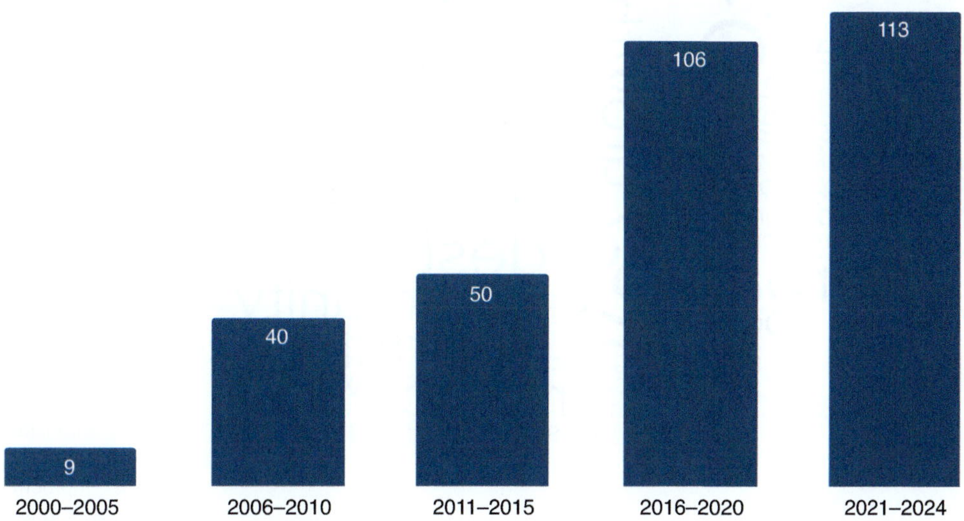

Fig. 1 Number of papers published on tourism and Latin America (2000–2024). (From Scopus database).

Navarrete Escobedo, 2017, 2020; Valverde *et al.*, 2022) and rural areas (Lorenzen, 2021; Neef, 2021); or the opportunities for bottom-up development of community-based tourism (De la Rosa *et al.*, 2017; Vargas *et al.*, 2020), social tourism (Schenkel, 2018; Schenkel and de Almeida, 2020) or the value of 'Buen Vivir' as cements for tourism policies (Fischer, 2019); the unfolding to international financing on tourism inclusiveness in Latin America (Carrillo-Hidalgo and Pulido-Fernández, 2019); participation and autonomy of local communities on tourism policy making (Ziritt Trejo *et al.*, 2022); labour precarity in tourism industry (Hack-Polay *et al.*, 2022; Porto and Garcia, 2022), and so on. To better understand the key areas explored in previous publications, we did a word cloud exercise. Fig. 2 shows the frequency of the keywords in the 318 papers published and available on the Scopus list. There is a clear focus on economic aspects but also includes perspectives on development, heritage, culture, sustainability, Indigenous people and in communities.

More research is needed to grasp the real participation of Latin American scholars on tourism production. Researching and publishing on Latin America should not be an exclusive task for authors from the region, however, the

Fig. 2 Word Cloud based on the cases published in the *CABI Tourism Cases, Special Issue, Tourism in Latin America*.

more scholars based in the region make their reflections available to broader international audiences, the better the epistemologies of the South can be comprehended.

Latin America and Tourism Studies: A Closer Look and a Call for Action

In the previous section we pointed out the (originally) limited but (currently) growing literature on Latin America that has been published in English. But, looking with sharper eyes, specific contents can tell a relevant story.

In 2012, Gui Lohmann and Dianne Dredge published in English the edited book *Tourism in Brazil: Environment, Management and Segments*. Most of the authors of chapters were from Brazil and based in Brazil. This was not the first book of the Routledge series on Contemporary Geographies of Leisure, Tourism and Mobility to focus on Latin America. In 2004 David Duval had published an edited collection on *Tourism in the Caribbean* but mostly written by academics from and based in North America, Europe and New Zealand even if many are renowned for their work in the Caribbean.

More recently, in 2020, Sophia Rolle, Jessica Minnis, and Ian Bethell-Bennett published another edited book as part of the same series on *Tourism Development, Governance and Sustainability in the Bahamas*. This book also counts with 16 chapters mostly written by researchers from and/or based in the region. Before this, Alexandre Panosso Netto and Luiz Trigo (2015), based in the University of São Paulo (Brazil), had edited a 13 chapters volume with 11 cases of success on tourism in Latin America, having 100% of authors from the region. The book was published in English and, to date, has been accessed 27,000+ on the Springer platform.

If there is evidence of growing literature in English focusing on Latin America, it is also true that the literature in other languages such as Spanish and Portuguese is already extensive. In order to demonstrate the open space for research in other languages, the *World Leisure Journal* created the Global South Leisure Perspectives that has published papers in their original language as well as translated into English. Three of the four papers published came from Latin America. The first one from Rodríguez (2023) focused on intellectual colonialism in leisure research while the latest one from Monterrubio *et al.* (2024) explores the landscape of tourism, sport, and leisure research in Mexico. The opportunity to submit their work in their own language, be peer-reviewed by other researchers who are fluent in that language and getting it translated into English seems to be an important step in allowing their research to be better understood and appreciated by academics in other parts of the world (Carnicelli and Uvinha, 2023).

It is important to emphasize that countries in Latin America have their own academic journals and have been publishing research for many years. Moreno-Gil *et al.* (2020) have previously pointed out that the knowledge published in Spanish and Portuguese seems to be mostly ignored and rarely cited by English-written research. In their analysis of seven tourism journals based in Latin America, they found 2145 articles published between 2006 and 2018. According to the authors' analysis within the Brazilian journals, the contributions by Brazilian authors are higher than 85%, followed by far for the authors from Spain (4.5%). When analysing the journals from Mexico and Argentina, Brazilian authors are still leading (46%), followed by Mexican (25%) and Spanish (10%) authors. These results show how Brazilian authors are dominating the dissemination of tourism research in Latino-American journals (Moreno-Gil *et al.*, 2020, p. 555).

This is not surprising if we take into consideration the data published by Pimentel *et al.* (2018) showing an analysis of tourism courses and research centres in Argentina, Paraguay, Uruguay, Brazil and Venezuela. In their research Brazil appears with 347 higher education programmes in tourism as well as 57 research centres, while Argentina had 60 tourism degree programmes and 9 research centres, and Paraguay, Uruguay and Venezuela combined had 80 tourism programmes and 2 research centres. These discrepancies have also been seen in this edited volume with 25 out of the 57 authors based in Brazil, followed by 9 in Argentina, and 4 in Colombia. But in this volume we also have authors who are based in Peru, in Mexico and in Ecuador demonstrating the diversity of research and tourism knowledge developed in the region.

So the time has come to continue the process of diversification of voices in tourism literature. This special issue originated from our beliefs that other regions in the world could learn from some of the excellent practices taking place in Latin America as well as understand some of the struggles in the region.

Beyond the diversity of authorship, this edited book also approaches a variety of tourism products, topics, theories and concepts that are relevant to tourism researchers, organizations, policy makers and businesses in Latin America, but also in other parts of the world. For example, the concept of 'collectives' as a community organization is characteristic in Latin America and has been represented in some of the cases. Quilombolas, communities of Afro-Brazilian people who descended from escaped slaves, are also part of the cases presented in the chapters. Community-based tourism, a theoretical framework heavily used in Latin American research also appears frequently in this book. Brazilian carnival, Indigenous practices, as well as elements of colonial heritage have been explored demonstrating that the epistemological constructions in Latin America are complex but also extremely rich. Finally, the authors of the chapters presented in this volume are not only academics. The format of the *Tourism Cases* allowed policy makers, business representatives, community leaders and other agents involved in the communities to present their stories and share their experiences and learnings. Widening the voices in tourism has been at the forefront of this project. From opening opportunities to authors in Latin America to share their cases, to allowing perspectives beyond academia.

Structure of this Edited Book

This volume is a follow up from a special issue of *Tourism Cases* that has published more than 20 cases (and counting!) disseminating some best tourism practices in Latin America. The cases published are focusing on different parts of Latin America and the authors are mostly based in this region. Even if the majority of cases are from Brazil (as well as the editors of this book), the biggest country in the region, a significant number of articles are focusing on countries such as Argentina, Colombia and Mexico. We have also published cases from Peru, Cuba and Ecuador. We hope that such diversity has contributed to enrich the tourism literature and the understanding of tourism practices in Latin America.

This volume is divided into five sections.

The first six chapters are part of the section 'Community Empowerment and Social Engagement in Tourism'. This section was purposively developed to demonstrate the depth and breadth of research focusing on local communities and their participation in tourism decision-making processes. Community-based tourism and sustainable local development have been part of public governance in destinations such as in the Yucatan Peninsula (Chapter 1) and in Paraty/Brazil (see Chapter 5). The social engagement of collective community groups such as the Brazilian Comunidade Cultural Quilombaque (Chapter 2), Grupo Ururay (Chapter 3), and 'Pretos que Voam' (Chapter 6) have been essential to tackle injustices and inequities in the tourism development in the country. These cases demonstrate the power of such collectives in transforming realities and dealing with some of the worst aspects of tourism, social exclusion and environmental degradation. Such debate leads us to the second part of this volume.

The second section of the book is made up of four chapters under the theme 'Sustainability Challenges and Nature-Based Responses in Tourism'. Latin America is rich in natural resources, unique ecosystems and exotic flora and fauna attracting the attention of the world for many centuries. From the far south of Latin America we have a case of Patagonia (Chapter 8) with its national parks and lakes, a fragile and unique ecosystem that has seen significant increase in tourism in recent years. We have cases from the Galapagos Islands (Chapter 9), immortalized by the studies of Charles Darwin who visited the island in 1835. We also present a case from the different coastal regions including the Caribbean city of Barranquilla in Colombia (Chapter 10) and the fauna – birds – of the Pacific coast of Mexico (Chapter 7).

Chapters 11 to 16 move away from the natural environment in Latin America to focus on aspects of 'Culture and Heritage in Tourism'. This section presents a diverse set of geographies with cases from Peru, Brazil, Argentina, Colombia, and Mexico. The book presents cases from some of the World Heritage Sites in the region, such as Rímac in Peru (Chapter 11) and Qhapaq Ñan, or the Andean Road System (Chapter 15), that includes six South American countries (Peru, Bolivia, Ecuador, Colombia, Chile and Argentina). Elements of heritage also permeate traditional food such as the Argentinian Dulce de Leche and their touristification as debated in Chapter 13 and the experience of visitors at boutique and non-boutique heritage hotels in Mexico (Chapter 16). Such management of experiences leads us to Section 4.

Five chapters are part of the section 'Destination Management and Tourism Marketing'. Brazil and Ecuador are the settings for this section that look into tourism and policy that contributed to both the management of attractions, destinations, but also their marketing and brand. Chapter 17, for example, focuses on tourism branding at a national level. It specifically looks into the initiatives of hosting mega events in Brazil to leverage its brand awareness as well as the issues with widespread corruption and the inappropriate response to the COVID-19 pandemic as affecting the image of the country negatively. Political ideologies and nation-building is also a theme in Chapter 18, a case study from heritage railways in Ecuador. Chapters 19, 20 and 21 take us from national level to a more regional/local level looking at cases in specific areas of Brazil and how it evolved in partnership with local policy makers, universities and community organizations. Such examples of destination management also explored some of the economic opportunities that tourism provided in small and rural communities.

This takes us to the final five chapters under the theme 'Economic Development and Policy in Tourism'. This final section has case studies from Argentina, Peru, and Cuba and explores elements such as fiscal incentives (Chapter 22), subsidies (Chapter 24) but also economic development and infrastructure (Chapters 23 and 26). These chapters demonstrate the diversity of approaches to tourism policy and governance in Latin America. The chapters also demonstrate the struggles to develop an infrastructure for tourism and the legislative changes needed to encourage the development of a private sector. Public and private partnerships are also discussed in the chapters with a focus on the support given by some countries to enhance small business opportunities.

References

Babb, F.E. (2010) Sex and sentiment in Cuban tourism. *Caribbean Studies* 38(2), 93–115.

Babb, F.E. (2012) Theorizing gender, race, and cultural tourism in Latin America: a view from Peru and Mexico. *Latin American Perspectives* 39(6), 36–50. https://doi.org/10.1177/0094582X12454560.

Brown, D.O. (1998) German and British tourists' perceptions of African, Latin American and Caribbean travel destinations. *Journal of Vacation Marketing* 4, 298–310.

Carnicelli, S. and Uvinha, R. (2023) Leisure, inequalities, and the Global South. *Leisure Studies* 42(3), 328–336. https://doi.org/10.1080/02614367.2023.2168032.

Carrillo-Hidalgo, I. and Pulido-Fernández, J.I. (2019) Is the financing of tourism by international financial institutions inclusive? A proposal for measurement. *Current Issues in Tourism* 22(3), 330–356. https://doi.org/10.1080/13683500.2016.1260529.

Chambers, D. and Buzinde, C. (2015) Tourism and decolonisation: locating research and self. *Annals of Tourism Research* 51(1), 1–16. https://doi.org/10.1016/j.annals.2014.12.002.

Costa, F. and Bauer, T.G. (2001) Latin America tourism: an Australian travel industry perspective. *International Journal of Hospitality and Tourism Administration* 1(3–4), 23–42. https://doi.org/10.1300/J149v01n03_02.

De la Rosa, F.J.U., Tur, A.A. and García-Andreu, H. (2017) Community-based tourism and the political instrumentalization of the concept of community: a new theoretical approach and an ethnographical case study in northeastern Brazil. *Anthropos* 2(112), 467–486.

Díaz-Carrión, I.A. and Vizcaino, P. (2022) Mexican women's emotions to resist gender stereotypes in rural tourism work. *Tourism Geographies* 24(2–3), 244–262. https://doi.org/10.1080/14616688.2020.1867886.

Duval, D.T. (ed.) (2004) *Tourism in the Caribbean: Trends, Development, Prospects*. Routledge, London.

Fisher, J. (2019) Nicaragua's Buen Vivir: a strategy for tourism development? *Journal of Sustainable Tourism*, 27(4), 452–471. https://doi.org/10.1080/09669582.2018.1457035.

Francisco, R.A. (1983) The political impact of tourism dependence in Latin America. *Annals of Tourism Research* 10(3), 363–376. https://doi.org/10.1016/0160-7383(83)90062-2.

Gibbons, J.D. and Fish, M. (1984) Changes in the composition of Mexico's international tourists, 1970–1980. *Journal of Travel Research* 22, 6–13.

Gormsen, E. (1988) Tourism in Latin America – spatial distribution and impact on regional change. *Applied Geography and Development* 32, 65–80.

Hack-Polay, D., Dal Mas, F., Mahmoud, A.B. and Rahman, M. (2022) Barriers to the effective exploitation of migrants' social and cultural capital in hospitality and tourism: a dual labour market perspective. *Journal of Hospitality and Tourism Management* 50, 168–177. https://doi.org/10.1016/j.jhtm.2022.01.012.

Herold, E.S. (1992) Canadian tourists and Dominican Republic sex workers. *AIDS and Society* 4(1), 7.

Hudman, L.E. (1978) Tourist impacts: the need for regional planning. *Annals of Tourism Research* 5(1), 112–125.

Khoo, C., Yang, E.C.L., Tan, R.Y.Y., Alonso-Vazquez, M., Ricaurte-Quijano, C., Pécot, M. and Barahona-Canales, D. (2024) Opportunities and challenges of digital competencies for women tourism entrepreneurs in Latin America: a gendered perspective. *Journal of Sustainable Tourism* 32(3), 519–539. https://doi.org/10.1080/09669582.2023.2189622.

Krohn, F.B. and O'Donnell, S.T. (1999) US tourism potential in a New Cuba. *Journal of Travel and Tourism Marketing* 8(1), 85–99.

Lohmann, G. and Dredge, D. (eds) (2012) *Tourism in Brazil: Environment, Management and Segments*. Routledge, London.

Lorenzen, M. (2021) Rural gentrification, touristification, and displacement: analysing evidence from Mexico. *Journal of Rural Studies* 86, 62–75. https://doi.org/10.1016/j.jrurstud.2021.05.015.

Melgaço, L. (2017) Thinking outside the bubble of the Global North: introducing Milton Santos and 'the active role of geography'. *Antipode* 49(4), 946–951. https://doi.org/10.1111/anti.12319.

Meyer-Arendt, K. (1991). Geographic research on tourism in Latin America, 1980–1990: conference on Latin Americanist geographers. *Benchmark* 17/18, 199–207.

Mings, R.C. (1981) Tourism development in Latin America and related research needs. *Proceedings of the Conference of Latin Americanist Geographers* 8, 336–346.

Monterrubio, C., Dashper, K., Wadham, H., Chávez-Dagostino, R.M., Martínez Moreno, O.C., Pérez, J. and Silva, S. (2024) Navigating the landscape of tourism, leisure, and sport studies in Mexico. *World Leisure Journal* 66(3), 323–333. https://doi.org/10.1080/16078055.2024.2347982.

Moreno-Gil, S., Parra-López, E., Picazo-Peral, P. and Díaz-Domínguez, C. (2020) The dissemination of tourism scientific research in Latin American journals: a bibliometric study. *Anatolia* 31(4), 549–564. https://doi.org/10.1080/13032917.2020.1795892.

Mowforth, M., Charlton, C. and Munt, I. (2008) *Tourism and Responsibility: Perspectives from Latin America and the Caribbean*. Routledge, London.

Navarrete Escobedo, D. (2017) Turismo gentrificador en ciudades patrimoniales: exclusión y transformaciones urbano-arquitectónicas del patrimonio en Guanajuato, México. *Revista INVI* 32(89), 61–83. https://dx.doi.org/10.4067/S0718-83582017000100061.

Navarrete Escobedo, D. (2020) Foreigners as gentrifiers and tourists in a Mexican historic district. *Urban Studies* 57(15), 3151–3168. https://doi.org/10.1177/0042098019896532.

Neef, A. (2021) *Tourism, Land Grabs and Displacement: The Darker Side of the Feel-Good Industry*. Routledge, London.

Nuñez, T. (1963) Tourism, tradition, and acculturation: *Weekendismo* in a Mexican village. *Ethnology* 2(3), 347–352.

Panosso Netto, A. and Trigo, L.G.G. (2015) *Tourism in Latin America: Cases of Success*. Springer Link, Cham, Switzerland.

Pimentel, T.D., Carvalho, F.C.C. and Oliveira, M.B. (2018) Centros de pesquisa em turismo no Mercosul: distribuição, ênfases e possíveis interações como fator de desenvolvimento. *Revista GUAL* 11(1), 319–342.

Piñeros, S.T. (2017) Tourism gentrification in the cities of Latin America. In: Gravari-Barbas, M. and Guinand, S. (eds) *Tourism and Gentrification in Contemporary Metropolises: International Perspectives*. Routledge, Abingdon, UK, pp. 75–104.

Pizam, A. (1999) The state of travel and tourism human resources in Latin America. *Tourism Management* 20(5), 575–586. https://doi.org/10.1016/S0261-5177(99)00026-6.

Porto, N. and Garcia, C.I. (2022) Unveiling the truth about tourism labour precarity in Argentinian cities. *International Journal of Tourism Cities* 8(2), 342–356. https://doi.org/10.1108/IJTC-10-2020-0237.

Rodríguez, A.D.R. (2023) Trazas del colonialismo intelectual en los estudios de ocio y recreación en Latinoamérica. *World Leisure Journal* 65(4), 447–453. https://doi.org/10.1080/16078055.2023.2268484.

Rolle, S., Minnis, J. and Bethell-Bennett, I. (2020) *Tourism Development, Governance and Sustainability in the Bahamas*. Routledge, Abingdon, UK.

Ruschmann, D. (1992) Ecological tourism in Brazil. *Tourism Management* 13, 125–134.

Saarinen, J. and Rogerson, J.M. (2021) *Tourism, Change and the Global South*. Routledge, Abingdon, UK.

Schenkel, E. (2018) Turismo social en América Latina: la conquista de un derecho desigual. *Lurralde* 41, 93–113.

Schenkel, E. and de Almeida, M.V. (2020) Social tourism in Latin America: regional initiatives. In: Diekmann, A. and McCabe, S. (eds) *Handbook of Social Tourism*. Edward Elgar, Cheltenham, UK, pp. 33–42.

Schlüter, R.G. (1991) Latin American tourism supply: facing the extra-regional market. *Tourism Management* 12(3), 221–228. https://doi.org/10.1016/0261-5177(91)90006-F.

Schlüter, R.G. (1993) Tourism and development in Latin America. *Annals of Tourism Research* 20, 364–367.

Valverde, C., Levi, L.L. and Narciso, C.F. (2022) From degradation to gentrification and touristification of historical centers in Latin America and the Caribbean: a decolonial perspective. In: Gonzalez-Perez, J.M, Irazabal, C. and Lois-Gonzalez, R.C. (eds) *The Routledge Handbook of Urban Studies in Latin America and the Caribbean*. Routledge, New York, pp. 221–256.

Vargas, R.C., Zuñiga, C.H. and Alcantara, L.C.S. (2020) Turismo de base comunitária: uma perspectiva desde o bem viver como alternativa ao desenvolvimento na América Latina e Abya Yala. *Desenvolvimento e Meio Ambiente* 54, 167–183. https://doi.org/10.5380/dma.v54i0.68745.

Vizcaino-Suárez, L.P. and Díaz-Carrión, I.A. (2019) Gender in tourism research: perspectives from Latin America. *Tourism Review* 74(5), 1091–1103. https://doi.org/10.1108/TR-02-2017-0021.

Ziritt Trejo, G.Y., Moreno Freites, Z.E. and Castellano Caridad, M.I. (2022) Community participation in the formulation of tourism public policy: systematizing your legitimizing process. *Jurídicas CUC* 18(1), 497–518.

Section 1: Community Empowerment and Social Engagement in Tourism

1 Community-Based Tourism Initiatives as a Tool for Empowering Indigenous Communities: Evidence from the Yucatan Peninsula, Mexico

Stefan Lazic[1] and Maria Laura Gasparini[2]
[1]Department of Economics and Management, University of Trento, Italy; [2]Centre for Advanced Studies in Tourism, Department for Life Quality Studies, University of Bologna, Italy

Summary
Community-based tourism (CBT) is seen as an opportunity for equitable development and sharing of benefits among host community members. It is an alternative to the mass tourism phenomenon and a way to introduce sustainable management of natural resources. Tourism social enterprises often accompany CBT initiatives, emphasizing the social force of tourism as an activity. They enable communities to address social issues while developing market-based activities, deal with injustice in tourism, and respond to the capitalist development strategies applied in the so-called major destinations. In this work, we explore four indigenous community-based initiatives in the hinterland of the Yucatan Peninsula, one of Mexico's main tourist areas. Through a fieldwork case study approach, we examine the internal governance of these initiatives, the role of external actors and the community approach to developing conscious tourism activities that support local livelihoods, value cultural heritage, and safeguard the natural environment. We find that tourism creates a spillover effect by providing opportunities for the development of complementary economic activities, decreasing the dependence on a single sector. Concurrently, tourism development empowers women and youth members of the communities by creating the possibilities for them to take leading roles in different activities.

The Value and Interest of the Case Study

We analyse tourism social enterprises as a suitable model for tourism development within indigenous communities that can improve local livelihoods, complement traditional economic activities, and support environmental protection.

Introduction

Current development models in tourism mainly derive from the neoliberal capitalist agenda and tourism as a fuel of capitalism (Fletcher, 2011). Based on extractive approaches, mass tourism business strategies embrace growth trajectories aiming at continuously increasing tourist flows (Ryan, 2002) and generating little benefits for local host communities (Pearce, 1992) with limited contribution to battling social injustices and environmental degradation (Rozemeijer, 2001). Tourism activities built on these premises can easily create imbalances among the social groups within a destination and galvanize negative consequences for the natural context of a tourism place. Calls to revise development models have increased with the COVID-19 pandemic exposing various fragilities of the modern tourism sector when a sudden break in tourism arrivals has disabled income flows, exacerbating already existing issues (Hockings *et al.*, 2020).

Among various alternative proposals that try to tackle inequalities and introduce more responsible practices in tourism development, community-based tourism (CBT) has often been identified as a type of tourism that can bring social, environmental, and economic benefits to host communities and inspire innovative processes to activate local networks, capacities, and resources (Sheldon and Daniele, 2017). Community-led initiatives have capacities to solve social injustices and position tourism as a social force that drives local development while empowering host communities to manage tourism activity in their vicinities (Higgins-Desbiolles, 2006). At the same time, they can activate mechanisms for the conservation of nature and renewal of environmental context (Rozemeijer, 2001). CBT has different forms, and tourism social enterprise (TSE) is often identified as the form that can generate new job opportunities, additional social capital, financial support for education and other positive outcomes while enabling the local community to independently organize tourism activities (Sloan *et al.*, 2014).

In this case study, we analyse four TSEs developed within the rural Mayan communities of the Yucatan Peninsula in Mexico, a region where tourism has had a vital role in local economies for decades. Through collective organization, the communities of Ek Balam, Yaxunah, Nuevo Durango and Eugenio Echeverria Castellot 2have established a self-managed model of tourism, aligning it with local social and environmental peculiarities, respecting the cultural heritage and enhancing the livelihoods of the community on the whole. Through a fieldwork case study approach (Eisenhardt, 1989), we explore their actions and organization to understand how the creation of TSE made the management of tourism activities possible, paving the way for empowerment, respect, dignity and solidarity in rural indigenous communities. At the same time, we identify the critical challenges for the future of these communities and their tourism projects.

Community-Based Tourism and Tourism Social Enterprises

CBT refers to a tourism activity that involves community members as providers of tourism services and activities and as owners of enterprises in tourism and related sectors (Strydom *et al.*, 2019). Development and management of tourism activity in a responsible way within a community implies the involvement of all local stakeholders related to tourism to a lesser or higher extent (Jamal and Stronza, 2009). This process does not mean that all stakeholders should be involved to the same degree, as they have different positions and power. However, it does require that all interests are recognized and considered to create a balance between different social groups and achieve partnerships among them (Hardy and Beeton, 2001). Activation of stakeholder networks in local tourism development conveys empowerment for host communities through four lines of impact – economic, psychological, social and political (Scheyvens, 1999), enabling community members to actively bring about desired change by influencing factors that impact their lives. For Guia (2021), tourism managed by communities can inspire advocacy and solidarity of travellers with local communities in their efforts to solve injustice(s) and bring justice through tourism by activating support mechanisms on different scales.

In this context, tourism is considered a tool that can bring innovative solutions to potentially improve the social, environmental, and economic conditions in host communities by activating capacities, resources and social relationships that can positively influence a community's sustainable transformation (Higgins-Desbiolles, 2006; Sheldon and Daniele, 2017). CBT is often interlinked with entrepreneurship and the creation of TSE. It is considered an alternative to the mass tourism industry and even a need that facilitates a shift in tourism development in a capitalistic environment (Pollock, 2015). The interrelation between tourism and social entrepreneurship is evidenced by engagement in solving concrete social problems through market-based solutions and the production of benefits throughout the tourism value chain (Altinay *et al.*, 2016). Still, control mechanisms and parameters must be set to ensure that community-based initiatives remain under local control and that the external actors only serve as a channel to the market. Furthermore, Altinay *et al.* (2016) find additional challenges, such as gender inequalities in opportunities and pay, greenwashing and limited availability and access to resources, as some of the issues that might affect TSEs. Other authors, like Scheyvens (1999) and Sloan *et al.* (2014), also identify greenwashing (dishonest use of local communities and sustainability practices) as a possible pitfall of CBT, as it can be used merely for promotional and communication actions rather than actually being implemented in practice.

Research Context

Tourism in Mexico and the Yucatan Peninsula

Tourism is among the main economic activities in Mexico. Before the COVID-19 pandemic, the country was among the top 10 destinations by international tourist arrivals, with 45 million incoming tourists registered in 2019 (World Tourism Organization 2021). The national government has been planning and developing tourism since the second half of the 20th century, mainly through large-scale tourism projects under the *sun & sea* model, making Cancun an emblematic example (Guardado *et al.*, 2012). An isolated area until the 1970s, in a few years it became the number one tourist destination of Mexico and even of Latin America (INEGI, 2014), later extending to the rest of Quintana Roo littoral. In fact, the Yucatan Peninsula, formed by the states of Quintana Roo, Yucatan, and Campeche, is now the leading tourism destination in Mexico, mainly because of its pristine beaches along the Caribbean Sea, which have consolidated a model of mass (international) tourism development, with more than 90% of foreign direct investment belonging to Spanish capital (Guardado *et al.*, 2012). In the same period, the rural areas in the Peninsula's hinterland, which were already suffering from the crisis in the agricultural sector, saw much of its population emigrate to the coast in search of employment opportunities, thus abandoning rural livelihoods and traditional activities (Jouault *et al.*, 2022).

Community-based tourism in the Yucatan Peninsula

In the context mentioned above, tourism has been introduced in public policies for rural areas in Mexico as a poverty alleviation and local development tool, able to contribute to conserving the natural environment and preserving local culture, and prevent the rural exodus (Kieffer, 2019). In Latin America, CBT is understood as a tourism management model that enables rural families, especially indigenous groups, to develop tourism through collective action and maintain control over the activities and distribution of benefits (Cañada, 2015). In this process, throughout the continent, including Mexico, communities have been supported by international organizations and governmental institutions that provide funding to implement sustainable CBT programmes, although not always with successful results (Cañada, 2015; Kieffer, 2019). In fact, the support from these external actors is usually directed to infrastructure development, overlooking the commercialization aspect, which is one of the most common factors of failure in community enterprises in the region (Cañada, 2015).

The CBT model is considered an alternative to the Caribbean coast's sun and sea resort-based tourism in the Yucatan Peninsula. Mayan communities are often located in the vicinity of cultural heritage (more than 50 Mayan archaeological sites[1]) and natural heritage sites (39 Natural Protected Areas), and tourism activity enables them to link the safeguarding of the environment with valuing cultural heritage while generating

Fig. 1.1. Typical handicraft for sale at Nuevo Durango.

economic benefits for the community as a whole (Fig. 1.1). In the analysis of CBT enterprises in Yucatan, Garcia de Fuentes and Jouault (2021) find that 78% of 153 analysed businesses in 2013 had a form of cooperative and their existence was related to operations on communally owned agricultural land – *ejido*. According to these authors, some of the common characteristics of the cooperatives in Yucatan are: the link to previous forms of ownership of the land, incorporation of the inherited cultural forms into work relationships, the role of family relationships in the organizational structure of the cooperative, and the importance of paid work for cooperative's functioning and relevance of collaborating networks (Garcia de Fuentes and Jouault, 2021). At the same time, CBTs in Yucatan face some common threats, such as the indiscriminate territorial distribution of subsidies, the privatization of rural land (Garcia de Fuentes and Jouault, 2021) and the commodification of Mayan cultures (Poot *et al.*, 2022).

Methodology

With the aim of understanding the CBT model in practice, a fieldwork case study was performed between January and February 2019 (Eisenhardt, 1989). The fieldwork was organized as part of a collaboration project with Totonal Viajes[2], a social enterprise and tour operator that links rural cooperatives and small tourism enterprises with potential international markets. Since 2012 Totonal has developed partnerships with community-based initiatives in several states of Mexico to insert them in tailor-made itineraries (Cañada, 2015). By 2018, they directly impacted around 115 rural families, creating a spillover effect to other community activities, and were recognized by the UNWTO as one of the 12 best practices in tourism innovation (World Tourism Organization 2017).

During the fieldwork, the authors visited four communities and their tourism enterprises located in the Yucatan Peninsula – two initiatives in the state of Yucatan, one initiative in Quintana Roo and one in Campeche. Totonal Viajes suggested the four analysed social enterprises, as they were communities with whom the tour operator has already built a trust relationship, and their projects were already successfully implementing various best practices, with the potential to continue improving. The collaboration between

Totonal Viajes, the researchers and the four initiatives was based on mutual respect and reciprocal benefit for all parties involved.

Our work in the field included participating in the activities and experiences typically offered to tourists and providing capacity-building workshops in which the members of the initiatives were involved. Community members agreed to collaborate with authors and exchange knowledge valuable for the research. Data was collected through participant observation, semi-structured interviews with the coordinators of each initiative and personnel employed in the social enterprises (i.e. tour guides, chefs, craftsmen, women, etc.). In total, eight interviews were performed (two per initiative), which were complemented by informal conversations with members of the communities involved in tourism projects. The interviewees' identities have been kept anonymous, and only some demographic attributes have been disclosed to distinguish participants. We obtained their permission to take pictures and carefully selected those included in this study to ensure the participants' anonymity. Data were interpreted and analysed between 2019 and 2022 by both authors. A second round of online interviews was performed in 2023 with two local practitioners and one scholar familiar with the analysed context to assess the current situation of the communities and account for the COVID-19 pandemic impacts.

Case-Study Analysis: Community-Based TSE of the Yucatan Peninsula

We analyse four TSEs developed within the rural Mayan communities of the Yucatan Peninsula, where tourism has been relevant in socio-economic development since the early 2000s. Through collective organization and external support, these initiatives were able to establish self-managed models of tourism activity that are aligned with local social and environmental peculiarities, respect cultural heritage and enhance the livelihoods of the community as a whole.

1. U Najil Ek Balam (The house of the black jaguar in the Mayan language) is a cooperative located in the Mayan community of Ek Balam of about 400 inhabitants within Temozón municipality in Yucatan state. The initiative is well established, has gained considerable experience in providing tourism services since the early 2000s and has partnered with several private and non-profit initiatives to increase its reach, including the recently formed Peninsular Alliance for Community Tourism (APTC).
2. Yaxunah (First House or Green House) is a TSE located in the Yaxunah Mayan-speaking community of 700 inhabitants in the Yaxcabá Municipality, Yucatan state. Their tourism experience is shorter than the other three initiatives and derives from previous community projects related to their culinary traditions, which made them value their cultural heritage as a potential tourist asset.
3. A'aktun Jaaleb (The cave of Tepezcuintle or Paca) is a cooperative specializing in adventure activities, thanks to the rich geological past and underground caves and natural wells surrounding the community. The cooperative is located in Nuevo Durango, a Mayan-speaking community of about 250 inhabitants, part of the Lázaro Cárdenas Municipality in the Quintana Roo state. A'aktun Jaaleb is an example of a project involving the entire community, with 41 members, including almost all families in the village.
4. El Hormiguero is a tourism cooperative established in Eugenio Echeverría Castellot 2, a Chol community (an ethnic group closely related to Mayan people) located in the Municipality of Calakmul, Campeche state. As in other cases, the cooperative was formed with the support of government programs and has developed slowly since then. It has high potential due to its location in the buffer zone of the Calakmul Biosphere Reserve.

In Table 1.1, we summarize the main characteristics of every initiative and then discuss them according to their main strengths, common challenges and future perspectives.

Results and discussion

The dominant form of TSE in the four analysed communities is a cooperative. U Najil Ek

Table 1.1. Four indigenous TSEs and their main characteristics.

TSE name	U Najil Ek Balam	Yaxunah	A'aktun Jaaleb	El Hormiguero
Community	Ek Balam	Yaxunah	Nuevo Durango	Eugenio Echeverría Castellot No. 2
Origins	• Civil Association formed in 2004 • Bottom-up with external support from CDI[a]	• Early 2000s • Top Down, support by FHMM[b]	• Cooperative formed in 2004 • Bottom-up with external support from CDI	• Cooperative formed in 2004 • Bottom-up with external support from CDI
Governance model	• Cooperative (since 2006) • 13 members • Members divide most duties amongst themselves, but they have employees as well	• Tourism group formed by 3 existing groups: Lool K'uum, Compadres and Parador Turistico Yaxunah[c] • 10 members	• Cooperative • 41 members • Duties divided across four areas: kitchen, accommodation, tours and insect museum	• Cooperative • 10 members
Tourist services	Ecotourism complex: • Mayan style lodging • Restaurant • Mayan culture museum • Camping area • Temazcal • Bike rental Experiences: • Community tour • Milpa (Maize) tour • Artisanal and culinary workshops	Ecotourism complex: • Mayan style lodging • Restaurant • Camping area • Bike rental Experiences: • Milpa tour • The maize experience • Artisanal and culinary workshops Other facilities: • Yaxunah Cultural Center • Community museum	Ecotourism complex: • Mayan style lodging • Restaurant • Bike rental Experiences: • Guided speleology tours • Artisanal and culinary workshops • Insectarium • Community museum • Mayan ceremonies with a *j-men* (shaman) • Visit to Cenote Toj (natural well)	Ecotourism complex: • Cabins using local materials • Restaurant • Camping area • Bike rental Experiences: • Excursion by bike to El Hormiguero archaeological site • Guided walking and cycling trails in the forest • Culinary workshops • Tour of the organic orchards of the families
Other initiatives emerging from the TSEs	N/A	• Xmajana – rent of hammocks, towels and sheets • Káanbal Kuxtal – female weaving workshop • Bull horns handicraft	• Unit for environmental protection (UMA) for endangered animal species • The Maya Farm • Organic communal garden association	N/A

Cultural/Natural heritage nearby	• Ek Balam archaeological site • 4 hectares of communal land transformed into a nature reserve	• Chichen Itzá archaeological site • Yaxunah archaeological site and cenote Lol Ha	• Cobá archaeological site • Wildlife Protection Area 'Otoch Maax Yetel Kooh' (Ramsar site)	• Calakmul Biosphere reserve and archaeological site (UNESCO WHS) • Protection of 1005 hectares of forest • El Hormiguero archaeological site
Commercialization channels	• Several tour operators • International NGOs • Coox Mayab[d] • APTC[e]	• National tour operators	• Several national and international tour operators	• 1 national tour operator • Visit Calakmul DMO brand[f]
External support/ Partnerships	Public sector: • CDI • Ministry for the Environment and Natural Resources • National Forestry Commission NGOs: • Tourisme & Développement Solidaires (TDS)[g] Universities: • Universidad Autónoma de Yucatán • Cornell University Networks: • Coox Mayab • APTC	NGOs: • Fundación Cultural Yucatán (FCY) • FHMM facilitating other partnerships • International Maize and Wheat Improvement Center – CIMMYT (through FHMM) Private sector: • Traspatio Maya (social enterprise of FHMM)	Public sector: • CDI, Paraisos Indigenas programme	Public sector: • CDI, Paraisos Indigenas • National Forestry Commission (CONAFOR) NGOs: • Pronatura (Visit Calakmul) • FHMM Private sector: • Explorean Kohunlich Hotel (through FHMM)

Continued

Table 1.1. Continued

TSE name	U Najil Ek Balam	Yaxunah	A'aktun Jaaleb	El Hormiguero
Sustainability certifications/ standards	• 'Viaja seguro y solidario, viaja turismo comunitario'[h] (APTC) • 'Paraísos Indígenas' (CDI)[i] • 'Distintivo M Moderniza' (Ministry of Tourism)[j] • 'NMX-AA-133-SCFI-2006' standard[k]	N/A	• 'Paraisos Indigenas' (CDI) • 'Distintivo M Moderniza' (Ministry of Tourism)	N/A
Code of ethics	Yes	Yes	Yes	Yes

Translation of Spanish names into English and explanation of the acronyms are listed below:
[a]CDI – Comisión Nacional para el Desarrollo de los Pueblos Indígenas (National Commission for the development of Indigenous People), Programme of Alternative Tourism in Indigenous Areas.
[b]FHMM – Fundación Haciendas del Mundo Maya (Mayan World Estates Foundation) is a non-profit organization operating since 2002 in several Mayan communities in the Yucatan Peninsula with the aim of generating socio-economic opportunities for the local population. In Yaxunah and through their Comprehensive Community Development Programme, they have promoted artisanal workshops, culinary events, infrastructure, and other commercial initiatives to link the community with the tourism market.
[c]Culinary tourism cooperative was founded in 2005 by 10 women of Yaxunah with the support of the Yaxunah Cultural Center (CCY), a local initiative supported by non-profit organizations (Fisher and Ardren, 2020). Three groups comprise the TSE – one dedicated to the culinary experiences, one to hospitality and one to tourism services.
[d]Union of 12 cooperatives (as of 2023) committed to the practice of responsible, solidarity-based, fair and conscious tourism. It is part of APTC.
[e]APTC – Alianza Peninsular para el Turismo Comunitario (Peninsular Alliance for community tourism) is a network that brings together CBT cooperatives in all three states in the peninsula. Available at: https://www.viajaturismocomunitario.com/ (accessed 29 April 2023).
[f]Visit Calakmul is a network of five communities of alternative tourism in Calakmul municipality. Available at: https://www.visitcalakmul.com/ (accessed 29 April 2023).
[g]TDS – Solidarity Tourism and Development is a French non-profit organization that supports CBT initiatives around the world.
[h]'Travel safe and in solidarity, travel community tourism' is a certification provided by APTC to CBT initiatives as a response to the Covid-19 pandemic. Available at: https://www.viajaturismocomunitario.com/covid (accessed 29 April 2023).
[i]Paraísos Indígenas (Indigenous Paradises) is a brand created by CDI to promote and commercialize CBT initiatives developed by indigenous communities to national and international markets.
[j]Distinctive M for 'Modernization' – Quality assurance certification given by the Mexican Ministry of Tourism for micro, small and medium size tourism enterprises.
[k]Mexican sustainability standard for the tourism sector.

Balam was first constituted as a civil association, but soon they changed to the cooperative model to commercialize their services. The number of members fluctuates, with some cooperatives involving almost all families in the community (such as A'aktun Jaaleb). All members have the right to participate in decision making processes and organization. Regardless of the involvement of the members, the aim to generate direct or indirect benefits for the whole community is present as a guiding principle in all four TSEs. In Ek Balam, for instance, 10% of the cooperative's annual profits are given to the community as a contribution to using 4 hectares of *ejido* (collectively owned land). Even if members have main roles assigned, such as those in charge of the accommodation, the tour guides, and the chefs, everyone is able to perform multiple tasks.

Tourism was not previously the prevailing economic activity in any of the analysed communities but has taken on an essential role in the socio-economic development and communal life since the establishment of the TSEs. Traditional activities were implemented in the tourism offer, thus creating experiences that value local heritage, such as artisanal workshops, beekeeping, culinary workshops, traditional customs, and in-field activities like the *Milpa* tour (maize field) (Fig. 1.2). All these activities continue to be practised independently from tourism, thus enabling communities to avoid over-dependence on the tourism sector.

Cultural and natural heritage is an inseparable part of the identity of the Mayan livelihoods and is part of the different experiences offered to visitors. All four communities are located near Mayan archaeological sites that are also part of their tourism offer. Yaxunah, located near Chichen Itzá ruins, also manages its local archaeological site of the same name. In the vicinity of El Hormiguero is the Calakmul Biosphere Reserve (Fig. 1.3), the second-largest rainforest in the Western Hemisphere after the Amazon. It is recognized as a World Heritage Site. Maize is an element of social cohesion in these communities (and in Mexico in general) (Fisher and Ardren, 2020), and the culinary and cultural heritage based on this plant is shared with visitors. Degustation of maize-based drinks, workshops to prepare 'tortillas', and spiritual ceremonies related to the maize harvest are some examples of these experiences. All cooperatives recognize the value of tourism and its responsible management. A member of A'aktun Jaaleb confirms the complementing role of tourism in the overall socio-economic context:

> It is a complement to our [traditional] activities. Mayans have multiple activities, such as orchards, cornfields, beekeeping, and tourism is one more activity. The start of tourism helped us to avoid the migration of our young people. Thanks to the income from visitors, the youth can go to universities, something that did not happen ten years ago due to economic problems. Tourism strengthens our economy; it allows us to have a basic life where our children can go to school, have a career, and work here in the community. So that is what sustainable tourism helps us to do. (Male, early 40s)

Depopulation trends have been previously present in almost all the analysed communities. Although this phenomenon persists, the positive impact of tourism is easily observable. A young tourist guide from Ek Balam, who decided to stay in the community and work in tourism, explains:

> Tourism has helped a lot; before, young people had to migrate to Cancún, and the community was losing inhabitants. With the arrival of tourism, young people now have the possibility to work and be trained, so I think we have an opportunity to stay in the community and take care of our culture and roots. (Male, mid-20s)

Besides positive economic effects, responsible tourism development has impacted the socio-cultural context of the communities. A significant social shift is seen in the role of women in economic life and their position in the family and community. The process of socio-cultural change is especially seen in Yaxunah, where tourism is managed predominantly by female community members:

> Now is a time when women think differently. In the past, men thought that women were only good for raising children and being at home, but now that has changed because of the many organizations that come to Yaxunah and that have opened the minds of the women of the village. We see it in the whole community; women depend less on their husband's income and are involved in various activities like sewing cloth and napkins to sell, handicrafts and

Fig. 1.2. Milpa tour at Yaxunah.

> various other activities to help the family [...] The men have also changed. Previously, the husband would only go to Milpa, and that would be all his work, but now it is different. Males help their wives with the children, who once were women's responsibility, not the husband's, and now it is a mutual activity. (Female, mid-30s)

Establishing the cooperatives created a spillover effect in the communities, observable in the emergence of new initiatives related directly or indirectly to tourism. In Nuevo Durango, a few years after the establishment of the cooperative, several members started new projects, which they combined with the cooperative offer. The Unit for Environmental Protection (UMA) is one of those, a reserve established in 2006 by a member of A'aktun Jaaleb to provide space for breeding endangered animal species (deer, lowland paca, ocellated turkey and pheasant) in a protected environment. Another example is The Maya Farm project, whose founder (also a member of A'aktun Jaaleb cooperative) explains:

> Thanks to my experience in restaurants in Playa del Carmen (where I used to live), I started offering meals for tourists at my house. At the same time, we started to grow vegetables with my family to make the most of our land, have healthy food and take care of the economy. The idea of the gastronomic tour came after, adding the experience of picking up the vegetables from the garden and cooking the meals together to increase cultural exchange. (Male, mid-40s)

As is evident from Table 1.1, in all cases the tourism initiatives were born thanks to the support of external actors, with the communities willing to get involved in tourism to diversify their income and complement traditional activities. Three of the four TSEs were established in 2004 with the support in infrastructure provided by the CDI (National Commission for the development of Indigenous People). The support, as the cooperatives' members explained, was in the form of materials for building the accommodation, with the labour force provided by the members. Various public institutions, private companies, universities and NGOs provide them with

Fig. 1.3. Calakmul archaeological site near El Hormiguero cooperative.

possibilities for professional training, education and opportunities for commercialization and extension of their networks of collaborators. The role of external actors is visible not only in terms of financial support but also in providing capacity building to develop innovative and sustainable projects. A member from El Hormiguero explains the contribution from the forest authorities:

> Thanks to the forestry technicians, who come to give talks on conservation, we have focused on preserving 1005 hectares of forest. The forestry authorities pay the community for its ecosystem services, and then we collectively manage the reserve by doing maintenance and cleaning work. Through successful conservation, our children and grandchildren will be able to see the wealth left by their ancestors. (Male mid-40)

Finally, the interviews performed with experts in 2023 provided insights into the main challenges that threaten the survival of CBT initiatives in the Yucatan Peninsula. These include the lack of access to technological tools and links to the tourism market, inadequate capacity building, over-dependence on external funding, and the expansion of capitalist interests with significant investments in rural areas. The COVID-19 pandemic exacerbated these challenges and posed new ones since initiatives received little income from tourism for several months and had to think through their cooperative models. Nevertheless, they were able to overcome these challenges by reinforcing their cooperation mechanisms and all of them are still operative (Jouault *et al.*, 2021). Most notably, regional networks such as APTC and Visit Calakmul have played an essential role in strengthening the communities' position within the Yucatan Peninsula, not only by promoting them but also by raising the voice of cooperatives to influence public policies.

Conclusions and a Way Forward

The Yucatan Peninsula remains the leading tourism destination in Mexico and an emblematic example of the sea *&* *sun* model of all-inclusive tourism. Parallel to this mass-tourism reality, in the Peninsula's inner areas an alternative tourism model has been developing to contribute to improving local livelihoods while offering a different experience to travellers. This alternative model is provided through TSEs, formed by community members.

In this case study, we have analysed four examples of TSE within four communities that developed tourism activity in their surroundings responsibly and sustainably to contribute to their socio-economic development and safeguard the natural environment. U Najil Ek Balam, Yaxunah, A'aktun Jaaleb and El Hormiguero may be located in different states on the Yucatan Peninsula, but their socio-environmental realities are similar due to the shared indigenous roots and natural context in which they are settled. Their internal organization is based on collective ownership and decision making, and in three out of the four cases, communities chose cooperatives as a preferred model of social enterprises. This governance model allowed every member to be involved and participate in all the cooperative's activities.

We conclude that tourism contributes to generating economic opportunities for community members. It complements traditional productive activities, improves the quality of life of the communities and helps to prevent the rural exodus of people to the coastal cities and tourist destinations in search of job opportunities. Even though inequalities are still present, tourism empowers different social groups. It enables female members to take a leading role in a relevant economic activity that channels significant social change, such as the case in Yaxunah, and young people who decide to stay in their ancestral lands and work on tourism development, as seen in Ek Balam. In all communities, tourism activity values local cultural and natural heritage and includes them in provided experiences. External actors, such as foundations or public bodies, contributed to forming TSEs in all cases directly through a top-down approach or triggering a bottom-up process. Support from external initiatives cannot be neglected as an enabling factor for financial support, but over-dependence on external funding can also lead to the failure of CBT initiatives, as pointed out by Cañada (2015). In this case study, none of the four analysed TSEs depend entirely on subsidies, but they have used them to initiate community projects.

The COVID-19 pandemic has heavily affected the Mayan communities in the Peninsula, who also suffered from heavy storms and floods during the same period. However, as the study by Jouault *et al.* (2021) and our interviewees demonstrated, communities adapted relatively quickly to the health contingency, strengthening the Milpa for food security, exchanging products between communities on the coast and hinterland, and using joint savings. In this sense, the tailor-made certification '*Viaja seguro y solidario, viaja turismo comunitario*' was a proactive tool to restart tourism operations safely from the second half of 2020, relying on domestic tourism.

Besides these positive results, there are still several challenges CBT initiatives face in the region, including the privatization of rural land for large-scale projects (Garcia de Fuentes and Jouault, 2021) and the commodification of Mayan cultures (Poot *et al.*, 2022). While the former jeopardizes the communities' overall livelihood and creates a risk of a potential loss of the natural environment and habitat, the latter can lead to *over-commercialization* of the cultural heritage and banalization of its content. For this reason, the empowerment of local actors is vital so that they can work collaboratively and have a stake in public policies to support their livelihoods through CBT initiatives. Furthermore, partnering with external actors to commercialize the tourism offer of TSEs remains crucial since the absence of this support still results in failure to reach the tourism market. As evidenced by Cañada (2015), this constitutes the main reason for the collapse of CBT initiatives in Latin America and, therefore, is an issue that needs to be addressed in the future.

This analysis showcases various positive effects of TSE for indigenous communities that pursue tourism as one of the economic activities that can accompany their traditional livelihoods. We acknowledge the context-specific nature of some of the practices implemented in the analysed communities but, at the same time, identify their applicability in similar CBT enterprises in other geographical areas.

Notes

[1] Instituto Nacional de Antropología e Historia – INAH (National Institute of Anthropology and History). Available at: https://www.inah.gob.mx/zonas-arqueologicas (accessed 30 April 2023).
[2] As a consequence of the Covid-19 pandemic, in 2021 Totonal Viajes merged with Rutopia, forming a Destination Management Company that commercialises tailor-made itineraries involving rural communities in Mexico. Available at: https://rutopia.com/totonal/ (accessed 22 August 2023).

References

Altinay, L., Sigala, M. and Waligo, V. (2016) Social value creation through tourism enterprise. *Tourism Management* 54, 404–417. DOI: 10.1016/j.tourman.2015.12.011.

Cañada, E. (2015) La comercialización del turismo comunitario en Americal Latina. In: *Anuario de Estudios Centroamericanos*, Vol. 41. pp. 159–189.

Eisenhardt, K.M. (1989) Building theories from case study research. *The Academy of Management Review* 14(4), 532. DOI: 10.2307/258557.

Fisher, C. and Ardren, T. (2020) Partaking in culinary heritage at Yaxunah, Yucatán during the 2017 Noma Mexico pop-up. *Heritage* 3(2), 474–492. DOI: 10.3390/heritage3020028.

Fletcher, R. (2011) Sustaining tourism, sustaining capitalism? The tourism industry's role in global capitalist expansion. *Tourism Geographies* 13(3), 443–461. DOI: 10.1080/14616688.2011.570372.

Garcia de Fuentes, A. and Jouault, S. (2021) Las cooperativas turísticas en la Península de Yucatán: Una reflexión sobre un cuarto de siglo de turistificación de los espacios rurales. In: Ruiz de Leon, C.G. and Osorio Garcia, M. (eds) *El Desarrollo Turistico En Mexico*. Revisión general y casos de estudio. Universidad de Guadalajara. Puerto Vallarta, pp. 445–470.

Guardado, G.M., García de Fuentes, A. and Daltabuit Godás, M. (2012) *Turismo, globalización y sociedades locales en la península de Yucatán, México*, Vol. 7. PASOS, Tenerife.

Guia, J. (2021) Conceptualizing justice tourism and the promise of posthumanism. *Journal of Sustainable Tourism* 29(2–3), 503–520. DOI: 10.1080/09669582.2020.1771347.

Hardy, A.L. and Beeton, R.J.S. (2001) Sustainable tourism or maintainable tourism: Managing resources for more than average outcomes. *Journal of Sustainable Tourism* 9(3), 168–192. DOI: 10.1080/09669580108667397.

Higgins-Desbiolles, F. (2006) More than an 'industry': The forgotten power of tourism as a social force. *Tourism Management* 27(6), 1192–1208. DOI: 10.1016/j.tourman.2005.05.020.

Hockings, M., Dudley, N. and Elliott, W. (2020) Editorial essay: COVID-19 and protected and conserved areas. *PARKS* 26(1), 7–24. DOI: 10.2305/IUCN.CH.2020.PARKS-26-1MH.en.

INEGI (2014) *Anuario estadístico y geográfico por entidad federativa 2014*. Instituto Nacional de Estadística y Geografía, Aguascalientes.

Jamal, T. and Stronza, A. (2009) Collaboration theory and tourism practice in protected areas: Stakeholders, structuring and sustainability. *Journal of Sustainable Tourism* 17(2), 169–189. DOI: 10.1080/09669580802495741.

Jouault, S., Novelo Alpuche, C., Garcia de Fuentes, A. and Rivera-Núñez, T. (eds) (2022) *Cuando uno somos varios—Rostros y voces del turismo comunitario en la Península de Yucatán*. Alianza Peninsular para el Turismo Comunitario, Merida.

Jouault, S., Rivera-Nuñez, T., García De Fuentes, A., Xool Koh, M. and Montañez Giustinianovic, A. (2021) Respuestas, resistencias y oportunidades del turismo comunitario en la península de Yucatán frente al COVID-19 y las crisis recurrentes. *Investigaciones Geográficas* 104, 1–16. DOI: 10.14350/rig.60240.

Kieffer, M. (2019) Turismo rural comunitario en México: Apuntes para futuras investigaciones. *Dimensiones Turísticas* 3(5), 43–63. DOI: 10.47557/XSNY8857.

Pearce, D. (1992) Alternative tourism: Concepts, classifications, and questions. In: Smith, V.L. and Eadington, W.R. (eds) *Tourism Alternatives: Potentials and Problems in the Development of Tourism*. University of Pennsylvania Press, Philadelphia, PA, pp. 15–30.

Pollock, A. (2015) *Social Entrepreneurship in Tourism—The Conscious Travel Approach*. TIPSE, London.

Poot, M.A., Jouault, S. and Rodriguez Martinez, Y. (2022) Las vías de la mayanización: El Tren Maya, turistificación y representaciones de la mayanidad en la Península de Yucatán. *Maya America—Revista de Ensayos, Comentarios y Analisis* 4(2), 62–89.

Rozemeijer, N. (ed.) (2001) Community-based tourism in Botswana: The SNV experience in three community-tourism projects. SNV/IUCN CBNRM Support Programme, Gaborone.

Ryan, C. (2002) Equity, management, power sharing and sustainability—Issues of the 'new tourism. *Tourism Management* 23(1), 17–26. DOI: 10.1016/S0261-5177(01)00064-4.

Scheyvens, R. (1999) Ecotourism and the empowerment of local communities. *Tourism Management* 20(2), 245–249. DOI: 10.1016/S0261-5177(98)00069-7.

Sheldon, P.J. and Daniele, R. (eds) (2017) *Social Entrepreneurship and Tourism*. Springer International Publishing, Cham. DOI: 10.1007/978-3-319-46518-0.

Sloan, P., Legrand, W. and Simons-Kaufmann, C. (2014) A survey of social entrepreneurial community-based hospitality and tourism initiatives in developing economies: A new business approach for industry. *Worldwide Hospitality and Tourism Themes* 6(1), 51–61. DOI: 10.1108/WHATT-11-2013-0045.

Strydom, A.J., Mangope, D. and Henama, U.S. (2019) Making community-based tourism sustainable: Evidence from the free state province, South Africa. *GeoJournal of Tourism and Geosites* 24(1), 7–18.

World Tourism Organization (ed.) (2017) *Innovation in Tourism: Bridging Theory and Practice = La innovación en el turismo: acercamiento entre teoría y práctica*. World Tourism Organization, Madrid.

World Tourism Organization (ed.) (2021) International tourism highlights 2020 edition. World Tourism Organization, Madrid.

2 'We Are a Territory, a Territory of Struggles': Community and Resistance Tourism in a Territorial Museum on the Outskirts of São Paulo

Camila Cardoso[1] and Pedro Vianna Godihho Peria[2]
[1]Comunidade Cultural Quilombaque, São Paulo, Brazil; [2]Escola de Administração de Empresas de São Paulo da Fundação Getulio Vargas (FGV EAESP), São Paulo, Brazil

Summary

The case presents an interpretation of the actions of valorization of cultural heritage mobilized by the political-cultural collective Comunidade Cultural Quilombaque. Active in the neighborhood of Perus, in the north-western part of the municipality of São Paulo, since 2005, Quilombaque shows itself to be a pole of agglutination of several artistic and political manifestations and other groups in the territory. From the consolidation of the Territorial Museum Tekoa Jopo'i and the Queixadas Agency, they have elaborated a path of trails that go through the neighborhood's heritage that is marginalized by public preservation policies. Semi-structured interviews were conducted with central agents of the movement, in addition to participant observation on community tourism trails. The symbolic mediation promoted by the Museum educators brings out narratives of struggle eclipsed by the official policies of cultural heritage. The keys of solidarity and sharing operate by allowing a horizontal use of the riches organized by the Territorial Museum. The action of the Quilombaque Cultural Community transits in a fleeting limit between tourism and resistance, its tourism shows not only the resistance of the territory but is also organized with the community in such a way as to resist the tendencies of predatory tourism. On the one hand, the visits amplify voices silenced by hegemonic narratives. On the other hand, the very conception and organization of the trails are done in opposition to other models of tourism and predatory development.

The Value and Interest of the Case Study

We contribute to the reflection on how peripheral movements can act to value their collective memories according to the solidarity economy. Making resistance visible and thereby resisting, the Quilombaque agents call their practice resistance tourism.

Context

How do we value cultural heritage in peripheral areas of large cities? How to build tourism actions that aid the development of the community, economically, socially, intellectually, and culturally? How to fight against the gentrification of the territory, state violence, loss of community identity, and the constant pressure of market logic to the detriment of solidarity-based rationalities?

Processes of urban spoliation (Kowarick, 1979) such as these are experienced daily in contexts marked by precariousness.

Answers can only be creative and collective, and studies have called attention to relevant experiences in Brazilian peripheries (D'Andrea, 2013; Raimundo, 2017; Oliveira, 2021; Peria, 2022). By affirming the intrinsical relationship between tourism practices and power relations (Tribe, 2007), we link to critical perspectives in tourism studies. This standpoint allows us to broaden our questioning. As Crespo (2017, p. 531) states, we should 'consider tourism as a device that, beyond producing territorialities, involves forms of constitution, exhibition, and contemplation of who we are and are not'. We believe that the growing appearance of studies focused on Indigenous Tourism (Brandão *et al.*, 2014; Silvestre and Fontana, 2023) can provide rich insights into the complexity of understanding tourism as resistance. The idea of Indigenous-driven practices (Nielsen and Wilson, 2012), will help us frame one initiative that indicates a possible path to use tourism and a form of resistance.

Quilombaque Cultural Community is a non-profit organization that arose in 2005 from the initiative of a group of young people, residents of Perus, a peripheral district located in the north-western zone of São Paulo. This area concentrates on the worst socio-economic and cultural indices, where the most prominent victims are young people. Grouping the data from the 2010 Census, the Inequality Map, from Rede Nossa São Paulo (2017), shows that among the residents of the Perus district 40–49.9% self-declare as black or mixed race. In the Anhanguera district, this ratio rises to over 50%. In this district, as in the entire northern part of the municipality, we find a range of 0–0.99% formal jobs per inhabitant. In comparison, in the district of Pinheiros, a highly valued upper-middle-class neighborhood, this number rises from 2 to 5.99. In Campo Belo, South Zone, the average remuneration of formal employment in 2015 was more than R$10,000. The residents of Anhanguera and Perus received less than a quarter of that amount. Specifically in the Perus district, we know that between 15 and 29.9% of the population resides in slums, and the same percentage represents the portion of the neighborhood's population in a situation of high or very high social vulnerability (Rede Nossa São Paulo, 2017). While in the district of Jardim Paulista, an upper-class neighborhood, the average age at death was 79.4 years, in 2016, in the district of Perus, this ratio is 60 years (Rede Nossa São Paulo, 2017). In the COVID-19 pandemic neighborhoods like Perus, which has no hospital beds, had the worst indicators of contamination and death rates (Instituto Pólis, 2020).

The Quilombaque Cultural Community brings together several cultural actions and movements. It follows the strategies of other initiatives with similar purposes directed to cultural production and valorization of the peripheral territory in São Paulo (Raimundo, 2017; Oliveira, 2021). Among these initiatives, mapping places of memory and affection for the community subsidized the design of tourism trails. From this project, which began to take shape in 2011, the Territorial Museum of Culture and Landscape Interest Tekoa Jopo'í and the Queixadas Agency were born to expand knowledge about the territory itself and value the heritage that is often forgotten by official history. This operation generates a true counter-narrative (Noy, 2012; Lueg *et al.*, 2021), denouncing the dominant narrative that only denounces the precariousness of the *other* to announce the potentialities of this territory and its residents.

From the Indigenous presence in the urban environment to the historical railway heritage, to the struggles of workers and rural men and women, to the memory of the victims of state violence and reaching the most recent cultural manifestations, the Quilombaque Cultural Community weaves a path of difficult and relevant memories to understand the depth of violence and resistance in Brazilian society. Some principles govern the actions of the Quilombaque Cultural Community. One of them is sevirology (*'sevirologia'*, while impossible to exactly translate, it is a neologism that indicates the need to handle difficult situations), a systemic vision, a multidimensional methodology thinking about introducing knowledge and learning procedurally and permanently. So, what does this mean? 'If we have it, we do it, if we don't have it, we do it the same way; we turn around, but we don't stop doing it' (Souza, 2020). The 'turning around' is the ability to deal with precariousness dynamically and, therefore, it is not only a value, but it is the very reality of the cultural activities in that territory. It is a collective process aimed at facing challenges by doing, learning, remaking, and finding loopholes. In other words, 'it is the strength of we for us' (Camila Cardoso, interview, 2021).

To question, understand, and dialogue with these strategies, what follows is part of the product of a bigger research project about social movements and the democratization of heritage policies in Brazil. Interested in delimiting the narrative enunciated by Quilombaque participants, the collection of texts combined primary and secondary materials. The secondary materials were selected to incorporate into the analysis statements produced by Quilombaque itself or by its participants about the history of the organization and the nature of its work. The conducting of interviews followed the premise that the exercise of listening is central to understanding the narratives. The first interview conducted was with Camila Cardoso, a Quilombaque member, and coordinator of the Queixadas Agency (Quilombaque's arm directly linked to community tourism actions in the Territorial Museum Tekoa Jopo'i) since 2018. Another interview was conducted with Cleiton Ferreira de Souza, known as 'Fofão' and one of the founders and current coordinator of Quilombaque. The interactions were conducted virtually in mid-2021, being recorded in image and audio with the prior and express authorization of the persons interviewed, and lasted approximately 1 hour and a half. The interviewees approve and support the use of their speeches.

Agência Queixadas: Resistance Tourism, from the Territory to the Territory

Due to the damage caused by urban interventions, real estate speculation, and the great patrimonial heritage that exists in this region, a partnership was started between the Quilombaque Cultural Community, the Movement for the re-appropriation of the Perus Cement factory, and the Faculty of Architecture and Urbanism of the University of São Paulo (FAU-USP). It created a Free and Collaborative University that tried to point out with technical language all the particularities and especially the potentialities of this territory. The classes took place in the streets, alleys, alleyways, churches, terreiros, slums, and villages. The unfolding of these activities developed by the Free and Collaborative University was the TICP – Territory of Interest of Culture and Landscape.

The TICP proposes an educational, cultural, and collaborative space, recognizing the affective potential of knowledge production about the population's experiences, articulating cultural and natural heritage, places of memory, forms of participation, dialogues, and local cultural production. This instrument brings the dimensional recognition of the city alongside economic and functional issues, going against what traditionally guides conventional urban planning models. At this moment the process of 'hacking' city planning and architecture begins, developing its model of city management and its particularities.

In 2014, to preserve and properly develop the territory inclusively and sustainably, as well as the community and its histories, the TICP was transformed into law by the Strategic Master Plan of the City of São Paulo as a tool for participatory management and promotion of cultural education. However, until 2023, the year in which the city of São Paulo is reviewing the master plan, there were no regulations or investments by the public authorities for this new public policy. Considering all this context, there was the need to create a mechanism in which it would be possible to shape, articulate, and dynamically organize the local struggles so that these memories would be accessible to the community and the population in general. It is at this moment that for the second time, the collective went "hacking' the museum and tourism system. The Quilombaque Cultural Community started to develop two incubation projects: The Territorial Museum of Culture and Landscape Interest Tekoa Jopo'í and the Queixadas Agency of Eco-Cultural Touristic Development, which today operate independently.

The Territorial Museum of Culture and Landscape Interest Tekoa Jopo'í, named in the Guarani language, proposes dialogues, connects social movements, rescues the memory of the region, and disseminates the struggles that take place in the territory through living mechanisms

Fig. 2.1. Map of the Territorial Museum Tekoa Jopo'i. Photo: Queixadas Agency.

of collective actions and encourages local sustainability. Places of interest in history, affection, environment, and social struggles encompassed in the Perus, Anhanguera, and Jaraguá territories were mapped. Thus, the paths and narratives are given in the form of Educational Trails that enable different paths and educational and cultural perspectives through their attractions and dynamics. Fig. 2.1 shows a reproduction of the map made after the construction of the trail of memories of struggle and affection for the territory. At each point, there is a short explanatory text to help visitors read the map. The next section will discuss the entire route as suggested on the map.

To organize and streamline the Educational Trails promoted by the Territorial Museum and point out how to approach the TICP, the Queixadas Agency of Eco-Cultural Tourist Development was born. It offers a pedagogical proposal that provides a practical educational experience, providing values that account for multidimensionality when approaching the different possibilities of reading reality, the city, and new cultural making.

The Queixadas Agency has begun to develop a new method for tourism, the 'TOURISM OF RESISTANCE'. This is the mechanism that spreads the problematization brought by the Tekoa Japoí Museum of thinking about a change of reality and new possibilities of experiencing the periphery as a cultural source rich in forms, colors, knowledge, flavors, and recognition. It works directly on the self-esteem of the peripheral subjects as well as thinking about possibilities of territorial development and shared management of the whole territory. As the master, community leader of Perus, José Soró, used to say: 'Another city is possible and we are doing it'.

Trails of Resistance: Walking through the Struggles of Perus

Jaraguá Peak is the highest point in the municipality of São Paulo and, besides being one of the few remnants of the Atlantic Forest in the region, is home to the smallest Indigenous Land approved in Brazil: the Jaraguá Indigenous Land. It is located on the edge of the Jaraguá State Park, where around six hundred Indigenous people have lived in an area of fewer than 2 ha since 1987 (ISA, 2021). The trail called 'Jaraguá is Guarani' seeks to show the experience of the original people so that visitors can understand the relationship of the Guarani people within the land of Jaraguá. The trail makes a narrative return to the pre-colonial period to make visible the reluctant process of confrontation with colonialism. It aims to understand the suffering, publicize the struggles, and discuss the historical context of the relationship of slavery of both indigenous and black people, 'because we say that this heritage tells stories and forgets this process, which was a cruel process, but we have to highlight' (Camila Cardoso, interview, 2021). The process begins in the retelling of history from the Senzala's perspective – a thought and action born of the struggle of enslaved people, looking at Brazil's history through the eyes of those who resisted in the slave quarters. The images in Fig. 2.2 show part of the trail and a group dynamic in the Indigenous Land.

The second point of the trail makes a time jump to the coffee agro-exporting Brazil and the consequent expansion process of the railway infrastructure. The 'Perus-Pirapora Railroad' trail has as its material support the remains of a junction of the São Paulo Railway to take pilgrims from Perus to Pirapora do Bom Jesus. However, it was mostly used to transport raw material from the Gato Preto (Cajamar) quarries to Perus, where years later the Brazilian Portland Cement Company was installed.

Although the Perus-Pirapora railroad, which opened in 1910, has been little used, it is an important railway heritage. It is the only remaining 60 cm gauge railway in Brazil and contains several steam locomotives of various origins. However, they are in a very poor state of conservation. From these remnants, we see another face of 'progress', which produces structures and then abandons them. Fig. 2.3 shows a group of visitors visiting the trail and the railway remnants in the territory.

It was from this railroad construction that the first cement factory in Brazil was established. The Fábrica de Cimentos Portland Perus produced the cement that urbanized São Paulo and built Brasília. In the 1960s, after decades of neglect, negligence of labor laws, and the accumulation of terrible working conditions, the factory was the stage for a strike that lasted 7 years in the middle of the civil-military dictatorship.

The striking workers and their families, called 'queixadas' (bush pigs known for hunting in packs), applied the principle of active non-violence based on the inspiration of the French labor movement, and the heritage of Gandhi and Martin Luther King. Since the closure of the Portland de Perus activities, the Queixada movement demanded that a worker's memorial and cultural center be created. However, without the expropriation of the land what remains are the ruins and the cultural mediation work of the Tekoa Jopo'i Museum. The sensitization made during the trail 'Queixadas Memory' is focused not only on the struggles of the 1950s and 1960s but also the contemporary disputes to occupy the spaces, now empty, with all the political and cultural life that emanates from the community of Perus. In Fig. 2.4, we see the scene of the strike, in which workers and their families surrounded the entrances to the factory. In Fig. 2.5, with the factory ruins, we see a group getting to know the space through the mediation of the Queixadas Agency.

In the trail 'Dictatorship never again', the route goes to one of the most remarkable places of death in the city: the mass grave of the Perus Cemetery. The cemetery was built when Paulo Maluf was mayor, from 1969 to 1971. In the early 1990s during the first months of the government of Luiza Erundina, of the Workers' Party, an unidentified grave was discovered containing more than a thousand unidentified skeletons.

Some of the unnamed bodies are those of political activists arrested, tortured, and killed during the military regime. However, the educators remind us that history has many layers of violence. These skeletons are also those of children from the time of the meningitis outbreak in the 1970s, which the state tried to hide, and also of the death squad in the 1980s, which prowled the periphery.

Through these skeletons, the Queixadas Agency discusses the Genocide of Black Youth; something that is still going on in Brazil. Just for being black, peripheral, and manifesting Afro-Brazilian cultures, you were already a subversive. You died and were buried in a common grave, which still happens today. In this transit between the current context and that of the dictatorship, the movement has been discussing this issue of genocide, because the dictatorship continues for this black, poor, and peripheral population; it has only been veiled. After 30 years, after the comings and goings of the collection of mortal remains under constant neglect, only five skeletons were identified – all of them of

Fig. 2.2. Jaraguá is Guarani trail. Photo: Provided by Karen Siqueira.

Fig. 2.3. 'Perus-Pirapora Railroad' trail. Photo: Provided by Karen Siqueira.

Fig. 2.4. Striking workers, the queixadas. Photo: unknown, circa 1960.

young political activists. Observing the history of the dictatorship from the periphery, then, allows us to understand the violation of rights and denounce its continuity. In Fig. 2.6, we see part of the team from the Queixadas Agency and the Quilombaque Cultural Community in front of the cemetery with the person responsible for discovering the skeletons. In Fig. 2.7, we see a sign inside the cemetery that states: 'Martyrs' Hill, in this cemetery the military regime concealed corpses of politically persecuted people'.

The last points of the trail through Agência Queixadas are about the struggle for land and sustainable production. They are directly related to the appreciation of the cultural manifestations that take over the streets of the territory.

The trail 'MST (Landless Movement) Agroecology Field and City' takes place in the only settlement in the municipality created by the Landless Rural Workers Movement (MST) during the disputes around the proposal to take the debris from the silting up of Pinheiros River to Perus. Cleiton Fofão recalls that 'it has been 19 years producing food where the city's pinicão was going to be' (Cleiton Fofão, oral communication, 10/01/2021). The MST camp is home to more than 40 families. It has transformed the area into a space for the production of organic food from an agroforestry system. It serves as subsistence for families and is marketed through solidarity networks of consumption in the Perus district.

This trail is about understanding the agrarian reform process and the food process. It is also about bringing the merchandise that is produced in this settlement to market and making organic fairs for people to have better food within the territory. This trail acts as an example of the possibility of another form of production, capable of respecting both the ecosystem and the community. In Fig. 2.8, we see a group of visitors talking to residents of the MST settlement and getting to know the fields.

The trails 'Reappropriation and Ressignification of the Public Space' and 'Perusferia Grafite' were also created to enable

Fig. 2.5. Queixadas Memory trail. Photo: Provided by Karen Siqueira.

and foster the cultural manifestations of the territory. These trails allow us to see some of the processes of taking advantage of spaces abandoned by the public authorities in the neighborhood. They are manifestations that show some branches of Quilombaque's network of partners. These points along the way show that the management of public spaces can be done communally. For example, in only 2weeks, a reading house and a telecenter abandoned by the city hall were transformed into a theater and one of the main hip-hop venues in São Paulo. In this urban acupuncture, the lack of governmental equipment becomes a blank sheet of paper for the creativity of the several cultural movements that elected Perus as their territory of action.

In the sequence of images in Fig. 2.9, we see groups walking through the streets of Perus accompanied by mediators from the Queixadas Agency. We can see the presence of the hip-hop house and the graffiti on the city walls. By experiencing the territory and observing the landscapes, visitors can understand how the absence of public policies interferes with people's daily lives.

Conclusions

The Tekoa Jopo'i Museum is the lived experience, in general, of the residents of Perus and, specifically, of personalities who are difficult to dissociate from the trails they mediate. The creation of the Agency and the Museum does not conceive new histories but divulges those that already existed in the territory. These organizations, which this case study has been dedicated to describing, were created by the sharing of stories among neighbors. The innovation brought by the Museum and the Queixadas Agency is the agglutination of resistance narratives that, to a greater or lesser extent, were already known and valued by certain groups in Perus. It is, then, a platform for amplification of these narratives about the historical struggles of different social groups.

The boundaries between the Museum and the territory are very tenuous. If we can say that the former is the result of the work of signification on the latter, we could also imagine that the Tekoa Jopo'i Museum, as a form of work on the territory, also attributes new meanings

Fig. 2.6. Camila Cardoso, Toninho e Cleiton Fofão on Perus Cemetery. Photo: Personal archive by the author.

Fig. 2.7. Dictatorship never again trail. Photo: Provided by Karen Siqueira.

and creates new identities on that ground. In this sense, it is not a static museum, with a fixed address and collection, but is instead as dynamic as the territory of Perus, Anhanguera, and Jaraguá itself. More than a museum of territory, the praxis of the Museu Territorial Tekoa Jopo'i creates a museum that is territory and a territory that is a museum. In the reception of the Quilombaque, in the telling of the stories of each trail, and in the experience of this visitation we see a museum-territory in action.

The tourist dimension is a practical consequence of this valorization process. The keys of solidarity and sharing are at work when allowing a horizontal use of the symbolic riches organized by the Territorial Museum through the Queixadas Agency. The need to divulge other sides of history also becomes a profession, taking to its ultimate conclusion the sense of memory work. The conjugation of museological proposals, as amplification of narratives, and tourism as income generation is also placed under the meaning of resistance against an overwhelming development model, one which does not allow space for the self-sustainability of local networks. Similarly to the Museum, the action of the Queixadas Agency transits in fleeting boundaries between tourism and resistance. On the one hand, the visitation amplifies voices silenced by hegemonic narratives. On the other hand, the very conception and organization of the trails are done in opposition to other models of tourism and predatory development. At the same time that it narrates resistance, it resists. More than resistance to tourism, the praxis of the Queixadas Agency makes resistance into tourism, and tourism into resistance. In that museum-territory, we have, therefore, a tourism-resistance. We must remember: 'Here we choose to tell the stories that history chose not to tell' (Camila Cardoso, interview, 2021). These two dimensions of Quilombaque's and the Agency's actions manage to announce dreams: 'The Agency is maintained today because we also talk about dreams, a dream of an educational territory, a dream of Perus not being a dormitory neighborhood anymore, but also the dream of people feeling like studying, of people feeling like knowing another culture. The route as a whole proposes tourism based on the potentialities of the community while market trends do not see the richness in the territory' (Camila Cardoso, interview, 2021).

The Queixadas Agency is not afraid to shed light on difficult, sad, and dark memories. It exposes what Mignolo (2017) called 'the dark side of modernity' and what Benjamin (2020) qualified as 'documents of barbarism'. Looking

Fig. 2.8. MST (landless movement) Agroecology Field and City trail. Photo: Provided by Karen Siqueira.

Fig. 2.9. 'Reappropriation and Ressignification of the Public Space' and 'Perusferia Grafite' trails. Photo: Provided by Karen Siqueira.

at marginalized memories is not static and is not content with denouncing the pile of rubble bequeathed to the territory by 'progress', but invites perception of the disruptive potential of 'underground memories' (Pollak, 1989). As an Indigenous-driven (Nielsen and Wilson, 2012) initiative, the Territorial Museum Tekoa Jopo'i is part of a broad movement in the Global South to value the heterogeneity of traditional cultures and their relations with power structures that invisibilize them, as in Indian (Bandyopadhyay and Yuwanond, 2018), Panamanian, Mexican and Chilean (Bazan and Cabrera, 2019) cases.

Without hiding social ills, the Territorial Museum Tekoa Jopo'i and the Queixadas Agency weave a path of meanings that value different struggles, be they against colonialism, against torture, or the absence or abandonment of cultural equipment in the territory. The network of solidarity among residents and movements has organized itself to strengthen its bonds. The conjugation of multiple voices, even when telling the saddest and most visceral stories, creates an image of potency for the territory of Perus, as a territory that boils in constant agitation. This sequence of subjects narrating subjections occurs without losing the idiosyncrasies of each social group. It opens space for us to imagine what can be learned from examples of so many struggles. From 'Jaraguá is Guarani' to 'Reappropriation and Ressignification of Public Spaces', the tracks show broad processes of exclusion and resistance from local memories.

Receiving people from different locations as tourists made the residents of the Perus neighborhood resignify their relationship with the territory, not having a look that only highlights the problems, but an entrepreneurial look that transforms, prepares, and is proud of its peripheral condition. During and after the visitor hosting experiences many residents undertook significant changes. For example, extra rooms were built in the houses to attract more tourists, residents started to learn English, young people became interested in the tourist guide profession, and new ventures related to tourism in the region emerged. The headquarters of the Queixadas Agency was built and in it, courses such as English, Spanish, Guarani, software development, game development, applications, cultural production, Quebrada Tourism, and gastronomy were offered.

> Our tourism is the 'Tourism of Resistance' because we understand the complexities of living in an environment that can sometimes seem degrading, but that in the midst of all this, we can see the beauty of its stories and pass to tourists the best of our neighborhood. The Queixadas Agency has already received tourists from several places in the world, such as Brazil, Canada, Holland, South Africa, Cape Verde, Mozambique, United States, Hungary, Australia, Turkey, China, Korea, Romania, Mexico, Portugal, Spain, New Zealand, Peru, Bolivia, Colombia, Argentina, Uruguay, Japan, Venezuela, Chile, Italy, France, thus showing the importance of this innovative enterprise. (Camila Cardoso, interview, 2021)

We believe that the case of the Territorial Museum Tekoa Jopo'i indicates a path of investigation yet to be explored. We need to know more about examples of social movements that manage to collect a great diversity of subterranean memories and still create a consistent and coherent tourist route. Thus, we must ask ourselves what narratives make it possible to unite the heterogeneity of culture and the multiple forms of popular resistance in a single project.

References

Bandyopadhyay, R. and Yuwanond, P. (2018) Representation, resistance, and cultural hybridity of the Naga indigenous people in India. *Tourism Management Perspectives* 26, 164–171.

Bazan, C. and Cabrera, F. (2019) Turismo, pueblos indígenas y patrimonio cultural en México y Chile. *PASOS Revista De Turismo Y Patrimonio Cultural* 17(1), 53–64.

Benjamin, W. (2020) *Sobre o conceito de história*. São Paulo, Alameda.

Brandão, C.N., Barbieri, J.C., Junior, E. and João, C. (2014) Análisis de la producción científica internacional sobre el turismo indígena de 1990 a 2012: estudio bibliométrico y propuesta de una agenda de investigación. *PASOS Revista De Turismo Y Patrimonio Cultural* 12(4), 673–684.

Crespo, C. (2017) Contornos de lo decible, exhibible y pensable. Los pueblos originarios en las políticas turístico-culturales en el noroeste de Chubut (Patagonia, Argentina). *PASOS Revista de Turismo y Patrimonio Cultural* 15(3), 533–544.

D'Andrea, T.P. (2013) *A Formação dos Sujeitos Periféricos: Cultura e Política na Periferia de São Paulo*. 309 f. Tese (Doutorado). Programa de Pós-Graduação em Sociologia, Faculdade de Filosofia, Letras e Ciências Humanas da Universidade de São Paulo, São Paulo.

Instituto Pólis (2020) Quem são as pessoas mais afetadas pela pandemia? Available at: https://polis.org.br/noticias/quem-sao-as-pessoas-mais-afetadas-pela-pandemia/ (accessed 19 February 2021).

ISA (Instituto Socioambiental) (2021) *Terra Indígena Jaraguá*. Terra indígenas no Brasil, ISA.

Kowarick, L. (1979) *A Espoliação Urbana*. Rio de Janeiro, Paz & Terra.

Lueg, K., Bager, A. and Lundholt, M. (2021) What counter-narratives are: Dimensions and levels of a theory of middle range. In: Lueg, K. and Lundholt, M. (eds) *Routledge Handbook of Counter-Narratives*. Routledge, New York, pp. 1–14.

Mignolo, W. (2017) Colonialidade: O lado mais escuro da modernidade. *Revista Brasileira de Ciências Sociais* 32(94), 1–18.

Nielsen, N. and Wilson, E. (2012) From invisible to indigenous-driven: A critical typology of research in indigenous tourism. *Journal of Hospitality and Tourism Management* 19(1), 67–75.

Noy, C. (2012) Narratives and counter-narratives: Contesting a tourist site in Jerusalem. In: Tivers, J. and Rakic, T. (eds) *Narratives of Travel and Tourism*. Ashgate Publishing, Burlington, pp. 135–150.

Oliveira, D. (2021) Insurgências culturais e políticas e a emergência do intelectual periférico. In: Oliveira, D. (ed.) (coord.) *Periferias Insurgentes: Ações Culturais de Jovens Na Periferia de São Paulo*. Instituto de Estudos Avançados (IEA-USP), São Paulo, pp. 13–53.

Peria, P.V.G. (2022) *Museu-Território, Turismo-Resistência: política pública, ação pública e narrativas sobre o esquecimento e a memória*. Dissertação (Mestrado) – Programa de Pós-Graduação em Administração Pública e Governo, FGV EAESP. São Paulo. Disponível em. Available at: https://bibliotecadigital.fgv.br/dspace/handle/10438/31616 (accessed 12 February 2022).

Pollak, M. (1989) Memória, esquecimento, silêncio. Rio de Janeiro. *Estudos Históricos* 2(3), 3–15.

Raimundo, S.L. (2017) *Território, Cultura e Política: Movimento Cultural das Periferias, Resistência e Cidade Desejada*. 274 f. Tese (Doutorado) – Programa de Pós-Graduação em Geografia Humana, Faculdade de Filosofia, Letras e Ciências Humanas da Universidade de São Paulo, São Paulo.

Rede Nossa São Paulo (2017) *Mapa da Desigualdade 2017*. São Paulo.

Silvestre, R.P. and Fontana, R.de.F.. (2023) Turismo indígena en Brasil: una revisión bibliográfica de investigaciones publicadas en el período 1999–2021. *PASOS Revista De Turismo Y Patrimonio Cultural* 21(3), 487–501.

Souza, C.F. de (2020) *Sevirologia*: A Arte de Sobreviver e Construir um Território Educador. Trabalho de Conclusão de Curso apresentado como requisito para obtenção do título de Gestor Cultural, pelo Curso de Gestão Cultural do Centro de Pesquisa e Formação-SESC, São Paulo.

Tribe, J. (2007) Critical tourism: rules and resistance. In: Ateljevic, I., Morgan, N. and Pritchard, A. (eds) *The Critical Turn in Tourism Studies: Innovative Research Methods*. Elsevier, Amsterdam, pp. 29–41.

3 Tourism as a 'Tecnologia Social': The Case of the Ururay Group

João A. de Freitas[1] and Maurício Dias Duarte[2]

[1]Social Memory Graduate Program, Federal University of the State of Rio de Janeiro, Rio de Janeiro, Brasil; [2]Ururay Group, São Paulo, Brasil

Summary

The notion of social technology introduces a novel paradigm for technological advancement that emphasizes inclusivity, participation, and democracy. It is rooted in the idea of local citizens as empowered agents. Within this framework, we spotlight the initiatives of Grupo Ururay, a collective based in the East Zone of São Paulo. Through cultural, touristic, and educational endeavors, they have illuminated the pivotal role played by Black and Indigenous communities in shaping the city's history. The crux of our argument lies in portraying the collective's utilization of tourism as a social technology, effectively employing it as a catalyst for societal transformation. To substantiate this claim, we will delineate some of the group's initiatives and delve into the significance of revitalizing and celebrating counter-hegemonic histories within São Paulo's urban landscape.

The Value and Interest of the Case Study

The Ururay Group is a collective dedicated to activating the memory of a peripheral area in the city of São Paulo through various cultural activities. One of their initiatives involves conducting neighborhood tours that narrate the stories of the territory, aiming to shed light on the often overlooked contributions of the Black and Indigenous populations in shaping Latin America's largest city. The name 'Ururay' is derived from a region traversed by the Tietê River.

This case study explores the work of the Ururay Group and two other collectives in São Paulo that challenge the traditional notion of tourism. One of the authors is an activist within the group, while the other provides an external perspective on the actions of these collectives. Some members of the Ururay Group are researchers and encourage and become enthusiastic when other researchers engage in investigations around the collective's actions, which was crucial for this case study. The study examines how these initiatives provoke a reevaluation of the concept of tourism, emphasizing the importance of social engagement and critical reflection in understanding and experiencing a city.

São Paulo and Its History in Contrast

São Paulo, with a population exceeding 12 million, possesses a history intricately interwoven with its geographical features – rivers, hills, and plains (Fig. 3.1). However, the city's narrative is also heavily influenced by a foundational

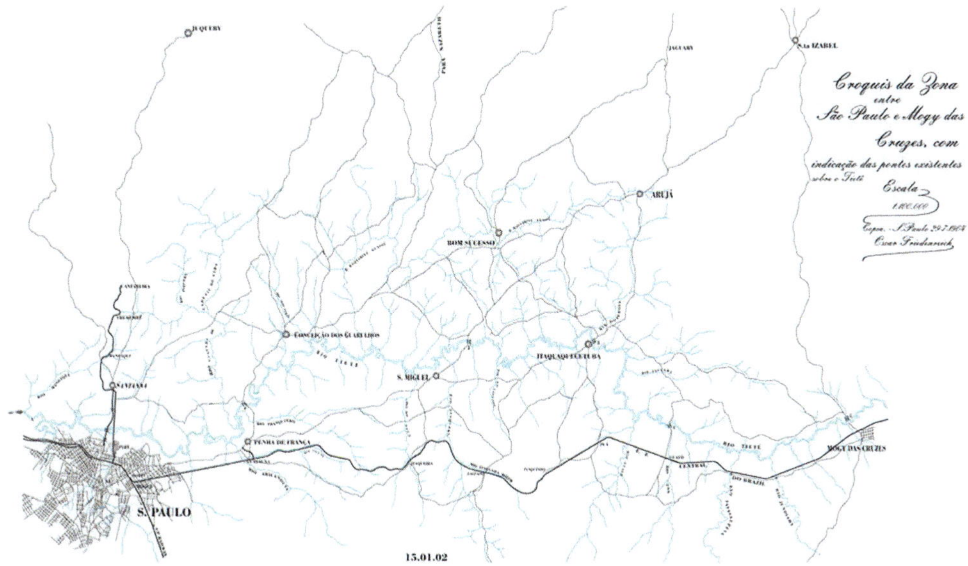

Fig. 3.1. Sketch of the area between São Paulo and Mogi das Cruzes, indicating the existing bridges over the Tietê River, copy: 09.07.1904 – The East region before population density increased. Source: São Paulo State Archive. Vectorization: Ururay Group. Courtesy of: Ururay Group.

colonial myth that romanticizes the genocide of Indigenous and Black populations under the guise of a European voyage of discovery. This myth distorts the historical events, portraying them as an adventurous exploration while concealing the tragic loss of lives, cultures, and societies (Monteiro, 2022). This narrative complexity underscores the need for a more accurate and inclusive retelling of São Paulo's past, one that acknowledges the harsh realities while fostering a deeper understanding of the city's development and its impact on various communities.

In the context of São Paulo, particularly concerning its Black communities, a series of significant events underscore the challenges faced. The demolition or expropriation of numerous Catholic religious sites with strong ties to the city's Black heritage is noteworthy. For instance, the historic Church of Rosário dos Homens Pretos, situated in the heart of São Paulo, fell victim to demolition during a period of urban expansion. Similarly, the former Cemetery of Aflitos faced expropriation to pave the way for a new avenue.

The struggles for the rightful occupation of the Brotherhood of São Benedito and the church of São Francisco have also shaped the narrative. These instances reveal the intricate battles for the preservation of cultural landmarks deeply rooted in Black history (Santos, 2021). Additionally, the demolition and subsequent displacement of the Church of Nossa Senhora dos Remédios, a space that provided solace and served as a hub for abolitionist activism, underscore the ongoing challenges faced by Black communities.

This historical process has led to the invisibility of the memory and contributions of the Black population in the city's development. Meanwhile, the material remnants associated with European populations continue to be preserved and celebrated. Numerous houses, factories, churches, and other buildings that adhere to the architectural canons of Spanish, Portuguese, and Italian traditions have been recognized and incorporated into public conservation policies and academic research throughout the city's history.

Within the Brazilian context, the diversity of situations regarding historical and cultural heritage and territories is always accompanied by a range of dilemmas, conflicts, and issues pertaining to politics, law, and social order. In the outskirts of cities, these challenges tend to be more pronounced. These areas are often further removed from the attention of expert communities such as universities, public authorities, and community mobilization efforts. Consequently, they are more vulnerable to neglect, the dominance of private and commercial interests, and the deterioration of public spaces.

Nevertheless, a burgeoning movement of community-based initiatives is emerging, aiming to reclaim the memory and heritage of these territories. These initiatives often prioritize social justice and acknowledge the diversity of cultural expressions. They encompass a range of activities, from preserving traditional practices and knowledge to fostering new forms of cultural production and artistic expression that challenge prevailing narratives and power dynamics.

The heritage, memory, preservation of environmental areas, and history of peripheral regions are far from being recognized as natural processes. Instead, they are overshadowed by a constructed notion of the city and its memory, predominantly centered around certain regions, particularly in the central and western areas. This narrative amplifies the stories of specific social groups, granting them a privileged role in shaping São Paulo's identity and establishing designated spaces for the celebration of culture, the city, and civic engagement.

Rolnik (2017) underscores these factors as fundamental to understanding the city's social geography, a phenomenon that gained momentum from the late 19th century onwards. The transformation she describes in the coffee-centric urban landscape undoubtedly involved the crystallization of more distinct spatial divisions – defined and separate regions catering to different activities and societal segments. This evolution unfolded through the establishment of working-class neighborhoods and upscale residential zones, the appropriation and renewal of the city's core by emerging elites, and the implementation of discriminatory practices via public investments and urban regulations, contributing to the shaping and persistence of these disparities.

Instances of this can be discerned in São Paulo's Eastern Zone, where already-established districts like Penha, São Miguel, and Itaquera were eventually be encompassed by areas impacted by new social housing policies. Although housing was provided by the state, these regions often lacked crucial infrastructural components such as transportation networks, sanitation facilities, schools, and healthcare centers. Unfortunately, this perpetuates stigmas associated with peripheral regions, wrongly portraying them as mere bedroom communities lacking historical significance and opportunities for meaningful social interactions.

On the contrary, it is in these peripheral regions that we can observe several positive examples of space activation and the fostering of a sense of belonging, knowledge, and community engagement through a diverse range of initiatives. Over the past 20 years, more than 100 groups have been instrumental in activating these spaces and promoting a variety of activities, such as cultural preservation efforts, the celebration and promotion of Black and Indigenous cultures, support for traditional popular cultural practices, hip-hop and samba communities, research endeavors, and other forms of strengthening local identities and tourism.

These initiatives have played a significant role in revitalizing these areas and bringing attention to the richness and diversity they possess. They have contributed to the empowerment of local communities, the preservation of cultural heritage, and the promotion of sustainable tourism practices. By embracing and celebrating the multifaceted identities and expressions found within these peripheral regions, it is possible to challenge the prevailing narratives that overlook their significance and instead recognize their vital contributions to the cultural fabric and vibrancy of São Paulo.

One of the milestones for the Ururay Group was the year 2013 when the Penha Cultural Movement, in collaboration with the Black Men's Rosary Community of France, organized the first *Seminar on the Historic Heritage of the Eastern Zone*. This event was followed by two additional seminars in 2014, bringing together residents and individuals interested in the preservation of heritage in the neighborhoods of the East Zone.

Tecnologia Social: A Latin American concept

The concept of *tecnologia social* focuses on promoting social change and aims to address social inequalities by redefining the relationships between social groups, tools, and methods used in everyday life, particularly in production and consumption. It presents an alternative to dominant perspectives on social innovation and incorporates decolonial lens by challenging notions of centrality and periphery (Souza and Pozzebon, 2020). *Tecnologia social* aims to create opportunities for political processes that question the traditional path of Western-centric 'technological development', which has often led to human impoverishment and economic dependency (Thomas, 2009). By promoting social change through the use of *tecnologias sociais*, it seeks to empower marginalized communities and foster a more equitable society.

One example of *tecnologia social* in practice is community-based tourism, which has emerged as a powerful tool for promoting sustainable development and social inclusion in many parts of the world (Pinheiro, 2016). Unlike traditional forms of tourism, which often rely on large-scale infrastructure and commercialization, community-based tourism seeks to involve local communities in the planning, development, and management of tourism activities. This approach helps to preserve not only local cultures and environments but also provides economic opportunities for marginalized groups, such as women, Indigenous peoples, rural communities, and Black populations. Additionally, community-based tourism can foster cross-cultural exchange and promote understanding between different groups, helping to build a more cohesive and peaceful society. Overall, *tecnologia social* offers a promising path toward creating a more just and sustainable society.

The main argument here is that this working methodology developed by the Ururay Group would be a *tecnologia social*. According to Dagnino (2014), the concept of *tecnologia social* proposes the construction of a new paradigm of technological development that is more inclusive, participatory, and democratic, based on the vision of local citizens as emancipated agents who make their own decisions about the technologies they use. In addition, as Pozzebon and Fontenelle (2018) encourage, we choose to use the term without translating it into English in order to uphold its status as a term coined by Latin American researchers and practitioners to describe an actual, powerful, and relevant intellectual social movement that merits more recognition.

Tourism as a Cultural Action – Ururay Premisses

Since its official establishment in 2014, the Ururay Group has been actively organizing 'walking tours' as a means to foster discussions about the history and heritage of the local region (Fig. 3.2). These tours serve as a platform for cultural initiatives and aim to introduce the concept of territory as a contested space. By engaging in these activities, the collective seeks to create opportunities for cultural actions and encourage a deeper understanding of the social and historical significance of the area.

The use of territory, rather than the territory itself, is what becomes the subject of social analysis. It is a dynamic and ever-evolving concept, constantly shaped by historical context. As Santos (2008) argues, the notion of territory is impure and hybrid, requiring continuous revision and reinterpretation. In this context, both the territory and its heritage demand ongoing reflection, always considering their current uses and relevance.

Material heritage serves as a catalyst for contemplating the present and understanding the historical formation of neighborhoods, as well as the role of public memory policies. Through the examination of material remnants within the territory, broader discussions emerge, encompassing topics such as the 'right to the city', community engagement, identity, and local development. These topics are consistently addressed in workshops organized by the collective for researchers, professors, and anyone else interested in delving deeper into the history of the region. These workshops provide a framework for understanding the fundamental role of territorial references. With the support of 13 listed heritage sites in the Eastern Zone, these discussions contribute to forging connections,

Fig. 3.2. Visitors from the city of Ribeirão Preto, in the interior of São Paulo, are welcomed by the Ururay Group and the Community of the Rosary of Black Men of Penha in 2023. Courtesy of: Ururay Group.

recognizing individuals within the context of history, and envisioning the future.

Cultural action is a process that lacks a defined endpoint or predetermined stages, as it does not aim to reach a specific final destination (Coelho Netto, 1989). This understanding has shaped the methodology of the Ururay Group, which continuously evolves to address the pressing demands of the cultural heritage field (Almeida, 2016).

One effective approach involves fostering collaboration between community members and heritage professionals. This collaborative effort empowers local stakeholders to assume responsibility for their cultural heritage while providing them with the necessary resources and support for preservation and promotion. Furthermore, there is a growing recognition of the significance of incorporating diverse perspectives and voices in the interpretation of cultural heritage, particularly those that have been historically marginalized or excluded. This necessitates ongoing dialogue and collaboration with underrepresented communities, challenging dominant narratives and power structures.

Preserving and promoting cultural heritage is an ongoing and dynamic process that requires flexibility, adaptability, and a profound respect for diverse histories and identities. It is a collective endeavor that engages community members, heritage professionals, and various stakeholders to ensure the vitality and inclusivity of cultural heritage practices.

For this purpose, the professionals in the field of cultural production act as valuable allies, continuously exploring improved strategies to stimulate both the social and economic dimensions of the territory, while actively engaging non-academic audiences. They identify the local reality and the diverse interests of the audience to develop awareness-raising actions. These actions may cater to various interests, such as crafts, religious themes, local development, or community-based tourism, among

Fig. 3.3. Tombamento Festival event, which featured a procession and a discussion on traditional cultures such as moçambique, folia de reis, jongo, samba de bumbo, and cordão carnavalesco, as well as providing space for new migrant groups, including the Afro-Bolivian Saya, 2019. Courtesy of Ururay Group.

others. Therefore, it should be emphasized that tourism is just one of several possible actions. The user's input underscores the importance of continuously reflecting on the territory and its heritage, particularly its material aspects, as triggers for contemplating the present and the historical constitution of neighborhoods and public memory policies. Fig. 3.3 is an example of a tour that is currently being developed in the territory.

Ururay's Tours

Architectures that teach

Do buildings teach us? Do the structures we navigate and inhabit shape us, influencing our thoughts and emotions? Is there coherence between architectural designs and current educational projects? How can we challenge and subvert the established norms within school spaces on a daily basis?

In this context, the itinerary is centered around the study of three schools built during different periods: 1910, 1950, and 1970. These schools represent distinct educational models, including early school prototypes, a modernist approach, and a military-influenced environment. The aim is to initiate discussions about the relationship between schools and communities, the evolution of pedagogical projects and practices over time, and the alignment between planned educational methods and their actual implementation within school environments.

This tour offers insights into how the subversion of entrenched norms within school spaces can manifest through various practices and actions. This includes adapting and reconfiguring physical spaces to better accommodate the needs and interests of students, introducing unconventional learning initiatives, empowering student participation and decision making,

Fig. 3.4. Visit to Padre Antão State School with a group of teachers and researchers from the city of São Paulo, 2018. Courtesy of Ururay Group.

fostering diversity and inclusion, and embracing innovative pedagogical approaches that transcend traditional structures. Subversion also involves questioning and challenging established norms, seeking new teaching and learning methods that align with the aspirations and requirements of students and the school community.

This tour was conducted in collaboration with three public schools in the neighborhood (Fig. 3.4). The participation of principals, teachers, and former students was coordinated to provide insights and perspectives. Additionally, the tour received support from a public Cultural Center and the Penha de França Cultural Memorial, both situated in close proximity to the tour location.

Historical Outeiro of Penha de França

The neighborhood of Penha has a rich history intertwined with its strong religious presence, particularly within popular Catholicism.

Moving away from clerical limitations, Brazilian popular Catholicism encompasses a broad range of faith experiences. These range from the exuberant celebrations and street processions to the irreverent practices of religious brotherhoods – with a significant influence from the Black population – and, in a lasting and remarkable way, the beliefs of private life (Abreu, 1994). This experience of religious reworking is significant in the formation of Brazilian social identity and has previously been described as rustic Catholicism (Queiroz, 1968). Brazilian popular Catholicism is characterized by being on the margins of institutional limits, historically conditioned by local customs and ways of life. The Catholic practice in Brazil has developed a popular dimension by submitting to the demands of the people rather than the Church, by translating and converting faith into the believer's world, often diverging from the scripture itself (cf. de Sousa, 2021).

Three significant churches serve as landmarks in the area: (a) the Eucharistic Sanctuary of Penha, (b) the Penha's Basilica, and (c) the Church of the Rosary of the Black Men of the Penha de França. Through this tour, the tourists can delve into not only the historical development of the neighborhood but also explore various themes such as the preservation of collective memory and the recognition of enslaved individuals and local elites as influential figures in the district's formation. We also examine the dynamics of Black and Indigenous presence, and the role of popular Catholicism in the region (cf. Almeida, 2019). Additionally, the tour engages with the contemporary urbanization movements and their effects on the neighborhood. These discussions encompass a wide range of aspects, shedding light on the cultural and social fabric of Penha.

The district of Penha has a rich and complex history deeply intertwined with the legacy of slavery and the influence of the local elite. Despite these challenges, popular Catholicism remains a vital and resilient aspect of the region's cultural identity. In recent years, the neighborhood has also witnessed a surge in urban occupation movements as residents strive to assert their rights and improve their living conditions.

The primary objective of the initiative is to recognize and celebrate the valuable work carried out by the Community of the Rosary over the past two decades. This community is dedicated to preserving and honoring the collective memory of Penha, emphasizing the territory's history as a site of resistance for over 200 years, particularly for Black and Indigenous people in the city of São Paulo. By highlighting these narratives of resistance and resilience, the initiative seeks to reclaim and reframe the historical significance of the neighborhood, promoting a deeper understanding and appreciation of its cultural heritage.

The research, of which this case study is a part, draws support from photographs sourced from the Penha de França Memorial, heritage preservation procedures of two schools endorsed by heritage institutions, and literature authored by historians specialized in educational history (Fig. 3.5). In their work, Duarte and Goulart (2018) delve into the specifics of the Ururay Group's undertaking, highlighting how investigating and coexisting with diverse architectures within the city present an opportunity for reflecting upon the city's history, public policies, heritage conservation policies, and the manner in which spaces are shaped and reshaped based on their local realities.

The standard audience of tours and the economic sustainability of these practices

Unlike what is commonly seen in traditional tourist circuits, these tours developed by the

Fig. 3.5. Tour through the streets of Penha de França with students from Padre Antão State School during a 3-day immersive experience as part of a training initiative, 2019. Courtesy of Ururay Group.

group do not intend to be the main economic activity sustaining the collective and its members. Instead, they are carried out in specific situations or events, constantly undergoing revisions or adaptations depending on the audience, age group, and visit objectives. However, each year the collective has organized between 10 and 20 tours in the territory.

One of the ways to enable these itineraries is through fundraising initiatives using municipal cultural incentive laws. Once secured, the collective collaborates with educational institutions and other partners to organize training processes. These incentives are crucial in providing free tours to schools in the region, which strengthens not only partnerships with educational institutions but also brings young audiences, teachers, and school communities closer to visiting and connecting with local stories. Typically, these activities are preceded by awareness-raising efforts, including one or two meetings before the tour.

Another aspect involves the demand from universities, technical schools, researchers, and cultural agents who are interested in getting involved in the actions carried out by the Comunidade do Rosário, the Memorial Penha de França, and other partners in the region. This presents an opportunity to strengthen the internal network, attract scholars of the subject, and stimulate local businesses such as restaurants, bakeries, and craftsmen. In both cases, the target audiences typically consist of residents, educational institutions from nearby neighborhoods, or individuals engaged in work within the region (Ferreira and Pacheco, 2022).

There is also occasional demand from external audiences, often associated with artistic residencies or services provided by the third sector, such as educational and training entities, as well as the SESC (Social Service of Commerce). In these cases, contracts may be established to cover the costs of the activities. These funds are then used to compensate the members of the group. Partner institutions also play a crucial role by providing spaces and supporting the implementation of these actions.

It is worth noting that tours targeting spontaneous audiences attract a diverse range of people, including individuals of different ages and economic backgrounds (Fig. 3.6). These add to the richness and inclusivity of the experiences.

The itinerary of the tours usually includes the screening of videos produced by the Cultural Movement Penha to the Community of Rosário, as well as studies and materials. Among these, the book *Territórios de Ururay* (2016), the Settlement Book of the Brotherhood from 1755 to 1880, from the Diocesan Archive of the Diocese of São Miguel Paulista, transcribed and republished in 2019, highlights the heritage preservation processes.

What Kind of Tourism Is This?

The actions carried out by the Ururay group primarily attract residents of the city of São Paulo. According to the conventional definition provided by the UNWTO (United Nations World Tourism Organization), these individuals would not be considered tourists as they do not leave their city of residence or stay at their destination for more than 24 hours. However, it is important to avoid solely adopting a positivist perspective when analyzing this phenomenon.

Urry and Larsen (2011) propose an alternative approach to understand tourism, focusing on the fundamental distinction between the ordinary and the extraordinary. Tourist activities often involve elements or aspects that provide pleasurable experiences, which deviate from everyday norms and routines. In this context, the tours organized by the Ururay group can be seen as offering extraordinary experiences to the participants, despite them being local residents. Richards (2017) argues that tourist strategies, like taking photos or admiring the landscape from a privileged viewpoint, can be utilized anywhere. Looking at Barcelona, he contends that the rise of tourism in the developed world has led to the adaptation of tourism practices from abroad to one's own domestic context.

Spatial segregation in São Paulo, coupled with staggering socio-economic inequality, creates invisible barriers (cf. Villaça, 2011). The working class crosses these barriers daily toward their workplaces, yet the middle and upper classes rarely make this reverse journey. For many of these individuals, it is easier to become familiar with a place on the other side

Fig. 3.6. Dialogue with the audience of the Social Service of Commerce (SESC), 2023. Courtesy of Ururay Group.

of the world than on the other side of the city. The tourism framed in this case study takes place when these borders are crossed, and people are willing to cast a different gaze upon their own city. By exploring the historical and cultural heritage of their own city, residents engage in activities that are distinct from their routine experiences. The tours provide an opportunity to discover hidden stories, explore different neighborhoods, and gain new perspectives, which can be regarded as extraordinary within the context of their everyday lives. Therefore, while the traditional definition of tourism may not strictly apply to these local participants, the concept of experiencing the extraordinary through these tours remains valid and relevant.

Indeed, in the case of tours organized by the Ururay Group, the experience can be equally extraordinary for both local residents and visitors from other states or regions. Regardless of familiarity with the physical space, a shift in perspective can provide an experience that resembles what is commonly associated with tourism. Even for residents of central São Paulo, for example, exploring different neighborhoods and uncovering alternative narratives can offer a fresh and extraordinary experience (Pais, 2009).

While these tours may not align with conventional forms of tourism, the participants are not conventional tourists either. People who engage in these tours are motivated by a desire to learn and discover the counter-hegemonic history presented by the Ururay Group. In the current scenario of political polarization that has intensified in Brazil, even leisure and enjoyment activities like these tours carry significant political meanings.

The activities organized by the Ururay group can be understood as a form of local tourism, focusing on exploring the city of São Paulo in a unique and alternative way. This type of tourism, often referred to as 'staycation' or 'hometown tourism', has been growing in popularity as people recognize the value of discovering the hidden treasures and cultural richness of their own cities.

Furthermore, local tourism promotes sustainable practices by reducing carbon footprints associated with long-distance travel and supporting local businesses and communities. It encourages responsible exploration and a more sustainable approach to tourism by valuing the cultural heritage, environment, and social dynamics of the local area.

By participating in these tours, individuals actively engage in a form of cultural and political resistance, challenging dominant narratives and exploring alternative perspectives. These experiences transcend mere entertainment and become opportunities for critical reflection, dialogue, and the promotion of inclusive and transformative practices.

Conclusions

The work of the Ururay Group encompasses various dimensions, including cultural, educational, touristic, and political activities. One of the key aspects that makes their work relevant is their mobilization of heritage and history in order to challenge established historical narratives and give visibility to the contributions of Black people in the East Zone. This multifaceted approach contributes to a broader social transformation by addressing historical injustices and amplifying marginalized voices.

The recognition of the Ururay Group's working methodology as a *tecnologia social* acknowledges not only its transformative potential but also emphasizes that it can be replicated by other groups. By examining the history of their own ancestors and assessing the fairness of their representation, other communities can engage in similar processes of reclaiming their heritage and empowering their narratives.

It is important to note that the classification of these actions as touristic practices is not solely based on their ability to attract tourists from other cities or generate revenue. Rather, it is grounded in their capacity for social transformation by bringing attention to overlooked heritage, stories, and people who may otherwise remain invisible. While sporadic visits from those traditionally labeled as tourists may occur, the primary focus is on the social impact and the promotion of a more inclusive and just representation of history.

It is imperative to underscore that the impacts of such initiatives are challenging to quantify, particularly within a relatively short timeframe. However, the paramount objective remains the pursuit of equitable avenues for representing the historical contributions of the Black and Indigenous communities in shaping São Paulo.

By shedding light on marginalized histories and perspectives, the Ururay Group's work contributes to a broader societal understanding and appreciation of cultural diversity and social justice. It serves as an inspiration for other groups and communities to engage in similar processes of reclaiming their own narratives and challenging dominant historical narratives.

References

Abreu, M. (1994) Festas religiosas no Rio de janeiro: perspectivas de controle e tolerância no século XIX. *Estudos Históricos* 7(14), 183–203.

Almeida, P. (2016) *Territórios de Ururay*. Movimento Cultural Penha, São Paulo, Brazil.

Almeida, P. (2019) *Contribuições para a história do protagonismo de negros e índios na Irmandade de Nossa Senhora do Rosário dos Pretos da Penha de França*. Movimento Cultural Penha, São Paulo, Brazil.

Coelho Netto, J.T. (1989) *O que E AçãO Cultural*, 1ª edição. Brasiliense, São Paulo, Brazil.

Dagnino, R. (2014) *Tecnologia Social: contribuições conceituais e metodológicas*. EDUEPB. Available at: https://books.scielo.org/id/7hbdt (accessed 31 December 2021).

de Sousa, E.J.F. (2021) O catolicismo popular brasileiro: notas em torno da sua invenção historiográfica. *Temporalidades* 13(2), 724–745.

Duarte, M.D. and Goulart, M. (2018) Arquiteturas que ensinam. *Revista E – SESC* (264). Available at: https://portal.sescsp.org.br/online/artigo/12107_ARQUITETURA+ESCOLAR (accessed 24 August 2023).

Ferreira, P.T.A. and Pacheco, R.T.B. (2022) Turismo, movimentos sociais e patrimônio cultural na cidade de São paulo/SP. *Turismo: Visão e Ação* 24, 340–359.

Monteiro, J.M. (2022) *Negros da terra: Índios e bandeirantes nas origens de São Paulo*. 2ª edição. Companhia das Letras.

Pais, J.M. (2009) Um Dia Sou Turista na Minha Própria Cidade. *CIDADES, Comunidades e Territórios* [Preprint] (18). Available at: https://revistas.rcaap.pt/cct/article/view/9287 (accessed 25 May 2023).

Pinheiro, R.C. dos S. (2016) Turismo de base comunitária como inovação social na gestão do turismo em espaços rurais. *Ponta de Lança: Revista Eletrônica de História, Memória & Cultura* 10(19), 146–165.

Pozzebon, M. and Fontenelle, I.A. (2018) Fostering the post-development debate: The Latin American concept of *tecnologia social*. *Third World Quarterly* 39(9), 1750–1769. DOI: 10.1080/01436597.2018.1432351.

Queiroz, M.I. (1968) O catolicismo rústico no Brasil. *RIEB – Revista Do Instituto de Estudos Brasileiros* 5, 104–123.

Richards, G. (2017) Tourists in their own city–considering the growth of a phenomenon. *Tourism Today* 16, 8–16.

Rolnik, R. (2017) *Territórios em conflito, São Paulo: espaço, história e política*. Três Estrelas.

Santos, F. (2021) Lugares do Catolicismo Negro na São Paulo do século XIX. *Estudos Avançados* 35(102), 279–294. DOI: 10.1590/s0103-4014.2021.35102.017.

Santos, M. (2008) *A Natureza do Espaço: Técnica e Tempo, Razão e Emoção*. 4a edição. Edusp, São Paulo, SP.

Souza, A. and Pozzebon, M. (2020) Práticas e mecanismos de uma tecnologia social: proposição de um modelo a partir de uma experiência no semiárido. *Organizações & Sociedade* 27, 231–254.

Thomas, H. (2009) Tecnologías para la inclusión social y políticas públicas en América Latina. In: Otterloo, A. (ed.) *Tecnologias sociais: Caminhos para a sustentabilidade*. Rede de Tecnologia Social, Brasília, DF, pp. 25–81.

Urry, J. and Larsen, J. (2011) *The tourist gaze 3.0*, 3rd edn. SAGE (Theory, Culture & Society), Los Angeles, London.

Villaça, F. (2011) São paulo: Urban segregation and inequality. *Estudos Avançados* 25, 37–58. DOI: 10.1590/S0103-40142011000100004.

4 Tourism Done by Indigenous Peoples and Traditional Communities in Brazil's First World Heritage Mixed Site

Renato de Oliveira dos Santos[1,2], Teresa Cristina de Miranda Mendonça[2,3], Edilaine Albertino de Moraes[2,4] and Sidnei Raimundo[1]

[1]School of Arts, Sciences and Humanities, University of São Paulo, São Paulo, Brazil; [2]TBC-REDE: Laboratory of Community-Based Tourism, Sustainability, and Networks, Brazil; [3]Department of Administration and Tourism, Federal Rural University of Rio de Janeiro, Rio de Janeiro, Brazil; [4]Department of Tourism at the Institute of Humanities, Federal University of Juiz de Fora, Minas Gerais, Brazil

Summary

Tourism as a sociocultural and economic phenomenon has been configured as an object of public policy actions, investments and research, mainly in countries located in the so-called 'Third World', emerging, developing and peripheral regions (aka the Global South). In the Latin American context, since 1970 tourism has been interpreted as a possible strategy for development and poverty reduction. However, the way in which tourism has been established historically is potentially exclusionary (both from the point of view of those who travel and those who operate activities related to the sector, as well as the receiving communities). In the analysis of popular participation processes in development projects in Latin America, it is possible to identify what is referred to as a 'silent majority', comprised the peripheral population. This population is also represented by indigenous and traditional communities that have consistently been marginalized and excluded from significant political, social, and economic processes. Given this context, we describe the case of the Nhandereko Community-Based Tourism Network, composed of indigenous people and traditional communities (maroons and 'caiçaras'), which had in July 2019, part of their territory (Paraty and the Ilha Grande Bay in Angra dos Reis–Rio de Janeiro) entitled a mixed world heritage site by UNESCO (i.e., both a cultural and natural site, expressing the richness of local diversity through living culture, associated with the way of life and cultural references of traditional communities). This can unveil and teach us, through community organization and tourism, other ways of valuing and safeguarding traditions, socio-economic development, governance, and political and social resistance.

The Value and Interest of the Case Study

What can we learn from the indigenous peoples and traditional communities who occupy and give life to the first World Heritage mixed site recognized in Brazil? What does it mean for tourism, locally called community-based tourism, to have its production and experiences proposed in World Heritage territory? What does it mean to become a world heritage site, when historically, native peoples and traditional communities have been neglected and silenced? How can tourism contribute to the construction of new meanings in this territory? This case study is the result of social research and presents reflections from a critical social perspective, both in practice and theory. It introduces the Nhandereko Community-Based Tourism Network, a collective composed of indigenous peoples, caiçaras, and maroons seeks to integrate into socio-economic practices related to tourism in a region considered one of the main tourism inducers in Brazil.

In this region, the economic, social, and political use of water (seas and rivers) and land, permeated by tourism, has driven the emergence of conflict-ridden relationships caused by serious social and spatial issues. The dispute over physical and symbolic space, and the ongoing struggle of indigenous peoples and traditional communities to remain in their inherited territory involves both physical and symbolic resistance. The potential of this region can inspire us and provide valuable insights and knowledge for tourism, as well as for understanding the relationship between this activity and its socio-environmental and cultural responsibilities in the contemporary world.

Tourism in Global South Countries: Perspectives on Community-Based Tourism

Tourism, as a complex phenomenon encompassing sociocultural and economic aspects, has gained prominence in the realm of public policies, financial investments, and academic research. Especially in nations derogatorily labeled as the 'Third World', emerging regions, developing and peripheral areas of the Global South, tourism has been recognized as a potential development strategy and a means of poverty alleviation, as seen in the case of Latin America. Since the 1970s, Latin America has been perceived as a rising tourism destination (Meléndez, 2000), while also being viewed as a space for experimenting with resistance against hegemonic influences (Lang, 2016). Therefore, it is possible to learn from the experiences of Latin America and other Southern countries 'to the extent that it is conceived as resistance to Northern domination and seeks in it what has not been completely disfigured' (Santos, 2004, p. 18). In this sense, the latent inequalities in the region must be acknowledged, making it an emblematic example to explore issues of global development, including the challenges, complexities, and uncertainties associated with tourism (Moraes, 2019).

Regarding tourism, it is recognized that its historical establishment has the potential to be exclusionary, for travelers, those involved in its activities and host communities, especially indigenous peoples and traditional communities who are recurrently marginalized and excluded from major political and social processes (Mendonça *et al.*, 2020). Often, tourism falls short of the promises propagated by official political-economic discourses and the media about tourism as a catalyst for job creation, income distribution, and local and regional development (Ouriques, 2015). This leads to various impacts and controversies in practice, questioning the hegemonic model of tourism.

Various studies, including Irving and Mendonça (2004), note community participation as an ethical element in the construction of tourism. They consequently highlight tourism as a tool for endogenous and local development. In this context, reflections on the various forms of economic organization among indigenous peoples and traditional communities stand out, established to ensure their socioeconomic, cultural, and political survival. This is achieved through actions that promote social inclusion and the preservation of their natural and cultural heritage, which are essential to their self-identity and cultural distinction across different territories. This perspective guides an approach to tourism known as community-based tourism (CBT).

CBT has gained prominence on a global scale, existing alongside the global expansion of the tourism sector. The unique and diverse CBT initiatives in Latin America reflect a comprehensive conceptual view, often associated with terms such as Community Tourism, Rural Community Tourism, and Peasant Tourism. These initiatives cover coastal, mountainous, and rural areas, encompassing regions of high biodiversity and their historical and cultural ties (Moraes, 2019).

Among the perspectives of CBT in Latin America, one can consider it as a practice based on a promising approach to improving the quality of life for vulnerable groups, indigenous peoples, and traditional communities (Bartholo *et al.*, 2009; Santos, 2020). This is achieved through socio-environmental responsibility (Lima and Coriolano, 2003), community participation (Irving and Mendonça, 2004), involvement in poverty alleviation discourses and policies (Coriolano, 2006), promoting cooperation and equity in labor and resource sharing. It is rooted in sustainable ownership and self-management of community resources. Additionally, there is an emphasis on the human and cultural dimension (Maldonado, 2009), as well as on community ownership, self-management, and community protagonism (Fabrino, 2013).

Furthermore, CBT strengthens the promotion of actions that value traditional practices and modes of organization. It highlights situated practices in diverse and close-range tourism (Zaoual, 2009), promoting social inclusion (Irving, 2009), and diversifying and reinforcing traditional/conventional activities rather than suppressing or replacing them (Cañada, 2010). Additionally, CBT fosters cultural exchange between visitors and local residents, integrating them into daily life and the local way of living (Moraes *et al.*, 2013).

It is also worth noting that CBT is not neutral and is often associated with circumstances of conflicts in and about the territory as well as in policies and allocation of public resources, due to the exploitation of nature as a resource, and imposition and patrimonialization processes. Thus, it also takes place in scenarios of resistance against exploitation (Cañada, 2010), as a movement to fight for staying in the local/territorial origin, transmission of knowledge, and political and territorial visibility (Mendonça *et al.*, 2021). Furthermore, it aims at safeguarding and communicating traditional knowledge and crafts (Santos, 2020), as a defense against real estate speculation, gentrification, and safeguarding of natural knowledge (Martins Fonseca and Rocha, 2021), as a tool for environmental education (Rodrigues *et al.*, 2022), and in the possibility of shared action and management in protected natural areas (Rodrigues and Botelho, 2023), among other perspectives.

In this scenario, CBT takes on a new strategic orientation as part of a broader process of resistance, mobilization, and collaboration. This is due to how these various grassroots expressions act in opposition to various forms of territory and natural resource exploitation. This encompasses everything from biofuels, industrialized agriculture, mining, and mega-infrastructure construction to real estate speculation, illegal land appropriation, overfishing, and large-scale tourism (Moraes, 2019).

However, CBT also faces challenges in terms of the disconnection of commercial strategies from the conventional tourism market in terms of marketing and integration into the conventional market. Both the inaccessibility and the limited management of the internet and social media by the involved communities is a potential problem (Cabanilla and Gentili, 2015).

Nonetheless, it is essential to highlight successful strategies of collaboration. When communities come together in collective actions, they have demonstrated the ability to introduce and sustain alternatives that incorporate distinct values and logics (Mendonça, 2004; Cañada, 2010). In this context, communities engaged in CBT initiatives, along with established partnerships to strengthen them, have played an active role in discussions and promotion of diverse actions in their territories (Moraes, 2019). This dynamic is facilitated through networks, as exemplified by experiences self-identified as CBT networks in Latin America, as identified by Moraes (2019): Indigenous Network of Communities of Alto Napo for Intercultural Coexistence and Ecotourism (RICANCIE) – Ecuador; Cooperative Consortium National Ecotourism Network (COOPRENA R.L) - Costa Rica; Garifuna Community Tourism Network (MUTU) – Honduras; Indígenas del Mapu Lahual (MAPU) – Chile; Indigenous Tourism Network of Mexico (RITA) – Mexico; Brazilian Network of Solidarity Community Tourism (TURISOL) – Brazil; Rural Community-Based Tourism Network (HUELLA GAUCHA) – Argentina; Nicaraguan Network of Rural Community Tourism (RENITURAL) - Nicaragua; Argentine Network of Rural Community Tourism (RATURC) – Argentina; Chiapas Ecotours and Ethnicities Tourism Network (SENDASUR) – Mexico; Bolivian Network of Solidarity and Community Tourism (TUSOCO) – Bolivia; Huataraco Suno Community Tourism Network (RETHUS) – Ecuador; Salta's Calchaquí Valleys Peasant Tourism Network (SALTA) – Argentina; Ceará's Community Tourism Network (TUCUM) – Brazil; Community Tourism Network of Lake Titicaca (APTHAPI) – Bolivia. Most of these initiatives were mobilized by the actions of the

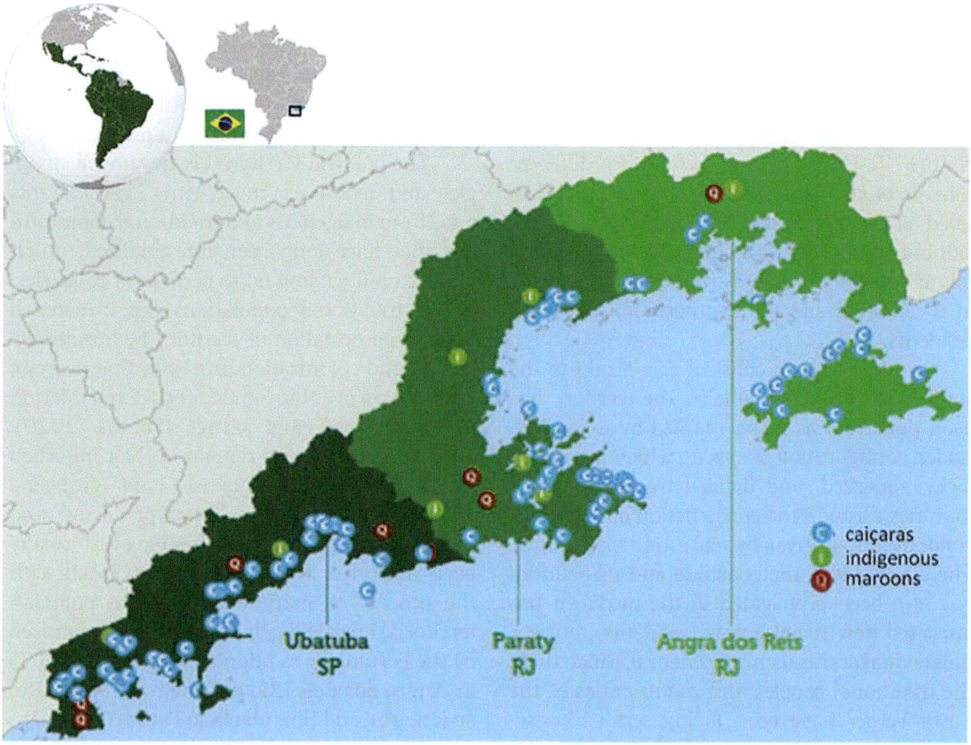

Fig. 4.1. Territory and indication of communities.

Latin American Community Tourism Network (REDTURS), driven by the International Labour Organization (ILO).

The perspectives and scenarios discussed echo and support the significance of the case to be described in this study, aiming to present the inspiring initiative of the Nhandereko Community-Based Tourism Network.

'Preserving Is Resisting' – Tourism in Defense of Traditional Territories

In the region located on the southern coast of Rio de Janeiro state (Angra dos Reis and Paraty) and part of the northern coast of São Paulo state (Ubatuba), territories integrated by historical cultural relationships, indigenous and traditional communities (Guarani indigenous people, maroons, and caiçaras)[1] (Fig. 4.1) have historically been immersed in a context of socio-environmental conflicts. This region is known for its stunning natural landscapes, breathtaking beaches, and lush nature, often characterized by dense vegetation and mountains near the sea.

The region is situated within the Atlantic Forest biome, often referred to as the 'Mata Atlântica' in Portuguese. This unique biome spans the eastern coastline of Brazil and is renowned for its remarkable biological diversity and richness in endemic species. It is characterized by lush tropical forests, diverse ecosystems, and a variety of habitats ranging from coastal areas, including a coastline with numerous beaches and islands, to rural hinterlands and mountains/ranges, harboring an incredible spectrum of flora and fauna. It is one of the most endangered biomes globally, one of the world's five key biodiversity hotspots. Historically, it covered a much larger area, but today it is greatly reduced due to human activities, becoming a crucial focal point for conservation efforts.

Among various issues, these communities have been and continue to be influenced by the arrival of tourism, the pressure of mass tourism, and the creation of Nature Conservation Units (UCs), the equivalent of Protected Areas or Nature Reserves, mostly belong to different spheres of public administration and are classified as integral protection units. With tourism gaining importance as a new economy, new real estate interests have also emerged, resulting in speculation and the expropriation of native residents.

From the north coast of São Paulo to the south coast of Rio de Janeiro, the territory has been portrayed as being marked by a conflict-ridden configuration of spaces, which has driven away, squeezed, and divided traditional communities through the construction of luxurious condominiums driven by real estate speculation. The discovery of the coastline and its beauties has also been propagated in the media, it has attracted new social actors (residents, tourists, holidaymakers, and entrepreneurs). Since then, the traditional peoples and communities of the Costa Verde[2] have been facing serious consequences due to the advance of mass tourism and the growing real estate speculation that continues to occur to this day.

Thus, tourism-related issues and nature conservation processes have brought about new sociocultural and economic dynamics at the local level. Consequently, these groups, namely, indigenous, maroons, and caiçaras, have been resisting to remain in their inherited territories, fighting against threats and violent processes generated and intensified by land grabbers, real estate speculation, the establishment of UCs, and large economic development projects.

The political-territorial issues unveil struggles and resistance movements. These fights arise from indigenous and traditional communities that have long been marginalized, much like the reality common to peripheral countries, where traditional populations are threatened by the appropriation of territory by hegemonic capital and colonization processes that exercise subjugation over the 'other', considered inferior, underdeveloped, and savage (Santos, 2020). These groups resist to remain in their inherited territories, fighting against land grabbers, violent real estate speculation processes, and 'development' projects that they have historically been neglected by. In this context, faced with the process of marginalization and political-territorial invisibility, these groups are organized by the Forum of Traditional Communities (FCT) in Angra dos Reis (Rio de Janeiro), Paraty (Rio de Janeiro), and Ubatuba (São Paulo) (Fig. 4.2).

The articulations of the FCT began in 2006 when community members started gathering for debates and discussions about the interests of the communities and organizing themselves as a social and political resistance in favor of their continued presence in the territory. Thus, the main theme of the FCT's work can be summed up in the phrase: 'Preserving is resisting,

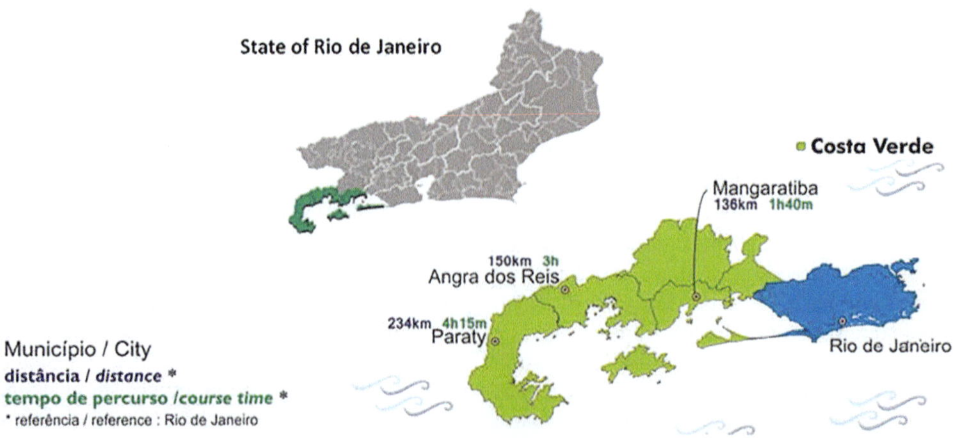

Fig. 4.2. Location of the Costa Verde tourist region, with a focus on Paraty and Angra dos Reis.

Fig. 4.3. 'The Sono caiçara community receives a Community-Based Tourism Sharing from the Nhandereko Network (Paraty)'. Source: Comunicação Popular FCT 18 July 2018.

resisting is conserving, conserving is knowing how to use, knowing how to use is the art of traditional communities'.

Among the strategies developed by FCT to strengthen its actions are activities referred to as 'partilhas' (sharing) (Fig. 4.3). This consists of meetings held among communities with the aim of promoting experiential learning, exchange of experiences, sharing of knowledge, and seeking to articulate, enhance, and replicate existing experiences, as well as drive the definition of action focuses in the territory (Oliveira *et al.*, 2019). The conducted 'partilhas' are linked to social technology incubators and inspired by the 'Campesino a campesino' methodology, which has been applied in rural territories across several countries in Latin America.

Among the actions of the Forum (FCT), in 2014, the campaign 'Preservar é Resistir – Em Defesa dos Territórios Tradicionais' (Preserving is Resisting – In Defense of Traditional Territories) was launched. Its main objective is to ensure residents' access to their original territories, advance the land regularization process, and raise visibility for traditional ways of life. Consequently, it aims to implement adequate public policies based on popular demands. According to the coordinator of the FCT and general coordinator of OTSS (Observatory of Sustainable and Healthy Territories), this organizational process constitutes a social movement, as described below:

> As a social movement, we fight organized for our rights and also bring to light the issues that are contested in this region on a daily basis due to the interests of different groups involved in environmental politics, real estate speculation, and large-scale projects (such as oil and energy exploration, highways, and predatory tourism, among others). All of these factors are transforming the relationship with the territory and affecting the daily lives of our communities, causing significant environmental and social impacts in our region. (Nascimento, 2019, p. 14)

In 2015, prompted by FCT, the I Community-Based Tourism Meeting of the Green Coast was held from 23 July to 25 July in Paraty. This provided a reflection on the territory of conflicts in which:

> 'With the arrival of the BR 101 Rio-Santos, the land became a scenario of increasing real estate speculation. This process intensified in the 1980s with the creation of Conservation Units and the installation of tourist ventures.... the dialogue about the territory must be contextualized within the conflicts and tensions experienced by traditional communities in the region,' as stated by the Quilombola leader of Campinho da Independência (Paraty). (Mendonça, 2015, p. 28)

As a result of a broad movement of articulation and organization, inspired by the phrase 'Without territory, there is no community-based tourism. For the protagonism and permanence of traditional communities in their territory', the FCT launched a community-based tourism map in 2015 (Available at: https://issuu.com/forumdecomunidadestradicionais/docs/folheto_tbc_fct_temporario_2015_liv, accessed 26 May 2023), indicating 15 nuclei/communities (guarani indigenous, maroons, and caiçaras) to be visited and which gave territorial visibility to the peoples and traditional communities of the region.

Through their collaboration with the Oswaldo Cruz Foundation (Fiocruz) and the National Health Foundation (FUNASA), the FCT established the Observatory of Sustainable and Healthy Territories of Bocaina (OTSS) as a shared strategy for management and planning oriented toward sustainable development, integrating regional, state, national, and global scales. They advocate for various causes and actions that permeate local life, including defending the territory, culture, health, ecological sanitation, agroecology, differentiated education, incubation of social technologies, community-based tourism, territorialized evaluation and monitoring of the United Nations' Sustainable Development Goals of the 2030 Agenda, and the establishment of international solidarity networks for cooperation with the Ibero-American Network of Sustainable and Healthy Territories, among others (Gallo and Nascimento, 2019).

One of the outcomes of this collaborative effort was the consolidation of the network for tourism organization, initiated in 2008, with the institutionalization of the already active movement, resulting in the 'creation' of the Nhandereko Community-Based Tourism Network (Fig. 4.4). Nhandereko means 'our way of being' in the Guarani indigenous Language.

It is worth highlighting that a part of the region where the Nhandereko Network is located (Paraty and the Bay of Ilha Grande in Angra dos Reis) has been officially declared as a World Heritage Site. In July 2019, the territory was recognized by the UNESCO World Heritage Committee as a mixed heritage site, encompassing both cultural and natural aspects, showcasing the richness of local diversity through living culture associated with the way of life and cultural references of traditional communities. According to the FCT coordinator and the National Coordination of Caiçara Traditional Communities (CNCTC), the designation brings hope that the communities, who make their culture come alive through their connection with nature, can be recognized and respected by all, as without this relationship between people and nature, the possibility of achieving such an important title would not exist.

According to Santos (2020), the designation came about through a change in the nomination projects, which previously focused solely on the historical and natural aspects of the territory. With the recognition and inclusion of the peoples and traditional communities that occupy and give life to this territory, the nomination was accepted, and it received the title of World Heritage Site, becoming the first mixed, historical, and natural site recognized in Brazil. The title bestows universal value upon the exceptional heritage of the territory and enables new measures of protection and preservation as an international commitment. Another inevitable impact is the international projection in various aspects/themes, including tourism.

The application for designation was carried out in collaboration between the Ministry of the Environment, the National Institute of Historic and Artistic Heritage (IPHAN), the Chico Mendes Institute for Biodiversity Conservation (ICMBio), the municipalities of Paraty and Angra dos Reis, the State Institute of Cultural Heritage (Inepac), the State Institute of the Environment (Inea), the Forum of Traditional Communities, and the Observatory of Sustainable and Healthy Territories of Bocaina (OTSS). The participation of the communities is evident in the text and also in the use of their own images, as can be seen in the following promotional video, recorded with

Fig. 4.4. Nhandereko Network logo representing the cultural elements of each group in the collective: gastronomy represented by the quilombola woman, the caiçara canoe, and indigenous craftsmanship. Instagram @Nhandereko (2023). Image of the logo of the Nhandereko Community-Based Tourism Network. It represents elements that embody the culture of the ethnic groups present in the territory. Source: Instagram: @redenhandereko.

community representatives, showcasing the images and beauties of their practices, knowledge, and territories.

Nascimento (2019, p. 21), the FCT coordinator and general coordinator of OTSS, emphasizes that 'our history as a social movement is always a story of resistance, struggle, critique, and confrontation, in search of a better quality of life and a more favorable process for the achievements of these communities.' Furthermore, in the current political context, all the achievements and progress of recent years are at risk: 'We are experiencing a scenario of losing these achievements, these social advances in the country ... We are confident that our project becomes very important to overcome this moment of crisis, coup, and political regression that the country is experiencing.' Nascimento's statement indicated the political context of the Bolsonaro government. Although we are now under a new left-wing government, there is still a significant contingent of far-right representatives in power who deny and attempt to obstruct social progress in the country.

The flag of CBT is the result of a broad movement of coordination and organization that firmly advocates the idea that CBT is only possible when traditional communities take center stage and remain in their territories. This assertion highlights the fundamental importance of the territory for the practice of community-based tourism. Through this movement, the goal is to ensure that traditional communities can play a central role in the development and management of tourism, valuing their ways of life, culture, and natural resources. By understanding this way of life as the ethical, political, and cultural foundation of solidarity-based modes of production, the FCT has embraced CBT as a productive arrangement capable of promoting sustainable, healthy, and solidarity-based development based on ethnic and cultural diversity. In this perspective, the FCT has been developing strategies for implementing CBT. These strategies value the cultural identity of traditional communities by coordinating their itineraries, products, and services, and organizing the desires of visitors to have a differentiated experience of encountering traditional communities (Oliveira *et al.*, 2019).

Lessons to Learn, Share, and Research

In the territory of Brazil's first natural and cultural World Heritage Site, the presence and contributions of Guarani indigenous people, maroons, and caiçaras, have often been overlooked and marginalized. However, these communities are an integral part of the rich cultural and natural heritage of the region. In recent years, there has been a growing recognition of the importance of their presence and knowledge in shaping the sustainable development and preservation of this unique territory. Through initiatives that promote CBT and emphasize the harmonious relationship between people and nature, these communities are reclaiming their political, economic, and territorial visibility and asserting their cultural significance.

Their deep connection to the land, traditional practices, and ancestral knowledge not only enriches the tourism experience but also serves as a reminder of the intricate relationship between humans and the environment. The inclusion of indigenous, maroons, and caiçara perspectives in tourism activities fosters cultural exchange, mutual learning, and a deeper appreciation for the region's natural and cultural heritage. By valuing and supporting the participation of these communities in tourism initiatives, we can ensure that their voices are heard, their rights are respected, and their contributions are acknowledged. This inclusive approach not only benefits the communities themselves but also enhances the authenticity and sustainability of tourism in the area, making it a model for responsible and respectful tourism practices in the context of cultural and natural World Heritage sites.

The Network Nhandereko of Community-Based Tourism claims visibility and space on the territorial and tourist map of the region as a means of cultural strengthening and an alternative source of income. Mendonça *et al.* (2020) emphasize that CBT represents a strategy of political and social struggle in defense of the inherited locality. CBT, in this case, also allows for the unveiling of a new network tourism map in the state of Rio de Janeiro and Ubatuba (SP), highlighting the memory and heritage that represent the indigenous, caiçara, and quilombola communities of the Bocaina territory, who seek to align themselves with the proposed political-economic project for the region.

Through conversations, local gastronomy (Fig. 4.5), traditions, music, Guarani singing (Fig. 4.6), dance, prayer, craftsmanship (Fig. 4.7), affection, the stories of struggle and resistance that permeate the itineraries or routes, informal conversations, sharing meals, or walking through the community, CBT plays a role in reconstructing past experiences and memories, revealing a new map of memories and identities for these groups. In other words, what is observed in the experiences of Nhandereko itineraries is that they are permeated by the past-present history, where tourism serves as a means of transmission and preservation of memory, safeguarding the inherited locality.

Between individual and collective memories, we find the elderly, the guardians of memory, the wise ones who narrate stories, as well as the younger generation and community leaders who explore various spaces of coordination and dialog in defense of their territory (Fig. 4.8). It is the memories of the past (Fig. 4.9), experiences, and current shared experiences that build networks of relationships with new social actors, the tourists. A past that unfolds into the present, a present that is constantly being reconstructed, a material and immaterial heritage of the country not found in history books but which finds in tourism a means to reveal and disseminate ancestral knowledge, cultural expressions, living heritage, and the everyday practices of life that have long been neglected and silenced.

This case may not fully capture the richness of actions and experiences within the Nhandereko network, but its purpose is to present and disseminate its existence to researchers in various fields of knowledge and our Latin American colleagues. Additionally, we hope to explore new questions and possibilities that may arise from these experiences. Among them, what does it mean for CBT to have its production and experiences proposed in the territory of the World Heritage Site? What does it mean to become a world heritage site, when historically native peoples and traditional communities have

Fig. 4.5. Caiçara from Praia Grande da Cajaíba (Paraty), receiving tourists and presenting cassava and the processes for flour production, December 2018.

Fig. 4.6. Guarani children's choir from Sapukai Village (Angra dos Reis), May 2016.

Fig. 4.7. Guarani indigenous craftsmanship and basketry, Araponga village (Paraty), June 2019.

Fig. 4.8. Roundtable discussion with fishermen from the caiçara community of Trindade (Paraty), July 2016.

Fig. 4.9. Grio, from the Santa Rita do Bracui Quilombo (Angra dos Reis), presenting the exhibition on the past, present, and history of slavery in the territory to university students/visitors, April 2016.

been neglected and silenced? How can tourism contribute to the construction of new meanings in this territory and strengthen 'nhandereko'?

Through the experience described in this case study, we hope that everyone can envision inspiring possibilities for better socio-productive and cultural practices related to tourism, embedded in very specific contexts involving indigenous peoples and traditional communities. In other words, we hope that what is presented and the questions raised can play a role and contribute to professionals and other readers in their specific realities. Looking beyond the locality, the Nhandereko Network is a powerful initiative that can inspire other indigenous peoples, traditional communities, and other vulnerable groups who are fighting for their living territories.

Notes

[1] According to the National Policy for the Development of Traditional Peoples and Communities these peoples and communities possess their own forms of social organization. They occupy and utilize territories and natural resources as conditions for their cultural, social, religious, ancestral, and economic reproduction, drawing upon knowledge, innovations, and practices generated and transmitted through tradition (Brazil, 2007).

[2] 'Costa Verde' is an expression used to refer to the coastal region of the state of Rio de Janeiro, Brazil. The term highlights the combination of lush green areas and the stunning coastline found in this region.

References

Bartholo, R., Sansolo, D. and Bursztyn, I. (eds) (2009) *Community-Based Tourism: Diversity of Perspectives and Brazilian Experiences*. Rio de Janeiro, Letra e Imagem. Available at: http://www.each.usp.br/turismo/livros/turismo_de_base_comunitaria_bartholo_sansolo_bursztyn.pdf (accessed 26 May 2023).

Brazil (2007) Decreto 6.040, de 07 de Fevereiro de 2007. Institui a Política Nacional de Desenvolvimento dos Povos e Comunidades Tradicionais. Available at: http://www.planalto.gov.br/ccivil_03/_ato2007-2010/2007/decreto/d6040.htm (accessed 26 May 2023).

Cabanilla, E.A. and Gentili, J.O. (2015) Características de las páginas de la internet de turismo comunitario en países de América. Una aproximación desde el análisis de contenidos y la cartografía temática. *PASOS. Revista de Turismo y Patrimonio Cultural* 13(1), 157–174. Available at: https://www.redalyc.org/pdf/881/88133268012.pdf (accessed 27 May 2023).

Cañada, E. (2010) Perspectivas del turismo comunitario: Cómo mantener vivas las comunidades rurales. *Revista Pueblos* 41.

Comunicação Popular FCT (2018) Comunidade caiçara do Sono recebe Partilha de TBC da Rede Nhandereko-Preservar é Resistir.

Coriolano, L.N.M.T. (2006) *Tourism in Discourses, Policies, and Poverty Alleviation*. São Paulo, Annablume.

Fabrino, N.H. (2013) Community-Based Tourism: From Concepts to Practices and from Practices to Concepts. (Master's thesis, Sustainable Development Program, Center for Sustainable Development, University of Brasília, Brasília, 2013).

Gallo, E. and Nascimento, W. (eds) (2019) The Pulsating Territory: Sustainable and Healthy Territories of Bocaina: Solutions for Promoting Health and Sustainable Territorial Development. Paraty, RJ, Fiocruz. Available at: https://issuu.com/otss/docs/o-territorio-pulsa-issuu-ok (accessed 26 May 2023).

Irving, M.A. (2009) Reinventing reflection on community-based tourism: Is innovation possible? In: Bartholo, R., Sansolo, D.G. and Bursztyn, I. (eds) *Turismo de Base Comunitária: Diversidade de Olhares e Experiências Brasileiras*. Rio de Janeiro, Letra e Imagem, pp. 108–121. Available at: http://www.each.usp.br/turismo/livros/turismo_de_base_comunitaria_bartholo_sansolo_bursztyn.pdf (accessed 27 May 2023).

Irving, M.A. and Mendonça, T.C.M. (2004) Community-based tourism: Participation as a practice in the development of tourism projects in Brazil-Prainha do canto Verde, Beberibe (CE). *Caderno Virtual de Turismo* 4(4), 12–22. Available at: http://www.ivt.coppe.ufrj.br/caderno/index.php/caderno/article/view/66 (accessed 27 May 2023).

Lang, M. (2016) Introduction: Alternatives to development. In: Dilger, G., Lang, M. and Pereira Filho, J. (eds) *Descolonizar o Imaginário. Debates Sobre pós-Extrativismo e Alternativas ao Desenvolvimento*, 1 edn. São Paulo, Fundação Rosa Luxemburgo, pp. 25–44.

Lima, L.C. and Coriolano, L.N.M.T. (eds) (2003) *Turismo e Desenvolvimento Social Sustentável*. Fortaleza, EDUECE.

Maldonado, C. (2009) O turismo rural comunitário na américa latina: Gênesis, características e políticas. In: Bartholo, R., Sansolo, D. and Bursztyn, I. (eds) *Turismo de Base Comunitária: Diversidade de Olhares e Experiências Brasileiras [Community-Based Tourism: Diversity of Perspectives and Brazilian Experiences]*. Rio de Janeiro, Letra e Imagem, pp. 25–44. Available at: http://www.each.usp.br/turismo/livros/turismo_de_base_comunitaria_bartholo_sansolo_bursztyn.pdf (accessed 27 May 2023).

Martins Fonseca, V. and Rocha, I.S.G. (2021) The marine extractive reserve of corumbau from the perspective of tourism in protected areas: The discursive perspective of the series Mar sem fim-conservation units. *Revista Latino-Americana De Turismologia* 7(Single), 1–17. DOI: 10.5281/zenodo.5781730.

Meléndez, A. (2000) Tourism in Latin America: Current situation and future trends turismo. *Visão e Ação* 2(5), 71–80. DOI: 10.14210/rtva.v2n5.p71.

Mendonça, T.C.M., *et al.* (2015) *Relatório do I Encontro de Turismo de Base Comunitária da Costa Verde*. UFRRJ/NEPET. Nova Iguaçu.

Mendonça, T.C.M., Santos, R.O. and Leite, M.C.S. (2020) Nhandereko network of community-based tourism: Caiçara, Quilombola, and indigenous resistance weaving a new tourism map in the Costa Verde (RJ). In: Alves, K.S. (ed.) *Social Dialogues in Tourism: Hegemonic and Counter-Hegemonic Elements*. Editora Dialética, pp. 163–180.

Mendonça, T.C.M., Santos, R.O., Souza, N.N.D.S. and Reis Andrade, S. (2021) Sapukai - the nhandereko mbya transforms into indigenous resistance tourism in the state of Rio de Janeiro, Brazil. *RITUR-Revista*

Iberoamericana de Turismo 11, 234–260. Available at: https://www.seer.ufal.br/index.php/ritur/article/view/12825 (accessed 26 May 2023).

Moraes, E.A. (2019) Follow the actors and their own actions: In the traces of sociotechnical controversies of community-based tourism in the TUCUM Network–Ceará–Brazil. (Doctoral Dissertation). Universidade Federal do Rio de Janeiro, Instituto de Psicologia, Programa de Pós-Graduação em Psicossociologia de Comunidades e Ecologia Social. Available at: http://pos.eicos.psicologia.ufrj.br/wp-content/uploads/2019_Dout_Edilaine_Abertino_de_Moraes-1.pdf (accessed 27 May 2023).

Moraes, W.V., Emmendoerfer, M.L. and Costa, N.M.C. da (2013) Las buenas prácticas del turismo de base comunitaria en el territorio de la Serra do Brigadeiro (Minas Gerais, Brasil). *Estudios y Perspectivas en Turismo* 22, 1074–1095.

Nascimento, W. (2019) The forum of traditional communities and the observatory of sustainable and healthy territories of Bocaina. In: Gallo, E. and Nascimento, W. (eds) *The Pulsating Territory: Sustainable and Healthy Territories of Bocaina: Solutions for Promoting Health and Sustainable Territorial Development*. Paraty, RJ, Fiocruz. Available at: https://issuu.com/otss/docs/o-territorio-pulsa-issuu-ok (accessed 26 May 2023).

Nhandereko (2023) A Rede Nhandereko tem uma grande novidade [Instagram post]. Available at: https://www.instagram.com/p/CsESJ9GrkDS/?igshid=MTc4MmM1YmI2Ng%3D%3D (accessed 26 May 2023).

Oliveira, F.J.R., Moço, E.B., Santiago, A. and Gallo, E. (2019) A Incubadora de Tecnologias Sociais (ITS). In: Gallo, E. and Nascimento, W. (eds) *O Território Pulsa: Territórios Sustentáveis e Saudáveis da Bocaina: soluções Para a Promoção da Saúde e do Desenvolvimento Sustentável Territorializados*. Paraty, RJ, Fiocruz, pp. 149–171. Available at: https://issuu.com/otss/docs/o-territorio-pulsa-issuu-ok (accessed 26 May 2023).

Ouriques, H.R. (2015) *The Production of Tourism: Fetishism and Dependency*. Editora Alínea.

Rodrigues, C.G.O. and Botelho, E.S. (2023) Partnerships in protected areas: Diversity of modalities, purposes, and involved effects. *Revista Brasileira De Ecoturismo (RBEcotur)* 16(3). DOI: 10.34024/rbecotur.2023.v16.15147.

Rodrigues, M.M., Santos, R.O., Raimundo, S. and Machado, G.C.X.M.P. (2022) The praia do Sono (Paraty, Brazil) as a gateway community to achieve the universalization of sustainable development goal 6 (water sanitation) in tourist areas. In: Slocum, S.L., Wiltshier, P. and Read, J.B. IV (eds) *Tourism Transformations in Protected Area Gateway Communities*. CAB International, Wallingford, UK. DOI: 10.1079/9781789249033.0009.

Santos, B.S. (2004) Do Pós-Moderno ao Pós-Colonial: e para além de um e de outro. Conferência de abertura do VIII Congresso Luso-Afro-Brasileiro de Ciências Sociais, Coimbra, 16 set. 2004, P. 45. Available at: https://estudogeral.sib.uc.pt/handle/10316/43227 (accessed 26 May 2023).

Santos, R.O. (2020) Traditional Knowledge and Occupations: Memory and Visitation in the Guarani Mbya Araponga Village - RJ (Master's Thesis, Master's Program in Psychosociology of Communities and Social Ecology, Institute of Psychology, Federal University of Rio de Janeiro). Available at: http://pos.eicos.psicologia.ufrj.br/wp-content/uploads/2020_Mestrado_Renato_Santo-1.pdf (accessed 26 May 2023).

Zaoual, H. (2009) From mass tourism to situated tourism: What transitions? In: Bartholo, R., Sansolo, D.G. and Bursztyn, I. (eds) *Turismo de Base Comunitária: Diversidade de Olhares e Experiências Brasileiras*. Rio de Janeiro, Letra e Imagem, pp. 55–75.

5 Public Territorial Governance and Community-Based Tourism for Sustainable Local Development: A Case Study of Paraty, Brazil

Thiago Chagas de Almeida[1], Magnus Luiz Emmendoerfer[1,2] and Marcos Eduardo Carvalho Gonçalves Knupp[2]

[1]Federal University of Viçosa, Viçosa, Minas Gerais, Brazil; [2]Federal University of Ouro Preto, Ouro Preto, Minas Gerais, Brazil

Summary

Community-Based Tourism (CBT) is a mode of tourism management that has several foundations articulated for Sustainable Local Development (SLD). Nevertheless, achieving its principles often depends on the interaction with other territorial actors, that is, on Public Territorial Governance (PTG). Consequently, this case study identifies how PTG mechanisms can involve CBT in the process of co-promoting SLD by focusing on the Paraty/Brazil Municipal Tourism Council (COMTUR). This was studied because it is a PTG space open to local CBT initiatives that discuss SLD-related agendas. We collected data through the document analysis of COMTUR minutes and the on-site observation of COMTUR meetings. Data were interpreted by the thematic content analysis technique. The results show that PTG can involve CBT in the process of co-promotion of SLD, through four main elements: Co-need, co-potentiality, co-effectiveness, and co-legitimacy.

The Value and Interest of the Case Study

This case study contributes theoretically to understanding the capacity of PTG mechanisms to help CBT co-promote SLD. This case study contributes theoretically to understanding the capacity of PTG mechanisms to help CBT co-promote SLD. Its findings show the importance of the CBT actor's presence in spaces of PTG. It considers the benefits of CBT in generating endogenous and sustainable development, which is an important developmental perspective in Latin American territories.

Introduction

Community-Based Tourism (CBT) was first identified in the 1980s in Latin American rural communities (Maldonado, 2009). According to Maldonado (2009) and Gómez *et al.* (2015), since its emergence CBT has been a way of managing visitation activities with principles that go against conventional mass tourism (which carries a prevailing market logic). These principles include giving decision making leadership to local actors, valuing the culture of traditional populations, performing activities that preserve environmental resources, and boosting economic gains for the local community from tourism (Graciano and Holanda, 2020; Almeida and Emmendoerfer, 2023).

Because of this, CBT is considered a tourism management approach that is associated with Sustainable Local Development (SLD). According to Almeida and Emmendoerfer (2023), SLD is a developmental perspective that aims, in an endogenous way (based on local resources and needs), to balance different factors, such as environmental, cultural, economic, political, and social.

Therefore, CBT is a practice that can help promote SLD, by harmonizing endogenous factors, something that is important for the management and execution of sustainable visitation activities (Burgos and Mertens, 2015; Pinheiro, 2016; Almeida and Emmendoerfer, 2023). These benefits are more likely to occur with the insertion of CBT in Public Territorial Governance (PTG), where it can have the support of other actors (Graciano and Holanda, 2020).

PTG mechanisms are initiatives, integrated by multiple representatives (governmental and non-governmental), aimed at discussing and managing public issues related to a specific territory (Dallabrida and Becker, 2003). They are essential for SLD, as they ensure the management of sustainable development is carried out by the local actors, not only for them (Fernández *et al.*, 2019). Through PTG, SLD is intentionally promoted, or rather 'co-promoted', by the collective effort to carry out actions based on the principles of this developmental perspective.

In addition, when there are representatives in PTG who defend responsible social and environmental principles, like those of CBT, fostering SLD becomes more feasible (Almeida and Emmendoerfer, 2023). Despite this evident relationship, the literature still needs to advance in the discussion of governance's ability to achieve CBT goals (Graciano and Holanda, 2020; Lindström, 2020; Dangi and Petrick, 2021), which are adherent to the theory of SLD. Based on this gap, this case study started with the following question: 'How can PTG mechanisms involve CBT in the process of co-promoting SLD?'

The investigation of this question was done through a case study on the COMTUR of the territory of Paraty, in the State of Rio de Janeiro, Brazil. We chose this COMTUR because Paraty has well-established CBT initiatives, compared to other Brazilian (Barros and Rodrigues, 2019) and Latin American locations. In addition, Paraty's COMTUR was chosen for being a PTG mechanism with the potential to promote SLD through the insertion of CBT representatives. Thus, the objective was to analyze how the PTG in Paraty can involve the CBT in the SLD process.

The study was carried out through analysis of the Minutes of COMTUR meetings. In addition, we did on-site observation of a meeting of that board, checking if the Minutes corresponded to what actually happened. Data were interpreted by utilizing the thematic content analysis technique, identifying in the categories of the documents that indicate how PTG mechanisms can involve CBT in the process of SLD co-production. In the next three sections, we present the theoretical framework that underpinned this research.

Origin and Meaning of Community-Based Tourism in Latin America

As mentioned in the Introduction, the first manifestations of CBT in rural communities in Latin America occurred toward the end of the 12th century (Maldonado, 2009). According to Maldonado (2009), CBT arose as an important income opportunity for part of the Latin American population that faced serious socioeconomic problems. In addition, CBT showed its relevance as a practice that would strengthen the relationship of belonging of traditional people with their territory. This sat alongside stimulating the local economy and performing actions compatible with the preservation of cultural and natural heritage.

Especially in Brazil, CBT has been discussed since the 1990s, but only since 2005 has been part of institutional agendas, through a Declaration prepared at the 5th World Social Forum, held in the city of Porto Alegre/RS (Neuhaus and Silva, 2006). CBT was included in this document as a form of tourism opposite to the logic of centralized management and alienation of the needs of local populations that is prevalent in conventional tourism.

In Paraty/RJ and the Costa Verde region (Brazil), CBT became more prevalent after 2007, with the establishment of the Traditional Communities Forum (Mendonça *et al.*, 2017).

The Forum was created as an important political space to strengthen the Costa Verde traditional people, mainly by their insertion in economic, cultural, and socio-educational activities, critical to the region and for valuing their identity. With that, CBT in Paraty was quickly seen as an essential practice to be fostered, by involving local indigenous people, *caiçaras*, and *quilombolas* (Mendonça *et al.*, 2017).

Over the years, the Paraty CBT was structured and consolidated (Cardoso, 2016), and on 16 November 2022, the Paraty Municipal CBT Policy was approved (Law no. 45, 2022). This law was the result of political articulations by local community actors, mainly through their participation in spaces for discussion and collective deliberation, such as COMTUR.

Knowing the origin of CBT helps better understand its main characteristics. According to Fabrino *et al.* (2012), Llupart (2022) and Sarabia-Molina *et al.* (2022), CBT takes place through self-management by community actors. The Chico Mendes Institute for Biodiversity Conservation (2018) defines CBT as:

> a visitation management model led by the community, generating collective benefits, promoting intercultural experience, quality of life, valuation of the history and culture of these populations, as well as the sustainable use of the Conservation Unit's resources for recreational and educational purposes. (p. 10)

Therefore, CBT is a particular way of managing tourism projects, aimed at enabling the community's leading role, appreciation of popular culture, decentralization of economic gains, and responsible use of natural resources (Graciano and Holanda, 2020; Almeida and Emmendoerfer, 2023). Because of this, CBT has been considered an adherent to endogenous and sustainable development, which is SLD (Burgos and Mertens, 2015; Pinheiro, 2016; Almeida and Emmendoerfer, 2023). Next, we present the main aspects of this developmental approach, as well as its relevance in Latin American territories.

Sustainable Local Development and its Relevance in Latin America

For a long time, developmental approaches – including sustainable ones, which have been widely discussed since the 1970s – disregarded the local nature of the territory (Fernández *et al.*, 2019). This only started to change around the 1990s, when it became evident that to provide quality of life for current and future generations, in addition to harmony between the social, environmental, and economic dimensions (which characterize sustainable development), it is necessary to consider the interests, resources, and culture of the territory (Olivera *et al.*, 2020).

Therefore, sustainable development should not only be for the local population but also conceived together with it. SLD is based on this understanding, that the elements which make up a territory should be included in the sustainable development process (Marín-González *et al.*, 2022; Sorea *et al.*, 2023). Thus, SLD is not just about sustainability but also about territorial development. This is not just a semantically micro-territorial development, because its benefits and influences go beyond the local territory. SLD should have a sense of nature, which comprises specific attributes of that territory, its local character (Tenório, 2012).

We highlight that 'local' can refer to any geographical delimitation with unique aspects, compared to another wider (or global) context, such as: Country (local) and world (global), state (local) and country (global), municipality (local) and state (global). Therefore, SLD is not territorially and geographically insulated, but endogenous, which considers the problems, needs, and resources characteristic of a territory.

Many scholars have emphasized the need to advance this developmental perspective, especially in developing countries, with great technological and economic dependence on those considered central and developed (Pacheco and Benini, 2018). Furtado (1974) was one of the developers of this idea, indicating that endogenous development would be the way for Brazil and other underdeveloped countries in Latin America to achieve real, meaningful, development – changing their relations with developed countries from 'dependent' to 'interdependent'.

To this end, interaction among territorially-local people is essential, combining their interests and creating a collective effort for strengthening the process of development in the territory. Hence, the existence of public governance mechanisms is important, in addition to actors committed to endogenous and sustainable principles – such as those of CBT (Almeida and Emmendoerfer, 2023). In the next section, the conceptualization of PTG and its relationship with the development of territories is further discussed.

Public Territorial Governance and its Relationship with the Development of Territories

Governance mechanisms have existed since ancient times (Tribunal de Contas da União [Federal Accounting Court], 2014), but they only gained more visibility in the corporate sphere in the 1930s (Álvares *et al.*, 2008). According to Álvares *et al.* (2008), corporate governance emerged and spread with the purpose of making company managers accountable to shareholders and owners – considering that many firms were no longer managed by their proprietors but rather by specialized professionals.

In the public sector, the concept of governance has been used more widely since the end of the 20th century (Tribunal De Contas Da União, 2014). Its goal is to ensure citizens have greater participation in public policies (Bovaird and Löffler, 2003; Bevir, 2011). Moreover, the mechanisms of public governance meet the principle of transparency, disclosing the acts of governments and their managers to the public (Silvestre, 2019).

PTG is a public governance tool for solving issues in a given territory. Hence, it is important to understand what is characterized as a territory. It is understood here not as a simple geographic delimitation but as a place appropriated, built, deconstructed, and rebuilt by social relations that are continuously established within it (Santos and Silveira, 2001; Dallabrida and Becker, 2003; Haesbaert, 2006; Saquet, 2007).

Thus, PTG is a mechanism for aligning the interests of territorial actors, to create strategies for improving relationships and lifestyles in the territory (Jørgensen and Fallov, 2022; Ravagnan *et al.*, 2022). Lima (2019) has a similar understanding, observing that PTG is a:

> process of actors' coordination, aiming to develop social, intellectual, political, and material capital, and coordinating territorial development based on the creation of sustainable territorial cohesion at different levels. (p. 133)

PTG is essential for SLD, especially due to its local nature (Fuini, 2014; Pires *et al.*, 2018). Among the different mechanisms of PTG existing in Brazil, we highlight the public policy management councils (such as COMTUR, which we analyzed), public hearings, and participatory budgets (Dallabrida and Becker, 2003).

Methodology

The case study addressed the COMTUR of the Paraty/RJ territory, showing the potential of PTG mechanisms for involving CBT in the SLD process.

Data were collected through an analysis of the Minutes of 31 COMTUR meetings, held from 1 January 2022 to 31 October 2023. The Minutes were accessed through the Council's official website (https://sites.google.com/view/comturparaty/, accessed 10 April 2024). According to May (2004), document analysis considers the circumstances of the material's production and the information it contains.

In addition, we did an on-site observation of a COMTUR meeting that took place on 17 May 2023. This data collection procedure captures the explicit and implicit relationships of a certain social practice (Gil, 2008). In this case, the observation sought to check if the information in COMTUR's Minutes corresponded to what happened in practice.

The analysis of COMTUR meetings is justified because this is a space for interaction between representatives of the CBT and the municipal public authorities (in addition to other local stakeholders) aimed at discussing and influencing territorial-public issues in Paraty – which is a place with very active and consolidated CBT (Barros and Rodrigues, 2019).

Document analysis was based on the thematic content analysis technique, extracting core issues related to the study goal (Bardin, 2018). The nuclei of issues were arranged by codes and categories. The codes are excerpts from the Minutes related to the objective of this research, and the categories are representations of several codes with common ideas. The

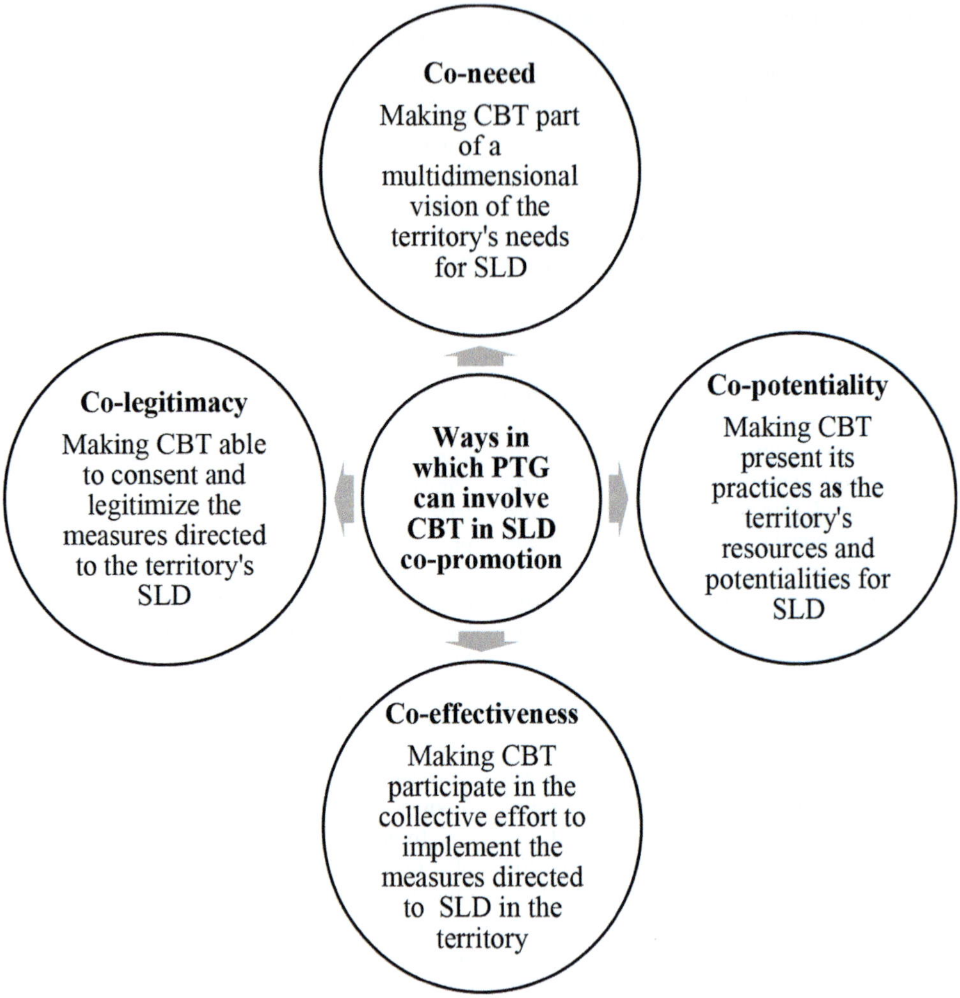

Fig. 5.1. Main ways in which PTG can involve CBT in the co-promotion of SLD.

categories emerged during data analysis and were not previously defined.

Results

Here we present and discuss the categories revealed by the analysis of the document data. They indicate how PTG mechanisms can involve CBT in the process of SLD co-promotion. Fig. 5.1 shows the categories identified in the research.

The analysis indicated that PTG mechanisms can involve CBT in the SLD co-promotion process through four main elements: Co-need, co-potentiality, co-effectiveness, and co-legitimacy.

These should not be understood as the only contributions from PTG to insert CBT in a collective and articulated action oriented for SLD but as the fundamental attributes of PTG perceived in the case of Paraty's COMTUR. Therefore, this case study does not enable or seek to generalize its findings. Instead, its purpose is to shed light on the topic as a whole, based on an example of the phenomenon that must be put into context.

The on-site observation that took place at a COMTUR meeting checked that the Minutes portray faithfully the dynamics of this space for discussion and collective deliberation. Therefore, we understand that the content of the documents related to the paper objective are codes that can be used. In the following sub-topics, the basic codes are mentioned, supporting their corresponding categories.

Co-need

Co-need is a category that shows that PTG can help CBT contribute to identifying problems and needs within the territory to be considered in the SLD process. Thus, PTG enables CBT to be part of a multidimensional territorial view. This relates to the conceptualization of PTG presented by Jørgensen and Fallov (2022), as an instrument for capturing different perceptions and interests from territorial actors.

The Minutes of 1 March 2023 demonstrate this element when the CBT representative indicates, along with the other COMTUR representatives, the need to implement a municipal project called 'Tourism Observatory' – which would be capable of raising data on the profile of visitors. At the meeting on 13 June 2023, COMTUR members expressed their opinion on the relevance of adjusting a Management Plan for an Ecological Station in the municipality, which dealt with monitoring actions to preserve the territory's biodiversity. In this case, the need to increase monitoring periods was raised, understanding that control needed to be more continuous. Co-need was also evident in other COMTUR meetings, like those on 1 January 2024, 14 April 2023, 20 April 2023, when collegiate representatives raised issues to improve the territory.

Therefore, we observed that PTG mechanisms can involve CBT in the co-promotion of SLD, so that it becomes an important actor in identifying the needs and priorities of the place. In PTG, this is done collectively, and CBT's vision does not always prevail, being just one of several positions.

Co-potentiality

The second category refers to the capacity of PTG spaces to enable CBT representatives to present their practices and knowledge to enhance the territory's potential for SLD. Different from the previous category, which is more related to the analysis of problems and priorities, co-potentiality brings more suggestions for the endogenous strategies to be defined. Pires *et al.* (2018) have previously highlighted this characteristic of PTG spaces for generating a collectively purposeful environment.

In the case of COMTUR this element was also evident. At the meeting held on 5 July 2022, for example, the representation of traditional people proposed that the council invest in the creation of the *samba-enredo* for the carnival of the Tatuapé Samba School in São Paulo (considering a proposal received previously). According to him, the school's theme could portray the culture, history, and biodiversity of Paraty.

Another meeting where co-potentiality appeared was on 1 February 2022, where some councilors suggested that the public authorities direct resources to revitalize trails in a peripheral potential location in the municipality. This would help to transform it into a sought-after destination by visitors to Paraty.

Therefore, we see that PTG can insert CBT in the SLD process, by enabling its actors to present strategies based on their conceptions and way of life. Many of them, as exemplified by the one addressed at the meeting on 5 July 2022, are based on the potential of tangible and intangible resources of the place.

Co-effectiveness

The third category shows that PTG can go beyond utilizing CBT to help identify the needs and potentials of the territory to also aid the implementation of SLD-oriented measures that are collectively deliberated. In addition to being an instrument for discussion, PTG has the role of mobilizing a joint effort for local development (Fuini, 2014).

This was noticed in the Minutes of 1 February 2023, when representatives committed themselves to helping to raise awareness against predatory tourism, informing tourists of basic rules for the appropriate use of the visiting spaces. At the meeting on 17 May 2023, the CBT representative and other participants made themselves available to help with the insertion of information about their respective locations for the Paraty tourist itineraries application that was being developed. This is valuable because, as residents, the representatives and participants have the best knowledge about the positive points and destinations to be visited in each location, making the application's descriptions more consistent with the services offered. The meetings on 1 September 2022, 1 November 2022, and 10 November 2022, also showed, in different ways the multiple participation of the council members to carry out the decisions taken.

Therefore, PTG can imply an active CBT intervention to carry out the actions for SLD proposed by the collegiate. This means that its actions for SLD should not remain only in the sphere of dialogue and decision making but contribute to implementing decisions.

Co-legitimacy

The last category that emerged from the data analysis was co-legitimacy. According to it, PTG can help CBT actors legitimize measures for SLD, by agreeing with the deliberations on these spaces. Lima (2019) emphasized this attribute of territorial governance mechanisms, as they result in decisions that are more territorially representative. This is different from an action stemming only from municipal government choices.

In the case of Paraty's COMTUR, the legitimacy of the territorial representations (legally provided for in the Council) is expressed when its members reach consensus on a particular meeting decision. The formalization of Paraty's registration as a SEBRAE (Brazilian Micro and Small Business Support Service) Intelligent Tourism Destination, which occurred at the meeting on 13 October 2022, is an example of territorial legitimacy that had the consent of CBT representation. According to the Minutes, this registration enables Paraty to seek alternatives to interact with tourists through technological resources, attracting more visitors and generating more income opportunities for the local population. Through the review of different local representations, the development of digital tools to promote tourism has become more accepted by the population of the territory.

At the meeting on 23 May 2023, co-legitimacy occurred when the council representatives approved the allocation of resources from the Municipal Tourism Fund for the maintenance of the Casa da Cultura de Paraty, which was experiencing financial difficulties and needed support to continue carrying out its important cultural actions. Other examples of legitimation of SLD-oriented measures through COMTUR can be identified in the Minutes of 14 March 14, 2023, February 1, 2023, and November 30, 2022.

Final Remarks

This case study has shown that PTG is an essential instrument for involving CBT (or other territorial actors) in SLD co-promotion processes through: Co-need, making CBT part of a multidimensional vision of the territory's needs for SLD; co-potentiality, making CBT present its practices as the territory's resources and potentialities for SLD; co-effectiveness, making CBT participate in the collective effort to implement the measures directed to SLD in the territory; and co-legitimacy, making CBT able to consent and legitimize the measures directed to the territory's SLD.

These categories show that PTG can, in different ways, involve CBT in the SLD process and potentialize the reach of its principles. It is considered that they should not be understood as the only possibilities of PTG mechanisms involving CBT in SLD but as the most evident in the studied context. So, this study generates reflections for studying the topic, mainly in the territories of Latin America, where CBT is quite noticeable (Maldonado, 2009).

For future studies, it is suggested that research be carried out in other Latin American territories based on the categories presented here, in order to verify more manifestations of them. In addition, theoretical studies on the subject are still necessary, which potentially develop new conceptions of the relationships between CBT, PTG, and SLD, and how they may be analyzed.

Regarding its contributions, this study advances the understanding of the capacity of governance to foster the achievement of endogenous-sustainable CBT goals, which is a gap in the current literature (Graciano and Holanda, 2020; Lindström, 2020; Dangi and Petrick, 2021). As a practical implication, the results show the importance of PTG involving key territorial players (like ones from CBT), and the need to think of collective strategies for the balanced development of territories based on their constraints and potentialities.

Funding

Coordenação de Aperfeiçoamento de Pessoal de Nível Superior, (Grant/Award Number: 404606/2021–0; 04193/2022–6; 20194/2022–3; 312764/2022–7; 403139/2022–8) Fundação de Amparo à Pesquisa do Estado de Minas Gerais, (Grant/Award Number: APQ-04471–22) Conselho Nacional de Desenvolvimento Científico e Tecnológico, (Grant/Award Number: 001).

References

Almeida, T.C. and Emmendoerfer, M.L. (2023) Turismo de base comunitária e desenvolvimento local sustentável: Conexões e reflexões. *Revista de Turismo Contemporâneo* 11(1), 2023.

Álvares, E., Giacometti, C. and Gusso, E. (2008) *Governança Corporativa: Um Modelo Brasileiro*. Elsevier, Rio de Janeiro.

Bardin, L. (2018) *Análise de Conteúdo*. Edições 70, Lisboa.

Barros, A.L.R. and Rodrigues, C.G.O. (2019) Differentiated education and community-based tourism in the caiçara territories of paraty (RJ). *Ambiente and Sociedade* 22, 1–20.

Bevir, M. (2011) Governança democrática: Uma genealogia. *Revista de Sociologia e Política* 19(39), 103–114.

Bovaird, T. and Löffler, E. (2003) Evaluating the quality of public governance: Indicators, models and methodologies. *International Review of Administrative Sciences* 69(3), 313–328.

Burgos, A. and Mertens, F. (2015) Os desafios do turismo no contexto da sustentabilidade: As contribuições do turismo de base comunitária. *Pasos Revista de Turismo y Patrimonio Cultural* 13(1), 57–71.

Cardoso, D.S. (2016) *Apontamentos Sobre o Turismo de Base Comunitária a Partir da Análise de Duas Regiões Díspares: Região do Cariri (Ceará) e Região da Fazenda de Santa Cruz (Rio de Janeiro).* Annals of the Seminário da Associação Nacional de Pesquisa e Pós-Graduação em Turismo, São Paulo, SP, Brazil, p. 13.

Dallabrida, V.R. and Becker, D.F. (2003) Governança territorial: Um primeiro passo na construção de uma proposta teórico-metodológica. *Desenvolvimento em Questão* 1(2), 73–97.

Dangi, T.B. and Petrick, J.F. (2021) Enhancing the role of tourism governance to improve collaborative participation, responsiveness, representation and inclusion for sustainable community-based tourism: A case study. *International Journal of Tourism Cities* 7(4), 1029–1048.

Fabrino, N.H., Costa, H.A. and Nascimento, E.P. (2012) Turismo de Base Comunitária (CBT): Elementos chaves para aferir seu desempenho na perspectiva da sustentabilidade. *Revista Brasileira de Ecoturismo* 5(3), 546–559.

Fernández, R.L., Calatayud, M.M., Peña, R.M. and Urquiza, D.E.P. (2019) Epistemological foundations that sustain a research in environmental education for sustainable local development, in the province of El Oro. *Conrado* 15(67), 282–287.

Fuini, L.L. (2014) Território, territorialização e territorialidade: O uso da música para a compreensão de conceitos geográficos. *Terr@ Plural* 8(1), 225–249.

Furtado, C. (1974) *O Mito do Desenvolvimento Econômico*. Paz e Terra, Rio de Janeiro.

Gil, A.C. (2008) *Métodos e Técnicas de Pesquisa Social*. Atlas, São Paulo.

Gómez, C.P., Falcão, M.C. and Castillo, L.A.G. (2015) Turismo de Base Comunitária como Inovação Social: Congruência entre os constructos. *Revista de Turismo y Patrimonio Cultural* 13(5), 1213–1227.

Graciano, P.F. and Holanda, L.A. (2020) Bibliometric analysis of scientific literature on community-based tourism from 2013 to 2018. *Revista Brasileira de Pesquisa em Turismo* 14(1), 161–179.

Haesbaert, R. (2006) *O Mito da Desterritorialização: Do 'Fim dos Territórios' à Multiterritorialidade*. Bertrand Brasil, Rio de Janeiro.

Instituto Chico Mendes de Conservação da Biodiversidade (2018) *Turismo de Base Comunitária em Unidades de Conservação Federais: Princípios e Diretrizes*. Instituto Chico Mendes de Conservação da Biodiversidade, Brasília.

Jørgensen, A. and Fallov, M.A. (2022) Urbanization and the organization of territorial cohesion – Results from a comparative Danish case-study on territorial inequality and social cohesion. *Journal of Organizational Ethnography* 11(1), 64–78.

Law no. 45, of August 22, 2022 (2022) Establishes the municipal community-based tourism policy within the municipality of Paraty and provides other measures. In: *Diário Oficial do Município*. Câmara Municipal, Paraty, RJ.

Lima, I.G. (2019) Governança territorial. *GEOgraphia* 21(46), 131–135.

Lindström, K.N. (2020) Ambivalence in the evolution of a community-based tourism sharing concept: A public governance approach. *Scandinavian Journal of Hospitality and Tourism* 20(3), 302–315.

Llupart, M.R.N. (2022) Theoretical model for the analysis of community-based tourism: Contribution to sustainable development. *Sustainability* 14(17), 1–41.

Maldonado, C. (2009) O turismo rural comunitário na américa latina: Gênesis, características e políticas. In: Bartholo, R., Sansolo, D.G. and Bursztyn, I. (eds) *Turismo de Base Comunitária: Diversidade de Olhares e Experiências Brasileiras*. Letra e Imagem, Rio de Janeiro, pp. 25–44.

Marín-González, F., Moganadas, S.R., Paredes-Chacín, A.J., Yeo, S.F. and Subramaniam, S. (2022) Sustainable local development: Consolidated framework for cross-sectoral cooperation via a systematic approach. *Sustainability* 14(11), 1–33.

May, T. (2004) *Pesquisa Social: Questões. Métodos e Processos*. Artmed, Porto Alegre.

Mendonça, T.C.M., Santos, R.O., Lopes, P.C.B., Andrade, S.R. and Moraes, A.P.V. (2017) Turismo de base comunitária na Costa verde (RJ): Caiçaras, quilombolas e indígenas. *Revista Brasileira de Ecoturismo* 10(2), 328–356.

Neuhaus, E. and Silva, J.S. (2006) *Um Outro Turismo é Possível! Reflexões Sobre Desigualdades, Resistências e Alternativas no Desenvolvimento Turístico*. FBOMS/Instituto Terramar, Fortaleza.

Olivera, C.M., García, K.A.M., Sánchez, L.A.M. and Velasco, A.E.M. (2020) Relation between public management and sustainable local development of the cities of Oaxaca, 2000–2017. *Revista Iberoamericana de Estudios Municipales* 22, 155–187.

Pacheco, A.P.D.C. and Benini, E.G. (2018) A economia criativa em época de crise: O desenvolvimento endógeno brasileiro na obra de Celso Furtado. *Brazilian Journal of Political Economy* 38(2), 324–337.

Pinheiro, R.C.S. (2016) Turismo de base comunitária como inovação social na gestão do turismo em espaços rurais. *Ponta de Lança: Revista Eletrônica de História, Memória and Cultura* 10(19), 146–165.

Pires, E.L.S., Fuini, L.L., Figueiredo Filho, W.B. and Mendes, E.L. (2018) A governança territorial revisitada: Dispositivos institucionais, noções intermediárias e níveis de regulação. *GEOgraphia* 19(41), 24–38.

Ravagnan, C., Rossi, F. and Amiriaref, M. (2022) Sustainable mobility and resilient urban spaces in the United Kingdom. Practices and Proposals. *Transportation Research Procedia* 60, 164–171.

Santos, M. and Silveira, M.L. (2001) *Território e Sociedade no início do século 21*. Record, Rio de Janeiro.

Saquet, M.A. (2007) *Abordagens e concepções de território*. Expressão Popular, São Paulo.

Sarabia-Molina, M.Y., Soares, J.R.R. and Lois-González, R.C. (2022) Innovations in community-based tourism: Social responsibility actions in the rural tourism in the province of Santa Elena–Ecuador. *Sustainability* 14(20), 1–20.

Silvestre, H.C. (2019) *A (Nova) Governança Pública*. ENAP, Brasília.

Sorea, D., Defta, M. and Popescu, I.M. (2023) Journeys to significant places in orthodoxy as a source of sustainable local development in Romania. *Sustainability* 15(7), 1–26.

Tenório, F.G. (2012) *Cidadania e Desenvolvimento Local: Critérios de Análise*. Fundação Getúlio Vargas, Rio de Janeiro.

Tribunal De Contas Da União (2014) *GovernançA Pública: Referencial Básico de GovernançA Aplicável A Órgãos e Entidades da Administração Pública e Ações Indutoras de Melhoria*. Tribunal De Contas Da União, Brasília.

6 Pretos que Voam – An Afrocentric Collective Project to Promote the Access of Peripheral Black People to Airline Crew Jobs in Brazil

Laiara Amorim[1], Cassiana Gabrielli[2], Natália Oliveira[3] and Gabriela Santos[4]
[1]Quilombo Aéreo, Rio de Janeiro, Brazil; [2]Departamento de Turismo, Universidade Federal do Paraná, Brazil; [3]Centro de Ciências Sócio-Organizacionais, Universidade Federal do Rio Grande do Sul, Brazil; [4]Departamento de Línguas e Cultura, Universidade de Aveiro, Portugal

Summary

This case analyses the 'Pretos que Voam' (Blacks who Fly) project, which aids the training of socially vulnerable black people to become flight attendants in Brazil. Since 2020, this project has been carried out by the collective Quilombo Aéreo, which brings together black crew members to aid the elimination of racism in aviation. One of the goals of this collective is to increase the representation of black people in the airline industry. Quilombo Aéreo (2022) conducted a survey of air crew members, using online forms based on self-identification, and found that black people do not reach 5% of active airline crews. The project is supported by crowdfunding and social partnerships like Fundação Tide Setúbal, Fundo Baobá for racial equity Rede de Mulheres Negras, Odabá Associação de Afroempreendedorismo, Movimento Negro Unificado and others. This made it possible to provide a complete course to become a flight attendant, a basic English course, air crew licences and a passport, and mentorship. In 2021, in Porto Alegre (RS), ten black people graduated in the first class – four women among them have already been hired by airlines. In 2022, the project raised funds to provide the first Afro-focused flight school in a distance learning model in Brazil. In 2023, they are working to start the operation of the school and crowdfunded a campaign for the second class of Pretos que Voam. Facing several forms of racism in Brazilian society and its airline industry, the project also dialogues with other forms of violence common in the sector, such as sexism and *machismo*. The case study examines concrete actions to transform the pain of its pioneers into fuel for structural changes, having as a framework the Afro culture and, as a scenario, a relevant mode of tourism transportation such as air travel.

This case study originated from research into empowerment and intersectionalities between racism and sexism through the Quilombo Aéreo (see Further Reading). Here we focus specifically on the Pretos que Voam project, one of the actions developed by the collective.

All authors are members of the research group *Gritus – Gender, Race and Intersectionalities in Tourism*, where this study was developed.

The Value and Interest of the Case Study

This case shows the power and resistance of a few black crew members in favour of changing the racist scenario identified in the Brazilian airline industry, together with the empowerment of young black peripheral people. It serves as inspiration for other minority movements to articulate themselves for change, even if gradual, of the inequalities structured in our societies and practised daily in the tourism sector, among others.

Contextualizing Black People in the Brazilian Airline Industry

In Brazil, the fight for racial equality is a constant battle. The search for positions of representation and opportunities for black youth has been a historical challenge. It is important to note that the term Black is used here to refer to black people, descendants of the African diaspora and their blending in Brazil. This use by the collective is a matter of political positioning and identity. In this scenario, an innovative initiative called 'Pretos que Voam' aims to empower and transform the lives of young blacks through aviation. This project is bringing innovation and hope, broadening horizons, and opening doors for a new generation of black dreamers.

In the Brazilian civil airlines, the collective Quilombo Aéreo (2022) estimated that only 5% of flight attendants and 2% of pilots are black. More than that, the collective noted that women represented the majority of flight attendant (66%) but just 2.5% of the pilots. No black women were identified among the pilots. In another survey, the Instituto Locomotiva identified, among public transportation workers, that 75% of bus drivers, 73% of uber drivers, and 65% of taxi drivers have witnessed situations of racism in their work. In addition, 39% of them have been victims of racially motivated crime in the workplace in Brazil (Geledés, 2022).

It is important to point out that coloured people (black people and other AfroBrazilians) represent about 56% of the Brazilian population (IBGE, 2019) and have lower income and education levels than the white population. In addition, black people are underrepresented in management positions and lead the ranking of victims of violence (IBGE, 2022). In this way, in 2018 two black women flight attendants met on social media and, with other black crew members, founded the collective 'Quilombo Aéreo', to promote discussion and change, mitigating racism and sexism in the national air industry. The main project by the collective was named 'Pretos que Voam'. So, the 'Pretos que Voam' project was born from the joint efforts of black aviation professionals and others who realized the lack of black representation in the civil aviation field. The initiative seeks to combat stereotypes and prejudice by providing opportunities for young blacks to enter the 'world of aviation' as pilots, flight attendants, mechanics, and other related areas.

The colonizing process is common to all Latin America countries. It structures these societies and is reflected in the current model of tourism development in the region, which is based on a Eurocentric model. Going in the opposite direction is a strategy to confront the matrix of power derived from colonization and its consequences. In the context of Brazilian tourism, Quilombo Aéreo's proposals dialogue with other practical initiatives that have been expanding inquiries about the means of production and consumption of tourism from a postcolonial and responsible point of view.

The term 'postcolonial', as highlighted by Boukhris and Peyvel (2019), is not just a historical concept, but takes on a much broader epistemological dimension and makes it possible to deconstruct 'the ideological discourses of modernity, which attempt to confer hegemonic "normality" on the uneven development and differentiated, and often unbalanced, history of nations, races, communities and peoples' (Bhabha, 1998, p. 239). It is a paradigm. In this sense, with a view to tourism, decolonial studies understand that tourism was – and still is – a powerful vector for the circulation of colonial thought structures. Therefore, they aim to understand all the implications of colonization for the way tourism works in colonized societies (past, present, and future) and, above all, the structures of colonial thinking, taking into account the practice, theory and centre-periphery/North/South/West/East attitudes that persist long after colonialism. In addition to criticizing the functioning of our contemporary tourist societies, the studies covered by 'decolonial tourism' point above all to a hope for reparation, reciprocity and sharing, in which the tourist economy can participate (Berger, 2006; Mbembe *et al.*, 2006).

Initiatives such as the 'Diáspora Black' (Diáspora Black, 2023) startup, which focused on Afro tourism, have promoted experiences and reflections for tourists and host communities with African culture in Brazil as a central element. There is also Bitonga Travel (Bitonga Travel, 2023), a collective of black women travellers who empower and help each other to travel in contexts that are still exclusionary, sexist, and racist. These are some initiatives, among many others, that have emerged as social innovations that intend to confront the intersections between racism and/or sexism and tourism in Brazil. While different they are all in opposition to colonizing tourism.

These initiatives interact with the limited academic research that deals with racism, sexism, or intersectionalities in Brazilian tourism. For example, França and Malta (2021) point to the reflection of racism in the tourist planning of Petropólis, Rio de Janeiro. Oliveira (2022a) examined the representation of black people in tourist luxury advertisements, discussed Afro entrepreneurship in tourism (Oliveira, 2022b) and studied racism in Brazilian tourism (Oliveira, 2020). The latter was also the main topic of Ferreira and Casagrande's work (Ferreira and Casagrande, 2018, 2020). The experiences of black women travellers has been looked at by dos Santos e Sá (2021), Oliveira *et al.* (2022) and Morais *et al.* (2023). Research has also been conducted on racism and sexism in the Brazilian airline industry (Oliveira *et al.*, 2022; Santos *et al.*, 2022).

In 2021 the authors of this study conducted qualitative research, in partnership with Quilombo Aéreo, to explore the barriers to and retention of black women in aircrew positions. This research was carried out with six women, including the founders of the collective, through online in-depth interviews. One of the creators and current manager of the project and founder of Quilombo Aéreo is one of the authors of this case study.

The 'Pretos que Voam' Purposes

The project's main objective is to promote the social inclusion and professional advancement of young black people, who often face racially motivated barriers and obstacles in their journeys. By offering technical training, skills building, and mentoring, the project aims to break the chains of structural racism and allow participants to achieve their dreams. The 'Pretos que Voam' project adopts a comprehensive approach, encompassing several stages, to train and prepare young people for the aviation job market. The activities include:

- *Technical training*: Participants receive training at specialized institutions, acquiring theoretical and practical knowledge about aviation. They learn about aircraft mechanics, flight procedures, aviation safety, and other relevant skills.
- *Mentoring*: Black aviation professionals act as mentors, sharing their experiences, offering guidance, and encouraging the young participants to move forward, even in the face of challenges. Mentoring is a key part of the project, providing emotional support and inspiration.
- *Internships and opportunities*: The project establishes partnerships with companies and institutions in the aviation industry to provide internships and job opportunities for participants. In this way, young people can put the knowledge they have acquired into practice and develop their professional skills.
- *Support network*: The project creates a network of support among the participants by promoting meetings, events, and discussion groups. This network strengthens community ties and offers mutual support, ensuring that the young people feel welcomed and safe to follow their professional journey in aviation.

The first course offered by the project took place in Porto Alegre (RS), aiming to train flight attendants, with an Afrocentric perspective, which means that in addition to the aviation content, it includes critical studies on race. It started in 2021 and was completed in 2022. The people who enrolled in the first class received, besides completing flight attendant course, other benefits necessary to be employed in the sector, recognizing this entails significant costs in the Brazilian context. These benefits included updated and adequate documentation, including a passport, which currently costs almost 20% of the national minimum monthly wage, and the Aeronautical Medical Certificate, which can cost up to half a minimum monthly wage. The participants also received support to gain the Technical Qualification Certificate (required to operate aeroplanes), complete an English language course (a requirement in airline recruitment processes), and to undertake the National Civil Aviation Agency (ANAC's) theoretical examination (which includes a significant enrolment cost). Participants were also provided with collective and individual career mentoring, psychological support, public transportation tickets to attend the course, and tickets to participate in job selection processes (which in general take place in different regions of the country), and food aid for activities out of state. Finally, other specific expenses, according to the needs of each student, were committed to by the project to facilitate their training and subsequent hiring in the sector.

It is important to note that in the academic research conducted by the authors of this case study, with black women crew members, about specific barriers found by them to achieve their job, five obstacles were identified: Cost of professional training, lack of career information, lack of representativity of other black women, unequal recruitment process, and non-acceptance of black bodies by potential employers (Oliveira *et al.*, 2022).

In order to discuss and overcome some of these barriers, intersectionality can highlight the specificities of the intersections of race, gender and class. Originating in Black Feminism, intersectionality deals with the correlation between different social markers that make up a dynamic matrix of oppression. As Akotirene (2019) points out,

> Intersectionality prevents mathematical hierarchical or comparative aphorisms. Instead of adding up identities, we analyse which structural conditions crossing bodies, which positionalities reorient subjective meanings of these bodies, as they are experiences shaped by and during the interaction of structures, repeatedly colonialist, stabilised by the matrix of oppression, in the form of identity. In turn, identity cannot refrain from any of its markings, even if not all of them, contextually, are made explicit. (p. 27)

The same author points out that 'Together, racism, capitalism and heteropatriarchy must be treated by intersectionality observing the identity contours of the diasporic anti-racist struggle' (Akotirene, 2019, p. 22). This shows the connection between the concept and the purpose of 'Pretos que Voam' which brings the theory to practise, promoting feedback since this theorization has its origin in the daily experiences of black women.

Practices and Results Achieved So Far

Managed by the collective Quilombo Aéreo, the project studied in this case study started in 2020 with a match funding campaign to raise funds to cover the costs of training the first Brazilian class of black flight attendants. This was part of the collective's counterpart for the support received from the Enfrente Collaborative Fund that aims 'unite people and institutions to cofinance Brazilian peripheric projects' since 2019, from a group of institutional partners (Benfeitoria, 2023). The project was one of 15 selected by the Enfrente Match Funding in 2020. This means that for 3 years, the project received the support and encouragement of the Enfrente Collaborative Fund to finance all the actions that were prepared for the campaign. As part of the requirement of the Enfrente Match Funding it was necessary to raise BRL90,000 over the 3 years, or BRL30,000 per year. This value was agreed to demonstrate the social and collective mobilization of Quilombo Aéreo's network. For every BRL1 donated to the project, the Enfrente Fund invested an additional BRL2, tripling the collaboration.

The first funding campaign was undertaken online between 12 November 2020 and 20 December 2020 (throughout the COVID-19 pandemic). It reached 230 donors, and generated a total of BRL91,485.39 (BRL31,485.39 having been donated and BRL60,000 coming from the Enfrente Fund). The second campaign reached 459 contributors between 18 November 2021 and 31 January2022. The total raised through donations on this occasion was BRL32,969, with BRL60,000 coming via match funding by Enfrente, for a total of BRL 92,969. The last funding campaign recently closed. It started on 15 January 2023 and ended on 20 April 2023. Although it lasted longer than the previous campaigns, it had fewer donations and the amount raised was less than the first collection. The final amount reached was BRL90,266.00, with 152 individual and two corporate donors. After 3 months, the goal had not been reached so Quilombo Aéreo covered the gap. Thus, BRL30,266.00 was added through donations (including that of the collective) and BRL60,000 through match funding.

The funds are managed by the Quilombo Aéreo collective and were invested in the training of a class of black flight attendants and in the opening of the first Afrocentric flight school in the country. For the first goal, a public selection process, with details of the application procedures, was advertised in printed newspapers, television breaks, and social networks such as Instagram and Facebook. For the first class, a traditional and active school was hired in Porto Alegre (RS). However, for this class, specifically, Afrocentric content was taught.

There were 136 applicants for the 12 vacancies offered. Applications for the project selection were open from 12 to 29 March 2021, and had the following pre-conditions: Self-declared black people, over 18 years old, in a situation of social vulnerability, availability to follow the course for 9 months in the evening, high school education carried out entirely in the public school system, with supporting documents.

Some requirements were declassified since the project is aimed primarily at black people and subsequently at other social minorities. The decisive criteria for selection were, if applicants were black peripheral solo mothers, black trans women, or men, indigenous and, *quilombola* women. The class started with 12 people (8 women and 4 men), with 11 in the classroom and one online. Ten people completed the training.

It is important to point out that the participants experienced several conflicts with the school. After the payment of the courses the contract was not followed. An Afro-focused teaching methodology should have been applied, as well as the instructors of the school to undertaking training provided by Quilombo Aéreo about anti-racist practices in the training of flight crew members, but both were refused. The collective accessed the legal sector of ODABA, an association of Afro-entrepreneurs and partner of the Quilombo Aéreo, to ensure compliance with the contract between Quilombo Aéreo and the school. In the school there were only two black instructors and one of them, who is also trans, was one of the leaders of the project.

The school was the only paid partner, the others took part on a pro bono basis. There were two black women engaged in the admission process, four black women helped in the selection of the students, and two white women provided legal assistance. In addition, one black man (technology issues) and three black women – one black woman pedagogical expert, one black woman psychologist and one black woman lawyer – all of whom are partners of the collective, provided support throughout the training.

The leaders of the project estimate that beyond the nine families directly related to the graduating students, the project positively impacted a further 21 black families. This included taking into account the black teacher and instructor at the school, photographers, seamstresses, and other collaborators contracted by the project on an AD hoc basis. These families were spread across nine different communities.

On 20 November 2021 the first class, with 10 black flight attendants, graduated (although one of the students failed the license exams and, therefore, cannot go into a professional career). This date is Black Awareness Day in Brazil, and alludes to the important *quilombola* leader Zumbi dos Palmares, a reference to the resistance of the black movements. Four of the black women were hired, with others going through selection processes with the airlines. With the proposal for the financial self-sustainability of the project after their third month at an airline the people trained by the project must return the money invested in them by the project to allow other people to benefit. This transforms the formative process of black crewmates into an organic one.

After the first class, all the efforts and funds of the project were turned into the creation of the Afrocentric flight school. The need for this relates to Borges's (2022) point that:

> We can say that a peripheral and afro centred black feminist research is characterised by the visibility of the entanglements, crossings and power relations present in the process of knowledge production, from references coming from the cosmo-perceptions of African matrix, from the experiences and ways of life of black women on the African continent and in the diaspora and from the dialogues and translations between different subjects and perspectives. (p. 117)

In the first Afrocentric school of aviation, this macro theme will be addressed in an interdisciplinary way. The choice of the macro theme and the other associated themes relates to the context where the graduates will act and the project's objectives of challenging racism and sexism in the airline industry.

The collective adopted the guidelines of law 10.639/2003 (Brasil, 2003), which establishes the mandatory teaching of 'Afro-Brazilian history and culture' within the subjects that are already part of the curriculum of elementary and high schools in the country. They used this reference to build the pedagogy. The content provided by Brasil (2003) is described *ipsis literis* below, followed by the examples cited by the creators for adapting to aviation issues.

- EM13CHS601: 'Relate the political, social, and cultural demands of indigenous and Afro-descendants in contemporary Brazil to the historical processes of the Americas and to the context of exclusion and precarious inclusion of these groups in the current social and economic order'. For example, the airport and the bus station as places marked by race and social class, through a glamourization constructed and constituted through the structure of society divided and hierarchized by whiteness.
- EM13CNT305: 'Investigate and discuss the misuse of knowledge of natural sciences to justify processes of discrimination, segregation, and deprivation of individual and collective rights to promote equity and respect for diversity.' For example, the study of geography, the second most important subject in terms of relevance for the evaluation and recognition of the African continent for the constitution of the other continents in the last five centuries as a result of a criminal system of slavery.
- EM13CHS205: 'Analyse the production of different territorialities in their cultural, economic, environmental, political, and social dimensions, in Brazil and in the contemporary world, with emphasis on youth cultures.' For example, the history of aviation from a decolonial perspective, through untold stories of important achievements of black people for the development of this market.
- EM13LGG303: 'Debate controversial issues of social relevance, analysing different arguments and opinions expressed, to negotiate and sustain positions, formulate proposals, and intervene and make democratically sustained decisions that consider the common good and human rights, socio-environmental awareness, and responsible consumption at the local, regional, and global levels'. For example, COVID-19 arrived in Brazil by plane, where its first fatal victim was a black female domestic worker who had never been on an airplane.
- EM13MAT202: 'Plan and execute sample research using primary data or material from different sources on current relevant issues, including or not, support of technological resources, and communicate the results through a report containing graphs and interpretation of measures of central tendency and dispersion'. Example: unpublished industry research and data that, prior to the questioning of the Quilombo Aereo, were not raised and discussed.
- EM13MAT102: 'Analyse graphs and sampling methods of statistical research presented in reports published by different media, identifying, where appropriate, inadequacies that may lead to misinterpretation, such as inappropriate scales and samples'. For example, current cases of aesthetic and racial violence, moral and sexual harassment, racism in airports, prioritizing a technical analysis of procedures adopted in the face of the reality of structural racism.

Beyond this specific focus, all the traditional content planned for the training of crew members (except for the practical load), will be taught online, aiming to reach a wide audience. The project plans to achieve financial sustainability in the following ways: the contribution of people hired after graduation with a daily rate determined through the category's labour legislation for 2 years; the sale of the course to an audience that is not socially vulnerable; participation in funding selections of employability and education for minority groups, direct donations; the sale of lectures by the Quilombo Aéreo for the tourism, airport and regulatory sectors. School operations are expected to start in October 2023. *Aquilombar* is needed!

Conclusions

This case study shows the possibilities of mobilization and transformation inside traditional scenarios, like the air industry in Brazil, which is noted as a racist and sexist entity (Oliveira *et al.*, 2022). In May 2023, two cases became notorious in the national media. A black woman passenger suffered racism by crew members on a domestic commercial flight (G1 Bahia, 2023), and a black man suffered racism by local airport agents (Panho, 2023). It is urgent to discuss these practices in a relevant modal of transportation, like the air industry. So, paying attention to the training content of crew members, as well as promoting greater diversity in the work and consumer environment becomes essential.

The main results achieved by 'Pretos que Voam' to date are:

- They graduated a class of black flight attendants from the socioeconomic periphery by the end of 2021, writing a page in the history of Brazilian aviation.
- Of 12 students, 10 reached the conclusion of the course. This is above the average even for courses without affirmative action, which can contribute to higher absenteeism. When opportunities are created with the tools of equity and properly applied, it is possible to achieve extraordinary results.
- That four out of the graduates in the first year of the program gained employment in the aviation industry during the COVID-19 pandemic is truly remarkable.
- Communication, involvement, and mobilization of different actresses and actors, prioritizing blacks and women is noticeable.
- This presents a contribution to the preservation and dissemination of the memory of other black women leaders.
- Production of knowledge and practices to fight racism and promote equity is encouraged.

The main author of this case study, Laiara Amorim, who alongside Kênia Aquino and Shirley Reis founded the collective, points out other subjective benefits. The following are some considerations from her on behalf of the Quilombo Aéreo Collective Gabrielli *et al.*, 2024: Santos *et al.*, 2022.

> We believe that for all the groups we have reached the issue of professionalisation and employability are the main direct factors of impact, benefit, and change. But we also need to mention subjective benefits such as self-esteem, which is certainly present in a different way in their lives. Being able to give our students their first air travel, seeing their eyes water during take-off, seeing their excitement was by far one of the coolest things we experienced. Another very important point was the graduation day (Fig. 6.1), seeing the eyes of the black mothers from the periphery filled with tears of joy to see their children in toga and graduating made all our work worthwhile.

Fig. 6.1. First Class of 'Pretos que Voam' graduating, 2021. Courtesy of Kadu.

Further Reading

Gabrielli, C., Oliveira, N., Santos, G. and Borges, L. (2024) Quilombo Aéreo: Changing colour and gender relations in Brazil's sky. In: Swart, M., Cai, W., Yang, E. and Kimbu, A. (eds) *Routledge Handbook of Gender in Tourism: Views on Teaching, Research and Praxis*. Routledge.

Oliveira, N. de, Gabrielli, C., Santos, G. and Amorim, L. (2022) Intersectionality between racism and sexism in the Brazilian airline industry: Perceptions and strategies of Black Women crewmembers. In: Cembranel, P., Soares, J.R.R. and Perinotto, A.R. (eds) *Promoting Social and Cultural Equity in the Tourism Sector*. IGI Global.

Santos, G., Gabrielli, C., Oliveira, N. and Amorim, L. (2022) Lazer, gênero e sexualidade [livro eletrônico]: diálogos entre brasil e portugal. In: Isayama, H. and Baptista, M.M. (eds) *Mercado de Letras*.

References

Akotirene, C. (2019) *Interseccionalidade*. Pólen, São Paulo, BR.

Benfeitoria (2023) Benfeitoria Fundo Colaborativo Enfrente. Available at: https://benfeitoria.com/projeto/fundoenfrente (accessed 30 October 2023).

Berger, A. (2006) *Traversées de frontières*: postcolonialité et études de « genre » en Amérique. *Labyrinthe* 24(2). DOI: 10.4000/labyrinthe.1245.

Bhabha, H. (1998) *O local da cultura*. (ed.) UFMG, Belo Horizonte, BR.

Bitonga Travel (2023) Bitonga Travel Instagram. Available at: https://www.instagram.com/bitongatravel/ (accessed 1 March 2023).

Borges, L. (2022) Metodologia de Pesquisa Afrocentrada e Periférica. Marginália, São Paulo, BR.

Boukhris, L. and Peyvel, E. (2019) Os estudos turísticos frente a recepção do paradigma pós-colonial: uma exceção francesa, Via. [Online] 16. DOI: 10.4000/viatourism.4111.

Brasil (2003) Lei N° 1069, de 09 de janeiro de 2003. Altera a Lei no 9.394, de 20 de dezembro de 1996, que estabelece as diretrizes e bases da educação nacional, para incluir no currículo oficial da Rede de Ensino a obrigatoriedade da temática 'História e Cultura Afro-Brasileira', e dá outras providências. Diário Oficial da União. Available at: https://www.planalto.gov.br/ccivil_03/leis/2003/l10.639.htm (accessed 21 April 2023).

Diáspora Black (2023) Diáspora Black. Available at: https://diaspora.black/ (accessed 22 April 2023).

Ferreira, M. and Casagrande, L.. (2018) E Quem Disse Que Não É Seu Lugar? Por um Turismo Democrático e Inclusivo Para Negros e Negras. *Revista Mundi Sociais e Humanidades* 3(2), 1–21. DOI: 10.21575/25254774rmsh2018vol3n2665.

Ferreira, M. and Casagrande, L. (2020) Movimentos, Tecnologia e Pessoas Negras: É Possível um Outro Turismo? *Revista De Turismo Contemporâneo* 8(1), 149–167. DOI: 10.21680/2357-8211.2020v8n1ID19551.

França, W. and Malta, G. (2021) A Cidade Vista do Lado de Lá: O Turismo Como Potencializador da Imagem do Racismo em Petrópolis– Rio de Janeiro. Anais do XIV ENANPEGE.

G1 Bahia (2023) Mulher é Expulsa de Voo Após se Recusar a Despachar Mochila. Available at: https://g1.globo.com/ba/bahia/noticia/2023/04/29/mulher-e-expulsa-de-voo-apos-se-recusar-a-despachar-mochila-testemunha-acusa-companhia-aerea-de-racismo.ghtml (accessed 1 May 2023).

Geledés (2022) Racismo no Transporte Já Foi Presenciado por 72% dos Brasileiros. Available at: https://www.geledes.org.br/racismo-no-transporte-ja-foi-presenciado-por-72-dos-brasileiros/?gclid=CjwKCAjwhJukBhBPEiwAnilcNUlb6Kubel3Zs2rVBJRHRywNhm4_K30QGDg8Gn3mZplbWwTDOEEk9xoC9mMQAvD_BwE (accessed 20 April 2023).

IBGE (2019) Desigualdades sociais por cor ou raça no Brasil. Estudos e pesquisas: informação demográfica e socioeconômica. 41. Available at: https://biblioteca.ibge.gov.br/visualizacao/livros/liv101681_informativo.pdf (accessed 20 April 2023).

IBGE (2022) Desigualdades sociais por cor ou raça no brasil (2nd edn). *Estudos e Pesquisas: Informação Demográfica e Socioeconômica* 48. Available at: https://biblioteca.ibge.gov.br/visualizacao/livros/liv101972_informativo.pdf (accessed 20 April 2023).

Mbembe, A., Mongin, O., Lempereur, N. and Schlegel, J.L. (2006) Qu'est-ce que la pensée coloniale? *Esprit* 12.

Morais, I.A. de L., Santos, E.M. dos and Santos, I.B. dos (2023) Mulheres Negras no Turismo: Desafios Enfrentados Pelas Integrantes do Coletivo Bitonga Travel. *Revista Brasileira de Pesquisa em Turismo* 17, e–2722. DOI: 10.7784/rbtur.v17.2722.

Oliveira, N. (2020) Turismo Afrocentrado:Debates Iniciais. In: Mello, R. and Freitas, P. (eds) *Novos Olhares Sobre Turismo, Patrimônio e Cultura*. e-Publicar, São Paulo, BR.

Oliveira, N. (2022a) Representação e Representatividade dos Negros em uma Revista de Turismo de Luxo do Brasil. *Revista Brasileira de Pesquisa em Turismo* 16, 2325. DOI: 10.7784/rbtur.v16.2325.

Oliveira, N. (2022b) Afroempreendedorismo e Responsabilidade Social/Corporativa: Turismo e Cultura Afrodiaspórica no Brasil. *Revista Iberoamericana de Turismo* 12(2), 04–23. Available at: https://www.seer.ufal.br/index.php/ritur/article/view/13753 (accessed 20 April 2023).

Oliveira, N., Silva, P. da and Almeida, H. (2022) Mulheres Negras Viajantes: Experiências e Relatos de um Grupo de Facebook. *Caderno Virtual de Turismo* 22(1), 63–76. DOI: 10.18472/cvt.22n1.2022.1987.

Oliveira, N., Gabrielli, C., Santos, G. and Amorim, L. (2022) Intersectionality between racism and sexism in the Brazilian airline industry: Perceptions and strategies of black women crewmembers. In: Cembranel, P., Soares, J.R.R. and Perinotto, A.R. (eds) *Promoting Social and Cultural Equity in the Tourism Sector*. IGI Global, Hershey, PA. DOI: 10.4018/978-1-6684-4194-7.ch009.

Panho, I.A. (2023) Bando de Racistas Ignorantes', diz Renato Freitas, Deputado Revistado pela PF Durante Voo no Paraná. Estadão. Available at: https://www.estadao.com.br/politica/policia-federal-abordagem-racismo-voo-azul-renato-freitas-deputado-nprp/ (accessed 11 May 2023).

Quilombo Aéreo (2022) *Quilombo Aéreo*. Available at: https://quilomboaereo.com.br# (accessed 1 May 2023).

Santos, G., Gabrielli, C., Oliveira, N. and Amorim, L. (2022) Lazer, Gênero e Sexualidade: Diálogos Entre Brasil e Portugal. In: Isayama, H. and Baptista, M.M. (eds) *Mercado de Letras, BR*.

Santos, J. and Silva Coimbra de Sá, N. (2021) A Mulher Negra Vajante: Experiências e Estratégias de combate à sua (In)visibilidade no Turismo. *Revista de Turismo Contemporâneo* 9(2), 252–269. DOI: 10.21680/2357-8211.2021v9n2ID23584.

Section 2: Sustainability Challenges and Nature-Based Responses in Tourism

7 Birdwatching Tourism in Sustainable and Accessible Destinations: The Case of the Entrepreneurial Venture Explora Caloccita in the Mexican Pacific Coast

Kenia Ramos[1], Luis Rubén Pérez[2] and Narcís Bassols i Gardella[3]
[1]La Quinta Gran Bahía, Cuastecomates, Mexico; [2]Universidad Autónoma de Bucaramanga, Colombia; [3]Division of Quality Science, Uppsala University, Sweden

Summary
Bird watching is an activity that may prompt sustainable tourism development in remote communities. This case study lays out some lessons learned from the entrepreneurship case Explora Caloccita, a young venture offering bird watching to tourists in the Mexican Pacific coast. This case is based on descriptive research, laying out the practitioners' insights and analysing the supply of experiences related to bird watching. The innovative vision of the entrepreneurs, the financial and logistical backing by the local hotel La Quinta, and the support of residents and local tour operators allow for the conclusion that bird-watching tourism has been, in this particular case, a sustainable economic alternative. Furthermore, the entrepreneurship Explora Caloccita is embedded in a destination that strongly supports accessible tourism, sustainable local development and nature preservation. The venture and the place's sustainability are both possible thanks to the strong connections among all stakeholders and residents in the area, as locals also support the bird-watching routes and thus contribute to the conservation of the native and migrant bird species.

Bird Watching in Latin America and Mexico

Bird-watching tourism (also called 'avitourism') with artistic, cultural or academic motivations started being promoted in Mexico at the beginning of the 20th century, as some US scientific associations transferred their bird register and study techniques to Mexico. In the United States (USA), the bird counting method proposed by Frank Chapman in 1900, reported by media such as Bird Lore, as well as the foundation of the National Audubon Society around 1900, resulted in the passing of protection laws, which led to a decrease of Christmas bird hunting (Audubon Society, 2023). Bird watching with research and conservation ends became an alternative to hunting. As a novum, the Audubon Society accepted as members bird watchers who were not scientists, thus wide spreading the interest in bird watching and conservation. For their part, the Boy Scouts of America launched the Medal for the Study of Birds in 1911, which promoted environmental and conservational values. This became a prerequisite for members to climb up the organizational ladder (Gómez and Alvarado, 2010). These American associations may be seen as the first lobby groups who pushed for the creation of preserved areas to protect migratory birds, both in the USA and abroad, such as in Mexico.

In 1905, the American bird scientist William Beebe led an expedition to Mexico. The results were published under the title 'Two Bird Lovers in Mexico'. Furthermore, some of the USA's Christmas or winter countings would be published by the 'Audubon Field Notes' bulletin. Later in the 20th century, some works were published about Mexican birds, such as Emmet Blake's book *Birds of Mexico: A Guide for Field Identification* (1953), Ernest Edwards's *Finding Birds in Mexico* (1955), and *A Bird Watcher's Guide to Mexico* (1967) by Margaret Wheeler, a member of the Audubon Society. These books brought an increasing number of US bird watchers to Mexico. Since the 1950s, then, native species started to be differentiated from migrant species in Mexico. The availability of local guides, who were knowledgeable of the endemic species, was also an important factor in this growth. Hence, birds could be watched without needing to be hunted (Gómez and Alvarado, 2010). Since then, bird watching has been on the increase as a specialized practise carried out by biologists and birds scientists. They usually research bird flows in migration times, contrary to the amateur bird watchers who come during the holiday season when there are no migrations (Ortega *et al.*, 2022).

Mexico is crossed by two main bird migration routes. One goes from California to Colombia along the Pacific coastline, and the other from California to the Mexican Caribbean region along the coast of the Gulf of Mexico (Mexico - Secretaría de Medio Ambiente, 2017). While the latter is better known in the country, it is the former which boasts more migrant species that travel longer distances, making it of greater interest to bird watchers. Mexico hosts a large diversity of birds (some 1120 species), and worldwide it ranks sixth in terms of native bird species (Cantú *et al.*, 2020; CONABIO, 2021). Birds can be found in the country all year round in a variety of environments, including cities, deserts, beaches and protected areas. The selected study area is especially interesting as it includes some 210 bird species, 12 of which are endemic, making it a suitable destination for bird watchers.

In a number of Mexican destinations, this 'bird-watching scene' has sparked a series of entrepreneurial ventures, which enable locals to supplement their income. In this context, the tourist entrepreneurship Explora Caloccita, located in the State of Jalisco on the Mexican Pacific coast (see Fig. 7.1), is a successful case based on the conservation of the local biodiversity championed by the community, whose knowledge has become a significant part of the tourist product. This entrepreneurship is made up of recent tourism and biology graduates and is settled in an accessible tourist destination. The guided tours are set up considering accessibility rules and rely on fully engaged tourist operators who strive for sustainable development.

Bird-watching tourism is a growing, albeit still niche, form of tourism. As a result, there has been limited research on the topic. However, several works can be cited. Mohan and Athreya (2011) report on a case in India that underscores the importance of the community in building an avitourism destination, and Sari *et al.* (2011) see bird watching as a possibility to diversify coastal tourism. Both articles underscore the importance of community and diversification aspects and therefore support this case study. Much research, however, is still based on assessing the potential of avitourism, such as Li *et al.* (2013). Theoretical views of avitourism have been quite limited. Nonetheless, we would like to point to the recent work by Randler (2023) on the various motivational dimensions of bird watchers. Our case contributes to the other cases published to date, in the hope that it will stimulate more theoretical ideas on avitourism.

Focusing on the Latin American context, it is worth noting that Colombia is the country holding the biggest stake in bird-watching tourism in the region. Thanks to its huge biodiversity, Colombian stakeholders see this tourism typology as a development possibility in the post-conflict era the country is undergoing. The national government has supported this development with a series of documents devoted to bird-watching tourism. For example, Colombia, MINCIT (2017), is a booklet aimed at enterprises and outlining good practices in bird watching. The book by Moreno *et al.* (n.d) provides an update on the National Strategy for Bird Conservation until 2030, and Audubon Society

Map 1. Locating the case study: The beach town of Cuastecomates in Jalisco State, Mexico.

(2016) is a general presentation giving details on avitourism for interested entrepreneurs. In comparison, in Mexico, a recent guide for good avitourism practices was published jointly by academia and an NGO (García *et al.*, 2022).

For the last century, the protection of bird species and the promotion of avitourism have mainly rested on the side of NGOs in Mexico. In comparison in Colombia, avitourism has been recently pushed by the government, in cooperation with international bodies of experts such as the Audubon Society. This national drive for bird watching in Colombia has led to an interesting literature corpus by Colombian researchers (e.g. Ocampo-Peñuela and Winton, 2017; Maldonado *et al.*, 2018; Mora and Ramírez, 2019; Bonilla and Gutiérrez, 2022).

Colombia, followed by Mexico, are the countries that have produced the most research on bird watching tourism in Latin America, according to the reviews by Steven *et al.* (2015), Colombia, MINCIT (n.d.) and Ortega *et al.* (2022). This fact is also confirmed by the existing literature, which was researched by the authors of this case. Articles, theses and chapters from the years 2019–2023 were retrieved using the search engines of the databases associated with Scopus, Web of Science and Google Scholar. As a result, from a sample of 150 research works on bird tourism in Latin America from said period, it was revealed that 41% of the publications focused on avitourism in Colombia, 27% in Mexico, 14% in Brazil, 8% in Peru, 6% in Ecuador and the remaining 4% in Costa Rica, Bolivia and Argentina.

As for works related to our study area in Jalisco State, we would like to highlight the recent publication by Cuevas *et al.* (2018), which provides an overview of the general bird-watching industry in Mexico, with some comments on the Costa Alegre area, which is our study area. Also, Cuevas and Íñiguez (2017) compiled an inventory of an area in the State of Jalisco.

Bird Watching: General and Local Aspects

Bird watching is an activity that does not require heavy or complex gear. Neither is transportation expensive or long in avitourism destinations, as the focus is on conserving the species' natural environment. Walking along a route and watching the birds from the watching stations are the main components of this activity. Therefore, bird watching has traditionally been a means for community participation in tourism, as well as for complementary local offers to emerge. The low cost of both watching gear and logistics implies that the visitors may spend more on lodging, food service or guiding services during their stay. Thus, bird-watching packages are in high demand among tour operators located in avitourism destinations (Ramos, 2022).

In our study case, the community benefits from bird tourism watching as the locals open their doors to tourists who buy their crafts or food. Through this, visitors get to know their environment and potentially learn about the local wisdom, natural traditions, group beliefs, and knowledge the community members inherited from their ancestors. In some cases, a comparison of the bird counting and naming between locals and visitors may take place. This exchange of knowledge and experiences lead to a higher satisfaction level among visitors than in the cases of guiding by experts who are specialized in certain ecosystems or bird species. Learning about the birds' habits on the spot, practising bird counting or following up the birds' life cycle gives locals an unmatched degree of specialization, as many locals have been carrying out bird-related practices since their childhood (Ramos, 2022).

In the case of the local bird populations in our study, all of the aforementioned factors have resulted in both growing numbers of the protected migrant bird species and the numbers of bird-watching tourists and travellers. These visitors can enjoy publicly open ecosystems and can take handicapped family members or friends along with them as the destination is inclusive for visitors with any kind of impediment.

The data for the present case study (both about the destination's conditions for avitourism and the operations by the entrepreneurial venture Explora Caloccita) were gathered between June 2022 and June 2023. They were provided by the entrepreneurs themselves who agreed to disclose their names, business ideas and operations as set forth in this case.

Context: An Inclusive Destination on the Mexican Pacific Coast

Cuastecomates is a small town located on the central Pacific coast of Mexico, in the State of Jalisto, halfway between the well-known resorts of Acapulco and Puerto Vallarta (Map 1). It is surrounded by a deciduous forest and has a small bay in the shape of a horseshoe. Located on the stretch of the coastline known as Costa Alegre ['The happy coast'], the place is named after a tree that grows in Mexico and Central America. Its lush forests attract migrant birds, whose watching enhances the inclusive stay experience of the tourists. Since 2016, it has been 'Mexico's first inclusive town and beach, with special-purpose facilities, adapted streets and other services geared towards handicapped people as well as the elderly' (La Quinta Gran Bahía Hotel, 2022). On the Mexican Caribbean coast, similar initiatives have been replicated at the beaches of El Carmen and La Esmeralda, in Quintana Roo State and Miramar Beach, in Tamaulipas state (Jalisco State, 2016).

Cuastecomates' main innovations as an inclusive destination are the adaptation of the street furnishings and fixtures to allow for universal accessibility, adapted WCs and showers on the beach, and easy access streets with fixtures on the pavement for people with visual, auditive or mobility impediments – which allow for convenient mobility from the town's entrance to the beach. There are also designated areas for guide dogs. A local rule demands 'training tourist service providers to make them aware of the needs of disabled people' (Entornoturístico.com, 2016). This natural, urban and tourist setting improve the quality of life of the visitors and their families. Therefore, there is a healthy demand from tourists of different ages with different needs or impairments.

It is in this context that La Quinta Gran Bahía Hotel, the town's largest lodging facility, has developed. It is located in an area of woodland with direct access to the beach, and surrounded by forested hills. The beach offers calm waters, no deeper than half a meter in the first 50m from the shore. These waters allow for activities such as sports fishing, paddle boarding, snorkelling and kayaking which are aimed at all family members as they are performed in a natural and safe

Fig. 7.1. A white-throated magpie-jay (Caloccita formosa). Source: Wikimedia Commons.

Fig. 7.2. Bird watching for families with children. Source: Explora Caloccita.

environment. There are complementary activities in the destination, such as visiting a crocodile farm or cycling tours in the town and its surroundings. The pathways 'are adapted to visitors with impaired mobility, visual handicaps or elder visitors' (La Quinta Gran Bahía Hotel, 2022). While the area is protected by rules, the community is also actively engaged in conservation.

A New Entrepreneurship Venture in a Nature-Based, Inclusive Destination: Explora Caloccita

It was the potential for bird watching that led Kenia Ramos, a graduate in tourism from Guadalajara University in the State of Jalisco, to set up the entrepreneurship venture Explora

Caloccita. She was working as the Human Resources Manager of the La Quinta Gran Bahía Hotel when she saw the bird-watching tourism potential in the area and the opportunity to set up her own entrepreneurship venture (Ramos, 2022). The name of the venture is made up by the words 'explore' and 'caloccita', the latter being the Spanish name of the local flagship bird (Latin technical name: Calocitta Formosa; English name: white-throated magpie-jay, Fig. 7.1). Her idea was soon backed by the hotel and two more people joined her in starting off avitourism in Cuastecomates.

The offer of bird-watching tourism is a coherent complementary choice for a hotel settled in a natural destination focusing on sustainable practices. Explora Caloccita's offer is adapted to the local accessibility rules as they offer tours intended for families with children as well as older tourist groups. These two segments are offered 'easy' routes (i.e. routes with a short time frame and low physical requirements). Handicapped tourists may also take these routes, as adapted transportation is provided for them on land and sea (Jalisco State, 2022; Fig. 7.2).

The bird watching offer by the La Quinta Hotel required Explora Caloccita to be staffed by a multidisciplinary team. This was first set up by the founder of the venture, Kenia Ramos, and marine biologist Alan Manjarrez, who had also graduated from Guadalajara University. Later, they were joined by marine biologist Julián Villegas; a local who is passionate about bird conservation. Heading entrepreneurship in avitourism requires highly devoted people with a high interest in scientific bird study and in educating those who come to visit. A strong entrepreneurial and courageous spirit is also required to cope with the project's teething problems (Ramos, 2022).

The avitourism business plan of Explora Caloccita adopted international and state norms on the development of responsible and sustainable tourism, such as: monitoring the impacts caused by this activity, selecting and training the guides, setting up a register of local birds and editing printed materials, marking up the routes and the watching stations, carrying out actions to protect the birds and their habitat, promoting the product to the visitors via talks and visual materials, and disseminating good practices via web pages and social networks. Last, but not least, the business plan had to foresee the coordination of services such as lodging and food, with the services offered at the beginning or end of the routes by the local community (Amparán and Téllez, 2005).

Fig. 7.3. Educating different audiences about bird conservation. Source: Explora Caloccita.

Fig. 7.4. Hotel employees receive training about the venture's vision and products. Source: Explora Caloccita.

Before they are able to promote the birdwatching routes to the customers of La Quinta Hotel, the employees experience the product themselves. Employees are taught how to watch birds, as well as how to maintain and protect the area and all the species inhabiting it, including the turtles. Once they have received this training, the hotel staff also contribute to the quality of the experience. As the routes were launched, the hotel's staff were the first to evaluate them, making proposals for their improvement based on their local knowledge. They also became ambassadors for the avitourism product. An employee's son 'talked about this experience to his friends, and they all wanted to get to know it. In fact, locals notice a positive impact as they see they are treated the same way as tourists are' (Ramos, 2022; Figs 7.3 and 7.4).

The case of Explora Caloccita is particularly interesting because its members try to bring bird watching to a large public, with products aimed at the 'mass' market, instead of just a niche market. Thanks to competitive price levels (see Table 7.1) and the type of knowledge passed on to the visitors (which can be termed as 'essential' or 'non-specialized'), Explora Caloccita is set apart from other offerings in the region, aimed at the upscale or specialist market. While bird watching as a community-based or niche activity has been documented (Maldonado *et al.*, 2018; Bonilla and Gutiérrez, 2022), the same cannot be said for avitourism as a 'mass' activity with an entrepreneurial basis. This adds an interesting aspect to the present case. Furthermore, Explora Caloccita has an open mindset when it comes to cooperation with the local stakeholders and academia.

Strengthening Avitourism Products

The financial backing for the project's initial phase came from the La Quinta hotel, and this allowed Explora Caloccita's team to fine-tune the

Table 7.1. The three bird-watching routes supplied by Explora Caloccita and their main features and prices.

Route	Ardea alba	Coa citrina	Fragata real
Goals	Watching sea, beach and land birds on the Tule lagoon (36 species, 400 birds)	Experience Cuastecomates' attraction: the deciduous forest and its endemic birds, with a sight above the bay.	A tour on the Tule lagoon (RAMSAR listed) with boat through mangrove forests. Bird watching of marine and coastal birds in their beautiful landscapes.
Type	Land	Land	Mixed land and sea
Schedule	7:00–9:00; 17–19:00.	7:00–9:00; 17–19:00.	7:00–9:00; 17–19:00.
Capacity	1 to 10 people	1 to 10 people	1 to 10 people
Age	From 6 years of age	From 6 years of age	From 8 years of age
Prices	Adults: US$19	Adults: US$20, 21	Adults: US$32, 49
	Young people: US$16, 25	Young people: US$19, 20	Young people: US$29, 79
	Children: US$15, 16	Children: US$17, 68	Children: US$27, 08
	External guest: US$19	External guest: US$24, 25	External guest: US$32, 49
	Foreign guest: US$30, 31	Foreign guest: US$26	Foreign guest: US$37, 91
Price includes	Binoculars, specialized tour guide, booklet about bird identification, snack, water, compass, pen and land transportation	Binoculars, specialized tour guide, booklet about bird identification, snack, water, compass, pen and land transportation	Boat trip, binoculars, specialized tour guide, booklet about bird identification, snack, water, compass, pen and land transportation
Zoning	Tour goes through both public and private lands	Tour goes through communal lands	Tour goes through public lands

routes and activities and ascertain the financial viability of the products offered. Said financial backing also covered the administrative costs of the venture. After 6 months of operation, the entrepreneurship had paid off the initial investment, meaning the service was fully backed and promoted by the hotel. The growth in sales led to marine biologist Alan Manjarrez joining the initiative.

The guides of Explora Caloccita are able to evaluate the groups of people they accompany and rank them according to their skills and experience in bird watching, and their physical abilities. Therefore, guides are able to adapt the offering to the customer's needs. For instance, it is important to know whether a visitor is an occasional watcher, a regular one or a very frequent one. The same goes for devising routes. Unpaved pathways may be acceptable for most visitors but the elderly may find them uncomfortable or dangerous (Figs 7.5 and 7.6).

Three different routes are offered, each lasting for 2 hours, with an acceptable degree of difficulty, but each with a different focus (Table 7.1). Also common to the three routes is that, apart from bird watching on land and rivers, they also represent an opportunity to enjoy other natural and cultural assets present in the Cuastecomate's destination. All three routes sport nice landscapes and diverse birds and are conducted according to pre-existing standards to meet the needs of the visitors.

The environmental education before, during and after undertaking the route is key to Explora Caloccita achieving its sustainability goal, as talks are given by the guides about

Fig. 7.5. The routes offer different watching and counting points: From the lake shore. Source: Explora Caloccita.

Fig. 7.6. Visitors on a boat watching and counting birds. Source: Explora Caloccita.

protection and ecological awareness. As the participants of a tour are educated about how to behave inside protected areas and how to relate with the animals living there, resident species may live in peace in their nesting areas. Visitor safety and security are also enhanced as they are not exposed to dangerous animals or plants. They are also encouraged to keep well-hydrated and use suntan lotion. Further, they do not take any risks as long as they stay on the designated paths.

As for accessibility, it is worth highlighting that, although the destination strongly promotes it, only a few people with impairments make bookings. Still, extra support is provided in the bird-watching tours for three major detected

Fig. 7.7. Tourists attending the opening ritual of the Culazapa coffee festival: The intangible traditions of native communities are also part of Explora Caloccita's products. Source: Explora Caloccita.

impairments. Wheelchair-bound visitors are recommended to take the land part of tour 3, as it is the most accessible for them. For groups including persons suffering from Down syndrome, an additional staff member is available for support. Perhaps the most creative adaptation is for blind visitors. When a sight-impaired visitor joins a tour, it starts with an audio guide to the birds' songs, and all the participants are blindfolded. This raises awareness about the importance of all the body's senses when exploring a new place. It is also an opportunity to go deeper into more spiritually-focused values and to provide a particularly relaxing experience for some guests.

Explora Calocitta also impacts the local community at large. First, the hotel employees and the external providers must follow best conservation practices both on and off the hotel premises. Thus, the responsibility of conserving the local environment and promoting environmental activities ultimately falls onto the shoulders of the local community. This ensures local sustainable development, which in the end attracts visitors. If the local environment is to be 'sold' successfully, the residents must care for it.

Since last year, product development has diverged to develop products targeting more specialized markets and products related to the local intangible traditions. As for the former, these include the Christmas Counting, the visit to the Biosphere Reserve of the nearby Manatlán mountains, and the school-focused tours. The range of intangible traditions comes strongly into focus in the tour offers, such as the day tour to a native community of women coffee growers in the nearby village of Culazapa, a product which highlights the ecological and human connections between the Cuastecomates town and its hinterland (Fig. 7.7).

As for visitor numbers, the most recent figures from the first half of 2023 vary from 150 to 300 per month. As the tours must respect the environment's carrying capacity, only 10 people are admitted to each tour and there are only 2 or 3 tours on any given day. With these numbers, in December 2022, the peak holiday season, some capacity problems arise due to the high demand. Therefore, prices are increased during this period to make sure tours are only booked by interested visitors. As for the ages, the majority of visitors are adults, with only 5% under the age of 18. Only a couple of individuals have required support for physical impairments.

Despite all the positive developments, there are external challenges and threats. One of these challenges is the overall impact of uncontrolled visitor activity on the routes, which can directly threaten the birds' ecosystems.

However, this cannot be changed, as most of the routes are on public land, see Table 7.1. What is more concerning are the informal development projects to build second residences which are altering the area's natural landscape. This means that there is less food available for the birds and so there are less of them to see, and this affects Explora Caloccita's activities. Consequently, the venture has recently asked the municipality to improve control over informal building. This situation demonstrates a conflict of interests in the destination. However, there is an initiative to nominate the Tule lagoon as a protected area, which would mean better control over visitors but, unfortunately, there is a lot of misinformation in the community about this issue. In the long term, if the lagoon and the routes continue to suffer from growing human pressure, the whole Explora Caloccita project may cease to be viable. Apart from these external threats, consolidating the finances and operations of this young company via a sustainable growth model is a challenge on the internal side.

Conclusions

Avitourism contributes to the diversification of the local tourist offer, helps to mitigate visitor seasonality and puts quality before quantity in visitor management. In a very special way, the bird-watching tourism we have reported contributes to the education of visitors as they acquire knowledge about the conservation of the different local species and their importance to the ecosystem. Furthermore, the entrepreneurship venture Explora Caloccita has proved itself as a means to impact and strengthen the communitarian and inclusive tourism as it benefits locals who want to become guides or offer support along the routes. Future planned expansion includes other areas in the region, as well as nearby universities, by introducing bird watching in their campus gardens to motivate future practitioners.

As avitourism in the destination is not just a business for specialized professionals or scientists but rather an open initiative targeting all locals and visitors alike, it highlights the locals' knowledge as they have co-existed with the birds for generations. Furthermore, the revenue generated through this activity is spread across the place as residents participate in the tourism economy as food providers to restaurants or as goods or crafts providers to local shops. This allows for the tourism revenue to trickle down through the supply chain.

The challenges remain to preserve the natural environment and strengthen the young company. Another issue to face, which requires further research, is to understand why the destination's accessibility is met by such a low demand, given the fact that the place has clearly engaged with it. Conducting market research and implementing improved communication policies may help to attract an enhanced market segment to Cuastecomates. Finally, the present case of avitourism development may be seen as an example of good practice in terms of conservation, community participation and entrepreneurship which can spark similar initiatives in other destinations.

References

Amparán, R. and Téllez, J. (2005) El aviturismo en la zona rural. In: César, A. (ed.) *Desarrollo Rural y Turismo*. Universidad de Guadalajara-and Universidad de Buenos Aires, pp. 59–64.

Audubon Society (2016) Aviturismo: Conservación y desarrollo económico. Presentation. Available at: https://www.colombiaproductiva.com/getattachment/009f1b5b-be72-4265-9e99-cd350598f34e/Aviturismo.aspx (accessed 21 June 2023).

Audubon Society (2023) History of Audubon and Science-based Bird Conservation. Available at: https://www.audubon.org/about/history-audubon-and-waterbird-conservation (accessed 13 February 2023).

Bonilla, N. and Gutiérrez, L. (2022) Métodos para evaluar potencial del aviturismo en Santa María, Boyacá, Colombia. *Turismo y Sociedad* XXXI, 227–280.

Cantú, J.C., García, E., González, G. and Sánches, M.E. (2020) *Riqueza alanda: el crecimiento del aviturismo en México*. Defenders of Wildlife and UABCS, Mexico City (online report).

Colombia, MINCIT (Ministry of Trade and Tourism) (2017) Guía de Buenas Prácticas para la Actividad de Aviturismo en Colombia. Official Document.

Colombia, MINCIT (Ministry of Trade and Tourism) (n.d.) Red Nacional de Rutas de Aviturismo. Official Presentation. Available at: https://www.colombiaproductiva.com/getattachment/203a2d52-277a-4cc0-ad76-a4d600abb335/Red-nacional-de-rutas-de-aviturismo.aspx (accessed 28 June 2023).

CONABIO (2021) Sistema Nacional de Información sobre Biodiversidad (SNIB). Registros de ejemplares, versión 2021-03. México. Available at: http://www.snib.mx/ejemplares/docs/CONABIO-SNIB-Version202103.pdf (accessed 26 June 2023).

Cuevas, J. and Íñiguez, L. (2017) Aves del Puerto Interior Turístico Jocotepec, en el Lago de Chapala, Jalisco. *México. Huitzil* 18(2), 261–271.

Cuevas, J., Tello, I., González, J. and Palomera, C. (2018) ¡Sal a pajarear! Una mirada a la observación de aves en México. *Orama* 2(1), 29–33.

Entornoturístico.com (2016) *Cuastecomates, primera playa accesible para personas con discapacidad en Jalisco*. Available at: https://www.entornoturistico.com/cuastecomates-primer-playa-accesible-personas-discapacidad-jalisco/ (accessed 20 March 2023).

García, J.E., Lizárraga, A. and Cantú, J.C. (2022) *Guía de buenas prácticas para el aviturismo en México*. Defenders of Wildlife and UABCS.

Gómez, H. and Alvarado, E. (2010) Breve historia de la observación de aves en México en el siglo XX y principios del siglo XXI. *Huitzil* 11(1), 9–20.

Jalisco State, Sistema para el desarrollo integral de la familia (DIF) (2016) *Familias en Red: Cuastecomates Playa Incluyente*. [Vídeo]. Available at: https://www.youtube.com/watch?v=kSGfuzhT8pU (accessed 14 February 2023).

Jalisco State, Tourism Department (2022) Logros Pueblo y Playa Incluyente: Bahía de Cuastecomate. Available at: https://es.slideshare.net/VIMVIM4/plan-maestro-y-cuastecomates-final (accessed 14 February 2023).

La Quinta Gran Bahía Hotel (2022) *Conoce Cuastecomates, Jalisco*. Available at: https://www.quintagranbahia.com/cuastecomates-jalisco (accessed 7 February 2023).

Li, F., Zhu, Q. and Yang, Z. (2013) Birding tourism development in Sichuan. *China. Tourism Economics* 19(2), 257–273.

Maldonado, J.H., Moreno-Sánchez, R.P., Espinoza, S., Bruner, A., Garzón, N. *et al*. (2018) Peace is much more than doves: The economic benefits of bird-based tourism as a result of the peace treaty in Colombia. *World Development* 106, 78–86.

Mexico - Secretaría de Medio Ambiente (National Government Agency) (2017) México ofrece avistamiento de hasta 1096 especies de aves. Available at: https://www.gob.mx/semarnat/articulos/en-el-dia-internacional-de-la-diversidad-biologica-vamos-a-ver-aves-106971 (accessed 10 June 2023).

Mohan, D. and Athreya, R. (2011) Sustainable bird based tourism in India's remote North-East frontier. *International Journal of Innovation Science* 3(1), 23–28.

Mora, J.A. and Ramírez, N.A. (2019) Potencialidad del aviturismo para el desarrollo de iniciativas comunitarias en Cumara, Meta (Colombia). *Revista Internacional de Turismo, Empresa y Territorio* 3(2), 84–112.

Moreno, N., Maldonado, O., Falk, P., Carantón, D., Baptiste, M. *et al*. (n.d) Estrategia Nacional para la Conservación de las Aves de Colombia 2030. Instituto de Investigación de Recursos Biológicos Alexander von Humboldt. Official Document.

Ocampo-Peñuela, N. and Winton, R.S. (2017) Economic and conservation potential of bird-watching tourism in postconflict Colombia. *Tropical Conservation Science* 10, 1–6.

Ortega, R., Sánchez, L. and Figueroa, J. (2022) Del Río Bravo a Tierra del Fuego: perfiles, práctica, motivaciones, preferencias y percepciones de la comunidad de observadores de aves de América Latina. *Sociedad y Ambiente* 25(1), 1–25.

Ramos, K. (2022) *El Aviturismo como herramienta de conservación y sustentabilidad (Webinar)*. Extensión Universitaria-Universidad Autónoma de Bucaramanga.

Randler, C. (2023) Motivations for birdwatching: Support for a three-dimensional model. *Human Dimensions of Wildlife* 28(1), 84–92.

Sari, C., Oban, R. and Erdogan, A. (2011) Ornotho-tourism in Antalya. *Procedia Social and Behavioral Sciences* 19, 165–172.

Steven, R., Morrison, C. and Castley, G. (2015) Bird watching and avitourism: A global review of research into its participant markets, distribution and impacts, highlighting future research priorities to inform sustainable avitourism management. *Journal of Sustainable Tourism* 23(8–9), 1257–1276.

8 National Parks and Landscape Commodification: New Challenges for Nature-Based Tourism in Patagonia

Gabriela Landini
Instituto de Geografía, Facultad de Filosofía y Letras, Universidad de Buenos Aires/CONICET

Summary

Some of the most important tourism destinations in North Patagonia are located inside or adjacent to National Parks, especially around a circuit known as the Lakes Corridor that runs along the Andes through more than 600 km. Even though the main attractions are in the protected areas, accommodation and other services are provided in the towns or cities nearby. As a result of the intensification of tourism and recent inward migrations, the region has experienced urban growth and the emergence or increase of social disputes around land use, environmental issues, and local residential needs. In this case study, we analyze this process, focusing on destinations from the Lakes Corridor and paying special attention to the COVID-19 pandemic effects. The main objective is to understand the characteristics, actors involved, and tensions that emerge in urban development related to this specific kind of tourism based on the positive valorization of nature. This case study is part of an investigation regarding the configuration of tourist landscapes in North Patagonia. Methodologically, it involves the construction of data through observation and interviews, as well as qualitative analysis of different sources, like promotion material and public documentation regarding territorial planning.

The Value and Interest of the Case Study

Nature-based tourism is currently a very popular tourism modality. National Parks and other protected areas are important tourist attractions. Small towns or cities near these places concentrate the infrastructure and services for the activity. In addition, mountains have been historically identified, for different reasons, as 'healthy' places. Therefore, small or medium-sized urban centers in mountain areas experience significant growth due to a combination of increasing tourist activity and amenity migration (Moss, 2003; Moss *et al.*, 2008).

In the present, some of these dynamics have been intensified worldwide. The COVID-19 pandemic had an important influence on tourism, not only because of its first negative impact but because of the effects of the quarantine on people's desire to go out and the consequent increase in tourist activity in general once the distancing measures were over. In this context, open spaces and nature-based tourist destinations emerged as preferred places to disconnect, recreate, and relax in a 'safe' environment.

For all these reasons, some mountain regions with nature-based tourism, like Andean North Patagonia, are experiencing a significant increase in mobilities and, as an outcome, several problems linked to urban growth and intensive tourism. In this context, social tensions and environmental deterioration should be addressed as problems that threaten the development of tourist activities themselves.

Background: National Parks and the Beginning of Tourism in North Patagonia

At the beginning of the 20th century, the Argentinean state created its first protected areas following the example of the United States National Parks. Most of them were in southern Andean areas, with the objective of consolidating national sovereignty in a region belatedly incorporated into the national territory after its military conquest during the last decades of the 19th century. The jurisdiction of the protected areas was defined by the deliberate process of the selection and valorization of some specific characteristics, like the presence of lakes, mountains, and forests, which led to an aesthetic appreciation of Patagonian landscapes, considered, in this sense, as an important part of the Nation's *repertoire* of natural beauties and a reason for developing an elite-orientated tourist activity.

The creation of the Dirección de Parques Nacionales (DPN), a state institution for the administration of the protected areas, settled the origins of tourism policies in the country in the 1930s (Piglia, 2012). This led to specific spatialization dynamics characterized by the development of urbanization processes closely linked to the conservation of extensive portions of land as reservoirs of a – supposedly – unaltered nature. While the preservation of these lands was designed to secure them for leisure uses, towns were necessary for providing tourist infrastructure and services. The creation of these protected areas can be read as the conversion of these spaces as specific means of production. They are preserved as public spaces, out of the market, securing their specific use for leisure. On the other hand, complementary spaces are necessary to provide infrastructure and services (Sánchez, 1991). Therefore, urbanization was driven by the preservation of 'natural' spaces.

The consideration of human presence as a problem affected the indigenous people who lived in those places before the arrival of the Argentinian state. From that moment, the Mapuche people – indigenous inhabitants of Patagonia – were seen as a threat to the forests, because of their livestock activity and use of wood for domestic purposes. However, in general terms, the increase in population was not frowned upon. On the contrary, this was desired in the context of boundary consolidation. The development of existing towns and the creation of new touristic *villas* were some of the main objectives of the National Parks administration during the 1930s and 1940s. The aim of this process was, in part, to integrate the touristic sub-region of North Patagonia around a line of protected areas along the Andes. It centered on the first National Park, Nahuel Huapi – created in 1934 – and its head city, San Carlos de Bariloche. This is, at present, one of the biggest and most important cities in the region.

State intervention in the territory included the creation or improvement of infrastructure. Some of the most important and urgent were directed to solve communication needs by opening roads and setting up ports, and providing constructions for the parks' administration – quartermasters, forest stations, personnel housing – and urban services like schools, hospitals, and post offices. These new buildings stood out in the small, recently founded, towns whose main activities until then had been self-subsistence activities, cattle raising, and logging. Architecture was an important instrument for homogenization and for acquiring a European aesthetic. In San Carlos de Bariloche and San Martín de los Andes, this is still an element of differentiation because later constructions followed in some way that style, giving these cities a particular appearance. One of the most iconic of these constructions is the Llao Llao Hotel, built in the 1930s. In early times, these places were compared to, and modified to resemble, the European Alpes, where the Argentinian high class used to vacation.

In summary, National Parks set the foundations for tourism in the region. From that moment, the organization of this activity continued being driven, partly, by policies designed and implemented by the national state and provinces, as well as private businesses that emerged in the period. The outcome has been the predominance of nature-based tourism in the region as its main economic activity at present. Because of the expulsion and displacement of the previous population, the process also set the foundations for social tensions that prevail at present.

Transformations in Nature-Based Tourist Destinations

Within the vast region of Patagonia – more than a million square kilometers, one of the most visited circuits is the Lakes Corridor that stretches for more than 600 km across three different provinces, Neuquén, Río Negro, and Chubut. The main destinations are small or medium-sized cities with access to national parks that also extend along this southern section of the Andes: Lanin, Nahuel Huapi, and Los Alerces. San Carlos de Bariloche plays a central role in this circuit. This city, with a history strongly linked with national parks, is one of the most populated in the region and very well known for its tourist activity among national and international visitors. Less than a 100 km north starts the most famous scenic road of the region. The Seven Lakes Road is a fraction of National Road n°40 that crosses the country from north to south. This fraction goes from Villa la Angostura – one of the tourist *villas* founded by National Parks on the shores of Nahuel Huapi Lake during the 1930s – to San Martín de los Andes (Fig. 8.1). Panoramic views play an important part in the touristic consumption of these spaces.

At present, tourism proposals incorporate an active attitude to the contemplative relation with nature. A wide range of outdoor sports and recreational activities are available in these destinations, including trekking, hiking, horse riding, kayaking, mountain biking, and fishing, among others. Winter sports are also an important attraction. There are currently five ski resorts along this corridor. Outdoor sports

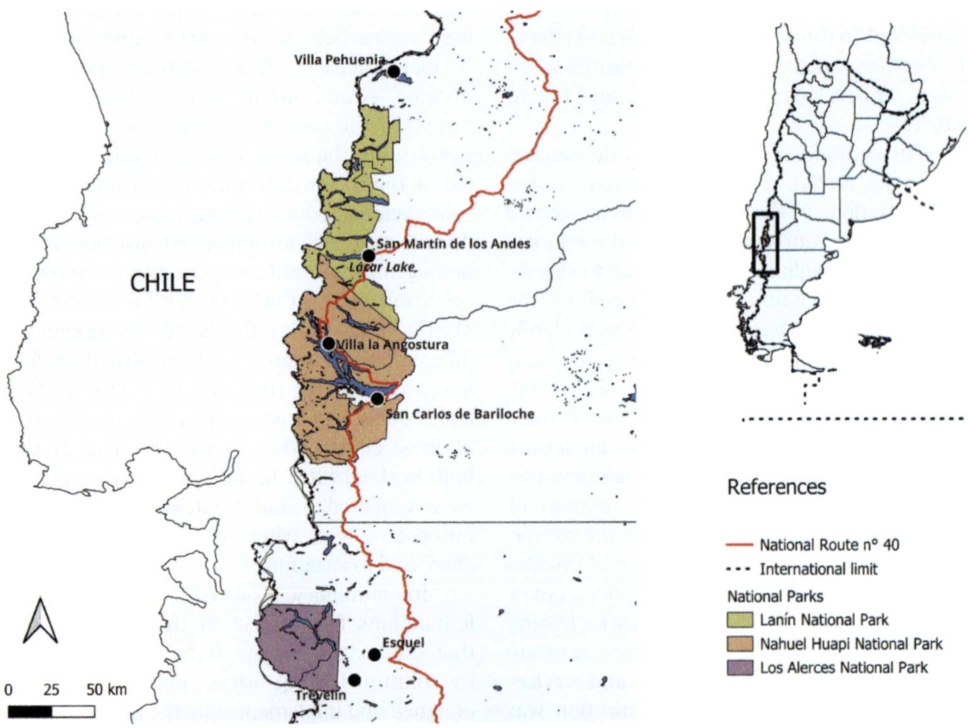

Fig. 8.1. National Parks in North Patagonia. The map shows location of Lanin, Nahuel Huapi, and Los Alerces National Parks and the main destinations of the Lakes Corridor. Author's elaboration.

competitions are important events organized during the year to attract tourists in different seasons. Trail and long-distance bike races are some of the most popular ones. Natural landscapes and the challenging mountain environment are the main attractions in the promotion of these competitions.

The imaginary of this region as a reservoir of 'untouched' nature has a strong influence on local and international tourism. In this sense, it is important to pay attention, to the visitors' desires and expectations, rather than the landscapes' real or perceived characteristics, in order to understand how this imaginary works and what consequences it brings in relation to commodification processes.

Different representations are built, through a process of images that are constantly changing and relating to each other. It is worth noting that tourism imaginaries are not independent of other imaginaries. Attractiveness is, therefore, related to other imaginaries of Patagonia that circulate in the market and in the cultural field. For instance, Patagonian landscapes, even before their exploitation by tourism, had been constructed as a border region, with low human impact, very vast, and 'mysterious'. At present, the region has relatively high notoriety, not only in the tourist field. Patagonia is a brand with a strong power of evocation in this nature-related sense, which explains the extended, and in some cases disputed, use of its name for selling different products beyond tourism, such as outdoor-specialized dressing, 'natural' food, and drinks with specific territorial characteristics like wines and 'craft' beer. Tourists look, therefore, to experience a 'return to nature', to a pristine space, while visiting these destinations. Patagonian protected areas are the most visited in the country. According to official information, during January 2022, more than 70% of visits to national parks in Argentina were registered in Patagonian parks (Ministerio de Turismo y Deportes, 2023).

From tourism to amenity migrations: Contemporary mobilities in Patagonia

The Lakes Corridor has been experiencing tourism growth during the previous decades. The increase in arrivals during the late 20th century and the beginning of the 21st is a visible fact, especially in small or medium-sized towns like San Martín de los Andes, which went from receiving 37,510 visitors during 1990 to 232,529 in 2019 according to local reports (MSMA, 2020). The COVID-19 pandemic introduced, at first, a deep crisis in tourism in the region. However, after this, Patagonian destinations have recovered relatively well.

In the context of tourism reactivation, in 2021, the Argentinean government launched a national program that consisted of the return of 50% of expenses made by domestic tourists in different categories like transportation, gastronomy, and hotels. Bariloche, Villa la Angostura, and San Martín de los Andes were among the most chosen places to travel in all four editions of this program. Although further proof should be presented in future research to confirm it, it is plausible to formulate the thesis that this can be, at least partly, related to the idea of mountain regions as 'healthier', especially when open spaces were recommended as safe contexts to be in and to practice recreational activities in during the pandemic. Another reason for the increase in tourist activity in these destinations has to do with difficulties associated with travelling abroad. People who used to go to other countries for vacations chose, instead, to travel to Patagonia. However, the first argument has more sense, considering that the region has also attracted people looking for 'better' places to live.

The pandemic had an important effect on immigration to these mountain areas. New temporary and permanent residents, particularly middle- and high-class people coming from big cities like Buenos Aires, have arrived motivated by the desire to undertake a radical change of lifestyle. This kind of migration is not new in the North Patagonian region. The combination of real or perceived high quality of life associated with nature-related imaginaries and landscape valorization has an important power of attraction for recent migrations. This has increased significantly since 2001, when a deep economic crisis affected Argentina. Insecurity drove middle- and high-class people towards the outskirts of big cities or to other smaller and more 'quiet' places in the country. Between 1991 and 2022, the population in the Bariloche

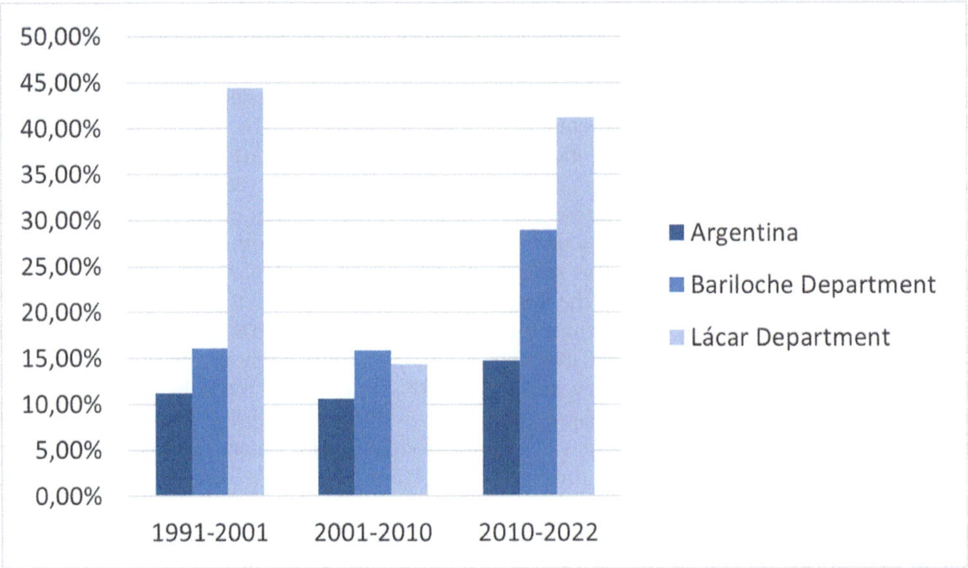

Fig. 8.2. Intercensal demographic variation expressed in percentage. Author's elaboration based on published results of national census (INDEC, 1991, 2001, 2010, 2023).

Department increased from 94,640 to 164,065 inhabitants, while the Lácar Department – where San Martín de los Andes is the head city – increased from 17,085 to 39,870 people. This means that in the last 31 years, these areas have increased their populations by 173% and 233%, respectively (e.g. Fig. 8.2).

This kind of mobilities, studied under the concepts of amenity migration (Moss, 2003; Moss *et al.*, 2008; Abrams *et al.*, 2012) or lifestyle migration (Torkingston, 2010; Benson and Osbaldiston, 2014), is characterized by those motivated by non-economic reasons, based on the differential appreciation of the quality of life in big cities and rural areas or small towns, especially in mountain areas. The appearance of these migrants is possible because of certain conditions, including an increase in discretional time and income, improvement in transportation and communication, and higher levels of comfort in historically remote places. The impacts of these migrations are varied, but include population growth, changes in services offered in the destinations and consumption habits, and social tensions produced by pressure over natural resources.

In Patagonia, especially Andean areas have been identified as destinations for amenity migrants. Researchers have verified the trend of amenity migration in the context of residential transition and conversion of tourism destinations, defined as post-tourism (González *et al.*, 2019). The real estate market has extended and gained a central role in local economic dynamics in traditional tourism destinations like Bariloche, Villa la Angostura, and San Martín de los Andes. The impacts of these new mobilities have produced new outcomes and deepened some pre-existing problems related to the way land and natural resources are seen and appropriated, on the one hand, and to inequalities based on access to housing, on the other.

Nature commodification: Disputes over land use and environmental conflicts

The creation of national parks set an important basis for the way the relation with nature is raised and represented in Patagonia. The consolidation of tourism during the last century implied a positive valorization of the region's landscapes and has led to a specific way of perceiving, occupying, and using the land. The concept of commodification (Prudham, 2009) is useful to understand this process of material and symbolic production of nature for its consumption in the market. The hypothesis is that, although this process was firstly driven by the tourist market, considering recent transformations in mobilities, the real estate business has emerged as an additional factor in nature commodification.

In both cases, the relation with nature is mediated by aesthetic appreciation of landscapes. Images have an important role in promotion and marketing. Usually built from a panoramic point of view, they tend to show still, mirrored lakes surrounded by bushy forests and snowy mountains in the distance. These 'natural', pacific, pleasurable, autonomous, 'untouched' landscapes are at the same time built and presented as 'given'. They are the main resource for tourism and real estate businesses in Patagonia.

The consumption of places in these terms needs infrastructure and services that allow tourists, residents, and temporal or permanent migrants to 'live' those landscapes. Therefore, places with certain 'natural' conditions and low populations – real or perceived – are especially sought for building new hotel complexes and gated communities. While population growth generally pressures the real estate market, this specific kind of place consumption seeks open spaces where those 'privilege landscapes' (Duncan and Duncan, 2004) can be built. As an outcome of this, different disputes emerge around land use and access to some spaces or natural resources. This often happens in lake or river coasts, which are used for recreational sports such as fly fishing – one of the most important activities in some of these destinations, but also highly valorized from an aesthetic point of view. Thus, lifestyle or leisure-driven migration joins the existing competition for the use of these places, deepening conflicts around them.

The increase in tourist services, buildings, and infrastructure also produces environmental degradation. Building or expanding ski resorts or hotel complexes in mountain areas has been a motive for public confrontations during previous decades in the region. Conflict involves different partialities among local communities, including environmental organizations, indigenous people, real estate agents, concessionaires, and business people. These ventures have been denounced as causing great harm to the environment, including deforestation and its negative effects on soil deterioration and flood vulnerability, and contamination of water that, in some cases, is consumed by people. Last, but not least, environmental deterioration impacts negatively on tourism itself, considering that, as has been noted, 'pristine' nature is the main attractiveness of the Patagonian region.

Urban growth and residential needs

As was pointed out in the beginning of this case study, National Parks are central to the construction of attractiveness in this Patagonian subregion. Although a great part of the recreational activities has been developed in those protected areas, they cannot hold the whole infrastructure and services like hotels. In addition, the attractiveness of these places has driven a wide range of spatial mobilities that go beyond traditional tourism. Since the late 20th century, Patagonia has been experiencing important demographic growth as a result of temporal and permanent mobilities driven by non-economic motives. This is a result of the combination of the idealization of rural, 'natural', and in this case, mountain areas and urban pressure. Therefore, urbanization and urban growth in the region have acquired specific traits, which have been defined by some authors as 'naturbanization' (Prados Velasco, 2012). This is explained as a product of the settling of people in protected areas, attracted by their desire to live, work, and recreate in a high-quality environment.

The main cities within the Lakes Corridor have extended in a particular way, as we have suggested earlier, with important participation of real estate projects that prioritize luxury offers for a high-purchasing power public. Consumers to which these products are orientated cannot be clearly categorized as tourists or residents. Some of them move to these places permanently, others are considered 'temporal residents' during vacation seasons, and some of them remain as long as their work and life in their place of origin allow them. This was a feature of the COVID-19 pandemic. Furthermore, some of these developments include recreational infrastructure like golf courses, and ski slopes, or even offer guide services for trekking, fishing, mountain biking, and other activities.

The location of these resorts, gated communities, or hotel complexes is defined by the preference for 'natural' spaces, on the outskirts of the city, with no neighbors in sight The outcome is a fast extension of suburban expansion, known as urban sprawl. Urban growth is currently driven by planned projects and land subdivisions which, in most cases, are driven before public services are insured. New population centers emerge in rural areas, often along lake shores, without potable water or sewer services, producing adverse effects on the environment, as it has been suggested in the previous section.

The pressure on rural and peri-urban lands is enhanced by the existence of protected areas which disable the possibility of developing extensions of these high environmental-quality lands. The convergence between these processes results in an important rise in land prices. At the same time, the supply of rental housing is very scarce, due to the convenience for landowners to offer their properties in temporary rent – over traditional contracts. Consequently, housing emergencies have been repeatedly declared in all these Andean cities in North Patagonia in recent years.

Conclusions

Nowadays, tourist destinations are going through transformation processes in relation to the emergence of new mobilities that cannot be read as traditional tourism or understood based on the dichotomic opposition between daily life and vacations. In this case study, we have seen how places near national parks in mountain areas of North Patagonia are currently points of attraction for heterogeneous mobilities that materialize in different kinds of residents, migrants, and tourists, all of which melt into a complex reality. These mobilities are driven by – and contribute to – processes of commodification of nature.

Tourism spatialization during the 20th century resulted in a paradoxical urbanization process in the region, driven by the creation of protected areas that sought to leave portions of space preserved from human impact. In that way, an important fraction of the territory was intended to be kept 'unaltered' through state regulation. The limits put on primary activities affected people who habited rural areas, mainly indigenous people who lived from cattle raising, while national parks were left as public spaces to visit and where an important part of the tourist activities is still practiced.

The imaginary of 'virgin' nature, mobilized through the Patagonia 'brand', still has a strong power of evocation. In addition, these imaginaries tend to link these pleasant images with the idea of certain places as 'healthier' and 'safer'. Mobilities driven by these imaginaries are having a great impact on tourist destinations in Patagonia. In the context of the COVID-19 pandemic, this effect intensified tourist arrivals and previous migration dynamics. These increases have led to disputes about land use and dispersed urban growth, which affects the environment through deforestation and contamination processes. It also places pressures on the land market, raising prices and increasing social inequalities based on difficulties in accessing housing. Finally, it is a factor of transformation of landscapes themselves.

In conclusion, all these consequences represent new challenges for local administration in nature-based tourist destinations, based on the threat of urban sprawl that pressures natural resources and, therefore, tourism itself.

References

Abrams, J., Gosnell, H., Gill, N. and Klepeis, P. (2012) Re-creating the rural, reconstructing nature: An international literature review of the environmental implications of amenity migration. *Conservation and Society* 10(3), 270–284.

Benson, M. and Osbaldiston, N. (2014) Understanding lifestyle migration. In: *Theoretical Approaches to Migration and the Quest for a Better Way of Life*. Palgrave McMillan, New York.
Duncan, J. and Duncan, N. (2004) Landscapes of privilege. In: *The Politics of Aesthetic in an American Suburb*. Routledge, New York.
González, R., Merlos, M. and Contreras, F. (2019) Post turismo en clave territorial. Una indagación teórica desde el diálogo posmodernidad-territorialidades. *Aportes y Transferencias* 7(2), 49–64.
INDEC (Instituto Nacional de Estadísticas y Censos) (1991) Censo Nacional de Población, Hogares y Viviendas 1991. Available at: https://www.indec.gob.ar/ (accessed 25 May 2023).
INDEC (Instituto Nacional de Estadísticas y Censos (2001) Censo Nacional de Población, Hogares y Viviendas 2001. Available at: https://www.indec.gob.ar/ (accessed 25 May 2023).
INDEC (Instituto Nacional de Estadísticas y Censos) (2010) Censo Nacional de Población, Hogares y Viviendas 2010. Available at: https://www.indec.gob.ar/ (accessed 25 May 2023).
INDEC (Instituto Nacional de Estadísticas y Censos) (2023) Censo Nacional de Población, Hogares y Viviendas 2022. Resultados provisionales. Available at: https://www.indec.gob.ar/ftp/cuadros/poblacion/cnphv2022_resultados_provisionales.pdf (accessed 25 May 2023).
Ministerio de Turismo y Deportes (2023) Sistema de Información Turísitca de la Argentina. Available at: www.yvera.tur.ar (accessed 2 August 2023).
Moss, L. (2003) Beyond tourism: The amenity migrants. In: Mannermaa, M., Inayatullah, S. and Slaughter, R. (eds) *Coherence and Chaos in Our Uncommon Futures: Visions, Means, Actions, Turku: Finland Futures Research Centre*. Turku School of Economics and Business, Turku, pp. 121–128.
Moss, L., Glorioso, R. and Krause, A. (2008) *Understanding and Managing Amenity-led Migration in Mountain Regions*. The Banff Centre, Banff.
MSMA (Municipalidad de San Martín de los Andes, Secretaría de Turismo) (2020) Evolución de la cantidad de turistas por año.
Piglia, M. (2012) En torno a los Parques Nacionales: primeras experiencias de una política turística nacional centralizada en la Argentina (1934–1950). *PASOS – Revista de Turismo y Patrimonio* 10(1), 61–73.
Prados Velasco, M.J. (2012) Naturbanización y patrones urbanos en los parques nacionales de Andalucía. *Boletín de La Asociación de Geógrafos Españoles* 60, 19–44.
Prudham, S. (2009) Commodification. In: Castree, N., Demeritt, D., Liverman, D. and Rhoads, B. (eds) *Companion to Environmental Geography*. Basil Blackwell, Oxford, pp. 123–142.
Sánchez, J. (1991) *Espacio, economía y sociedad*. Siglo XXI, Barcelona.
Torkingston, K. (2010) Defining lifestyle migration. *Dos Algarves* 19, 99–111.

9 Sustainable Tourism in Galapagos: The Perception of the Stakeholders

Lady Soto Navarrete[1,2] and Òscar Saladié[2]
[1]Faculty of Tourism and Geography, Universitat Rovira i Virgili, Vila-seca, Spain;
[2]Faculty of Social Sciences and Humanities, ESPOL Polytechnic University, Campus Gustavo Galindo, Guayaquil, Ecuador

Summary
The Galapagos Islands (Ecuador) are a living laboratory and a sensitive area where, among other activities, tourism is performed. The unique flora, fauna, and landscapes motivate thousands of people to visit the islands every year, generating economic benefits and various considerations regarding environmental issues.

The Value and Interest of the Case Study

The analysis of the perceptions of the different actors, especially the decision makers in the tourism and environment arenas, is crucial. The interdisciplinary nature of tourism makes it necessary to know what the stakeholders think to design future strategies to make better use of the natural resources of the archipelago and guarantee the sustainability of tourism activity.

Background

Tourism activity is important for the economies of many countries because the arrival of tourists implies the generation of foreign exchange, the creation of jobs, and also helps with infrastructure enhancements in destinations areas (Kumar *et al.*, 2015). Nevertheless, it can also negatively affect the environment and local communities (Theobald, 2012). Consequently, tourism must be planned under the parameters of sustainability to use resources appropriately without causing negative impacts (Fig. 9.1).

The adverse effects of tourism are aggravated on small islands as they are spaces that are more sensitive to anthropogenic activities than other destinations (Parra-López and Martínez-González, 2018). There is a confrontation between the expansion of tourism in island territories and the demand for resources, which are often scarce. Added to this is the concern for the generation and management of waste and residual water, which are not always adequately treated (Grabara and Bajdor, 2013). Protected areas on islands are unique tourist destinations due to their ecological and human development characteristics (Petrić and Mikulić, 2012). For these reasons, tourism management on these islands must be more exhaustive and stricter than elsewhere.

The COVID-19 pandemic affected tourism worldwide, but at the same time some authors stated that the pandemic offered the opportunity for the development of more sustainable tourism and the rethinking of the way in which we travel (Lew *et al.*, 2020). Responding to the need to

Fig. 9.1. Sardina Beach, San Cristobal Island.

adapt to new mobility conditions during the COVID-19 pandemic, several measures were implemented by public sector entities (Medina-Chavarría *et al.*, 2022). There is extensive literature available on the advantages that emerged from the adverse situation that COVID-19 produced, related to the transformation of tourism and the offering of activities. These incorporated corrective actions toward more sustainable tourism (Bernal Escoto *et al.*, 2021).

Taking into account the previous information, the aim of this case study is to analyze the perceptions of local stakeholders in the public and private sectors of the Galapagos Islands regarding the sustainability of tourist activities, including concerns of environmental economic sustainability, basic services, tourist attractions, and facilities in the archipelago, specifically in the three main islands.

Study Area and Methods

Study area

The Galapagos Islands, also known as the Enchanted Islands, are located about 1000 km from the Ecuadorian coastline (South America) in the Pacific Ocean (Fig. 9.2). It is here that Charles Darwin was inspired to write about the evolution of species.

The Galapagos archipelago includes 13 major islands and more than 115 small islands, islets, and rocks of volcanic origin, which form the archipelago (González and Cubero, 2010). Four islands are inhabited and three have been considered for the present study, San Cristóbal, Santa Cruz, and Isabela, as they are the most visited.

The archipelago houses, more than 7000 species of plants and animals (Figs 9.3 and 9.4), are both native and endemic. The Galapagos National Park was created in 1959, becoming the first protected area in the country and since then it has been part of the National System of Protected Areas of Ecuador. Due to its high ecological value, 97% of the terrestrial surface of the archipelago is categorized as a protected area, and the remaining 3% is the populated area (Parque Nacional Galápagos, 2023). It was also declared a Marine Reserve in 1998, and the Hermandad Marine Reserve was created in 2022 (WWF ECUADOR, 2022). Tourism is one of the primary activities that support the economy in the archipelago.

According to the Protected Areas Management Plan (2014), two models of tourism coexist in the Galapagos: cruise tourism and tourism in populated centers (Dirección del Parque Nacional Galápagos, 2014). The latter is more predominant on the islands. To manage the intrinsic value of ecosystems and their biodiversity ecotourism is integrated with the management of the Public Use of the protected areas of the archipelago, which recommends making a superior offer in line with the Galapagos protected areas' conservation principles This public use model allows visitors to natural areas to enjoy and appreciate natural and cultural heritage, with minimal impact equitable distribution of economic benefits for local populations.

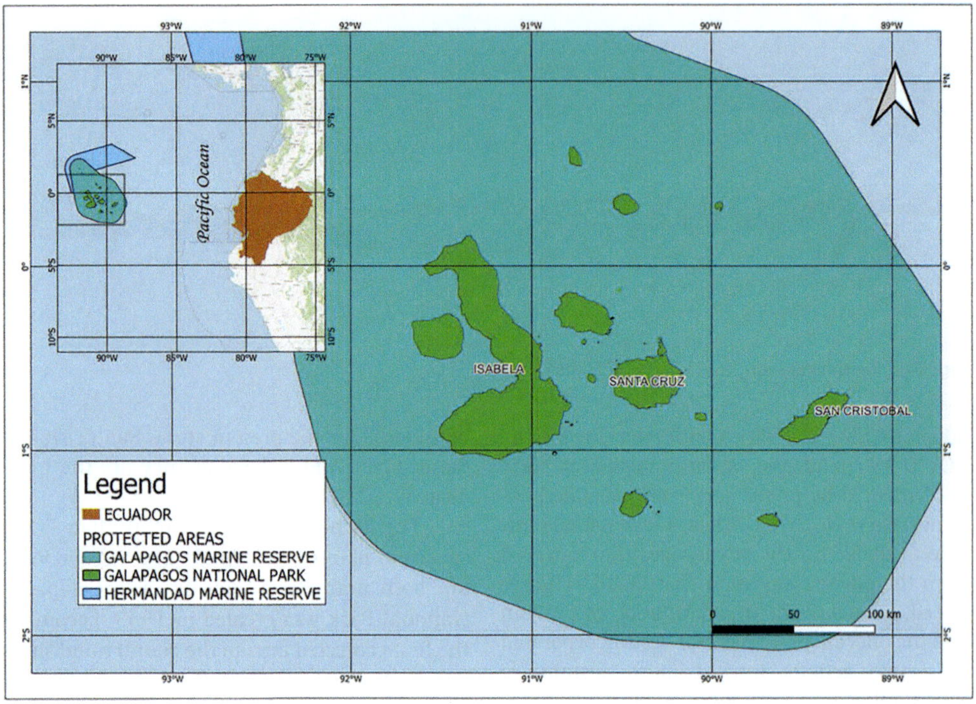

Fig. 9.2. Galapagos Islands.

As reported by the Population and Housing Census, 25,124 people live on the Galapagos Islands (INEC, 2010) and more than 267,000 tourists arrived in the archipelago in 2022. This compares to 271,000 tourist arrivals in 2019, the year prior to the COVID-19 pandemic, after 72,000 and 136,000 tourist arrivals in 2020 and 2021, respectively (Dirección del Parque Nacional Galápagos, 2022). Two-thirds of the visitors were foreigners before the pandemic but the number fell to two-fifths in 2021, due to the mobility restrictions and the different efforts of the tourism authorities that promoted internal tourism, especially in protected areas, since these natural spaces facilitated the social distancing that COVID-19 forced us to maintain. The non-Ecuadorians visitors in 2022 accounted for 54% of all visitors, once no mobility restrictions were enforced.

In the Galapagos tourist cadastre, 946 establishments that are part of the tourist offering are registered. These can be broken down as follows: 121 food and beverage establishments; 309 accommodation units, including tourist camps, guest houses, hostels, hotels, and lodges; 25 companies dedicated to tourist intermediation; 291 performing support operations; and 200 tourist transport companies.

Currently, 210 visitor sites exist where, according to the characteristics of each one, the following activities can be carried out: recreational diving, night diving, superficial diving (snorkeling), panga ride, kayaking, experiential fishing, surfing, hiking, and camping. The most visited places on Santa Cruz are Tortuga Bay, Turtle Route Fausto Llerena, the Breeding Center, Las Grietas, and Garrapatero Beach. The most visited places on San Cristobal are Gianny Arismendy Environmental Interpretation Center, Tijeretas Hill, the David Rodríguez Tortoise Breeding Center, Punta Carola, and Puerto Chino beaches. Attractions on Isabela include the Arnaldo Tupiza Breeding Center, the Wall Circuit, Sierra Negra volcano, Chico volcano, Pearl Shell, and the sulfur mines.

Fig. 9.3. Opuntia echios.

Fig. 9.4. Sea lion nursing.

The form of tourism that represents the greater risk to the integrity of the natural environment is mass tourism. It is a type of tourism that focuses on the sun and beach without taking into account the fragility of the ecosystem in the Galapagos Islands.

Methods

The data on which this case study is based was obtained through 13 in-depth, in-person, semi-structured interviews with key actors tied to the tourism and environmental arenas, from the public and private sector. In March of 2022 – 6 months after the COVID-19 restrictions to visit the archipelago were lifted – the interviews were conducted in Spanish. They had an average duration of 44 min. Each of the interviewees was informed about the research purpose and gave their informed consent.

In order to have the largest representative sample possible, actors from the public and private sectors and from the tourism and environmental arenas were considered, since these are all affected by tourism activity. All the decisions made in the archipelago about tourism are made in accordance with the environmental authorities at different levels, both provincial and local. Some actors, such as the national park officials, govern the entire provincial territory. However, decentralized decisions also need to be made, as a responsibility of each canton (i.e. the individual islands). All the previously mentioned actors have decision making power at different levels. Public sector actors are part of the public governance structure. They have key competencies for the development of tourism in accordance with the principles of decentralization and deconcentration that govern the country. Those of the private sector, when providing tourist services, live the reality that workers in this sector face in the archipelago.

Contact with the interviewees was made via email and in some cases via telephone. The list of interviewees could have been broader, but unfortunately, due to a variety circumstances, including people not answering our email, it was not possible to carry out all the scheduled interviews.

Seven actors from the public sector were interviewed (Table 9.1). Four of them belong to the local administration (three from tourism and one representative of the local community who does not belong exclusively to the environment or tourism spheres but we consider it appropriate to know their perspective). The other three belong to the Directorate of the Galapagos National Park (two are from the environmental arena and one from the tourism arena). Six stakeholders were from the private sector. Two of them were representatives of the hotel sector, three from travel services, and one from an environmental NGO. Other stakeholders were contacted to be interviewed but, unfortunately, some gave a negative response and others gave no response at all.

The questionnaire contained 16 open-ended questions to ascertain interviewees' perceptions of tourism activity in the Galapagos, including concerns and environmental issues, economic sustainability, basic services, tourist attractions, and facilities. A content analysis was undertaken on the basis of the information provided by the stakeholders.

Results

The following elements were analyzed, thus allowing a better understanding of sustainability in the archipelago. The elements were structured as follows: economic sustainability, tourist profile and environmental awareness, tourist attractions and facilities, basic services, and environmental issues.

Fig. 9.5 shows the visual scheme relating the topics covered in the interviews.

Economic sustainability

Both the public and private sectors recognize the importance of tourism contributions to the economy of the islands, it is, in fact, the basis of the local economy since it supports employment generation and helps small business growth. The value chain revolves around tourism. It is evident that there is interest in the population linked to tourism training and this has helped

Table 9.1. Public and private sector stakeholders interviewed.

	Private			Public	
	Hotel	NGO	Travel services	Local administration	National Park
Stakeholder 1	Tourism				
Stakeholder 2			Tourism		
Stakeholder 3				Tourism	
Stakeholder 4					Environment
Stakeholder 5				Community	
Stakeholder 6				Tourism	
Stakeholder 7			Tourism		
Stakeholder 8		Environment			
Stakeholder 9					Tourism
Stakeholder 10			Tourism		
Stakeholder 11				Tourism	
Stakeholder 12					Environment
Stakeholder 13	Tourism				

the hiring of more locals instead of foreigners or people who come from continental Ecuador.

> Stakeholder 2. 'Our agency opened in August 2019. We have been working for only 8 months practically and then the lockdown came. Boom! So we grew so much that I think we would give jobs to two more people, not counting the other guides, maybe 2 or 3 more jobs.'
>
> Stakeholder 9. 'Tourism is the basis of the economy, its entire value chain, in the case of large vessels, its supply centers generate transportation, hiring personnel. It's all a whole dynamic around that. Then, obviously, the tourists that arrive, everything related to the tourist operation itself, that is, the modalities, such as the tour of the bay, the experiential fishing tour, daily tours, and diving journals, are the operations that are here, plus the activity on land, the travel agencies and [tourist] guides. And from there, obviously, restaurants and souvenir shops, including supermarkets and bakeries – everything [the shops] depend(s) on tourism.'

We can appreciate how tourism represents an important engine driving the economy of the islands, both the services that are directly related to tourism and those services that in one way or another are available to tourists. This includes small shops and taxis, among others.

Although the economic factor is fundamental in terms of sustainability, it is necessary to consider the social and environmental spheres to achieve sustainability in a holistic way. These aspects are detailed in the following sections.

Tourist profile and environmental awareness

Tourism in the Galapagos Islands, considering the effects caused by COVID-19, is in the process of reactivation. Thanks to various efforts of public and private entities, taking place gradually.

It is observed that more and more tourists prefer to stay longer on the islands instead of arriving from cruise ships, highlighting the local-based tourist trend. The fact that a tourist prefers to stay longer on land has two implications. On the one hand, the circulation of foreign currency goes directly to the local population. On the other hand, staying on land contributes to reducing the tourists' carbon footprint because they have not opted for the cruise service.

Another point is that visitors have a greater environmental awareness. However, there are still several risks that the tourism activity could face, such as informality in the provision of the services, where visitors receive poor service, and not having learned the lessons left by the pandemic.

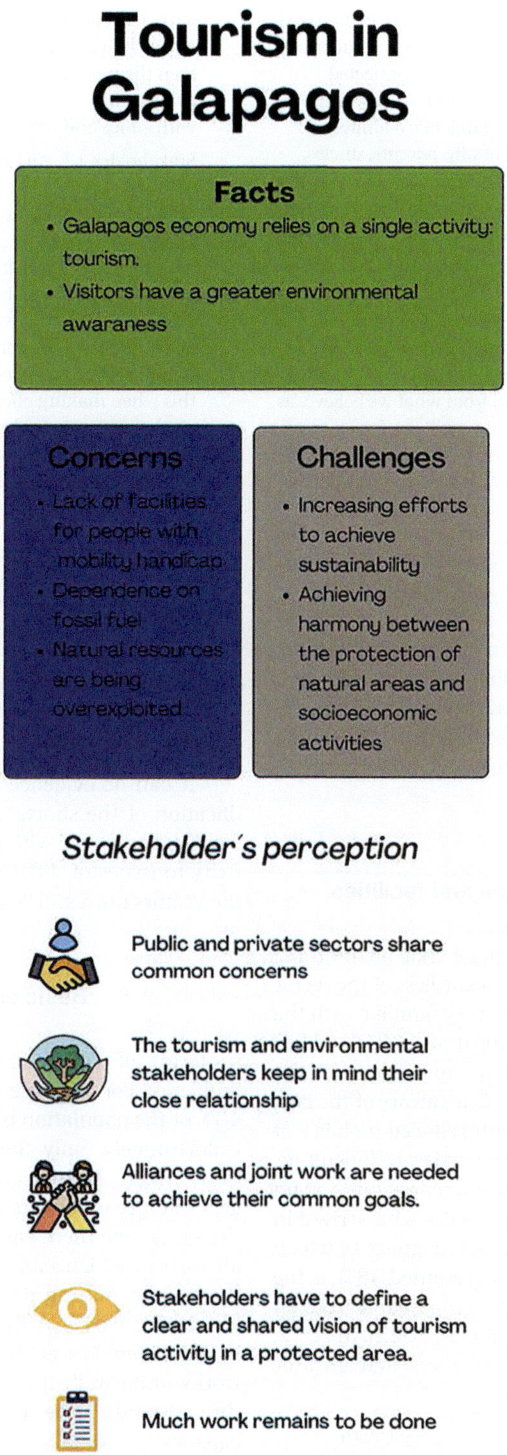

Fig. 9.5. Visual scheme relating the topics covered in the interviews.

> Stakeholder 6. 'We have a lot of people [tourists] who are looking for this issue of environmental awareness, people who are very connected to being very empathetic with nature, young people especially... The pandemic definitely brings us closer to our family, parents, uncles, so it's like an opportunity, there are promotions on the airlines... Let's go on a trip, I want to take my grandmother, my mother, for a walk, so it's something nice.'

> Stakeholder 13. 'The average services are still emerging incipient—not such good quality... it is in process. The majority of us had no knowledge... we are not studied in tourism, so services are provided from what we believe, as it is from tourism, but not really professional, so little by little that is going to change. We all want to give a better service... that will be gradual.'

Visitors are looking for awareness of the environment, as well as events linked to wellness tourism. Thus, they are able to integrate more with the community.

It is interesting to see how some stakeholders recognize the possible shortcomings that some tourist services have in order to achieve excellence. This can make it difficult to respond to the demands of tourists, relating mainly to environmental issues.

Tourist attractions and facilities

It is important to emphasize that in the three islands, the interviewees, regardless of the sector to which they belong, are very familiar with the different tourist attractions that each island has and all the activities than can be done by the visitors. But not all of them are aware of the lack of facilities for people with reduced mobility at certain visitor sites, which makes it difficult for visitors with this condition to carry out certain activities. Two-thirds of tourists who arrived in the Galapagos in 2019 were foreigners, of which those over 51 years old represented 38% of the total arrivals of foreigners in that year (Dirección del Parque Nacional Galápagos, 2022). In 2022, 54% of arrivals were foreigners and 45% of them were over 51 years old.

> Stakeholder 6. 'Especially for people with different abilities (handicapped) or for the same mobile limitation. We really need it everywhere (infrastructure)... I have seen many more people with disabilities or reduced mobility here, in wheelchairs or with canes. Well, we have always seen them with canes, but now, it is more evident that they have one—a difficulty walking with canes and with a crutch.'

> Stakeholder 12. 'All the visiting sites have accessibility and a good part of the sites have an infrastructure that already needs to be changed due to time... We must consider that a good part of the visitors are, what we now call, older adults, so the sites must provide those conditions. What must be clarified is that not all sites are suitable for everyone to visit, and it is important that operators are very clear about this when making offers on the island. It does not happen only in the Galapagos, but in any part of the world. There are sites that, due to site conditions and physical conditions or age, you could not access and you have to be aware of it.'

> Stakeholder 13. 'I think that some adjustments are needed in the tourist structure in the park areas, I am referring to visitor sites... they are still emerging [such as] the road to the sulfur mines, or sulfur volcano [because] they are a bit dangerous, so the infrastructure needs to be improved and no maintenance has been done after COVID.'

It can be evidenced that despite the identification of the shortcomings in the necessary facilities for people with reduced mobility, sensitivity to issues of differentiated abilities among the visitors must still be deepened.

Basic services

In terms of basic services, there are sewer problems. For example, on Isabela Island just 40% of the population has access to this service. Unfortunately, only San Cristóbal Island has fresh water. As indicated above, tourism can generate great benefits for communities but at the same time there can be negative impacts in all aspects which can affect the ecosystem of the islands, putting pressure on resources and generating waste. Although in the Galapagos Islands, there is a waste separation system that works quite well, the sanitary landfills of the three islands have reached their maximum capacity.

> Stakeholder 8. 'The positive impacts [of tourism] on the local economy are good, but on the other

hand, water, energy, and other resources must be managed to sustain the visitors, who can often be much more than the inhabitants of the Galapagos.'

Stakeholder 11. 'Of course, all these establishments, basically let's talk about hotels, have their reserve tanks because sometimes the water from the plant does not arrive because the pump is damaged, but this reserve for hotels doesn't last more than two or three days, and if the plant is damaged for five days, we have problems as they currently tell us. Hotels usually don't have these problems [of lack of water], but there are problems for the community that doesn't receive water, because the community has a small reserve tank in comparison with one of the tourist accommodation establishments. Sometimes the population doesn't have water all day because since they have a reserve and we open the faucet, we have water and we only realize it when that reserve runs out and that happens to us, since yesterday people [in the local community] have not had water.'

There is also the issue of dependence on fossil energy, although there are developing alternative energies, such as wind and solar. The population does not suffer from problems of electricity supply. Power outages do not lead to rationing electricity between locals and tourist establishments. This section allows us to have a better understanding of how worrisome the use of natural resources, which are consumed by both the local population and visitors, is.

Environmental issues

Tourist activity can generate negative impacts, especially in sensitive natural areas. The ecosystem of the archipelago is unique and also fragile. It has endemic biota, which is threatened by mass tourism without proper controls. Furthermore, the introduction of new species is considered invasive and affects biodiversity.

Stakeholder 12. 'The municipality should take measures regarding the capacity to offer services that are necessary. We need the local population to continue receiving income through tourism, but on the other hand, we have to be aware that this growth in tourism cannot grow without control... the territory would be affected by the demand for the basic services that it has to offer. This issue ends up harming all of us because uncontrolled growth leads us to an oversupply.'

Representatives of the Galapagos National Park are aware of the benefits obtained from tourism. However, they stress that limits must be established by the appropriate authority, in this case the municipality.

Tour guides play a very important role in the conservation issue, since they are the ones in charge of telling tourists about the rules of the park (about keeping their distance from the animals, not touching them, not feeding them, not taking flash photos, staying on the trails, and so on). Above all, they share information about the ecosystem of the islands with the tourist, so they can understand that this is *not a sun and beach destination* but a living laboratory, and as the saying goes: we love what we know and take care of what we love.

Stakeholder 10. 'but we as [tourist] guides have to give that message, that is, we are giving that message, right? To be more aware, with mother nature. Yes, because mother nature is our capital, right?'

Even when there is a harmonious relationship between the spheres of environment and tourism, it was observed that there is a feeling that more importance is given on some occasions to environmental care than to the population that inhabits the archipelago.

Stakeholder 2. 'We have realized that many negative things are being done against the human species here in the Galapagos in the name of conservation. Believe me, if they [Galapagos National Park] wanted to remove us humans, they would have removed us a long time ago. And that has been happening since the nineties. It's possible to reach a human-nature balance. However, here, on the other hand, the opposite happens than in other parts of the world. We feel that nature is covering us too much and they are also displacing us humans, within 3% [of the archipelago not included in the protected area]. Be careful, so much so that Leonardo DiCaprio comes, but he comes to deliver more money to whom? To conservation, but not to the human part of conservation.'

The discomfort of certain sectors can be verified when the local community feels displaced, so to speak, when attention is concentrated on environmental preservation, leaving the population aside.

Sustainability

The interviewees agree that efforts on the islands are moving toward sustainability. However, they are aware that there is still much to be done. It is necessary to maintain harmony between conservation and the different productive activities that are carried out to develop and maintain the local economy.

> Stakeholder 9. 'There are too many plans, we don't need more plans. What perhaps [needs to happen is] to return to the joint vision [between the different sectors independently on how to manage the national park].'
>
> Stakeholder 6. 'Institutionally, we can say that in recent years there has been a better relationship [between tourism and environmental arenas from the public and private sector]. We are also going to thank the pandemic, because for many decisions, obviously it was not possible alone. It had to be as a whole, but unfortunately, we do not understand that this connection between the community and the different public and private institutions, this interrelation, should not be lost and not only have it at the moment of the crisis, but just to have a clearer vision.'

The results of the study detail the positive aspects of tourism, such as economic sustainability and the tourist profile and environmental awareness, followed by the various concerns that revolve around this productive activity related to tourist attractions and facilities, basic services, and environmental issues. These aspects allow us to understand how sustainability can operate on the islands and also let us understand the perception of the stakeholders.

The stakeholders' perception

The context in which the interviews took place highlights the common thread between the concerns and requirements of both the public and private sectors regarding environmental and tourism issues in the Galapagos Islands. Although the purpose of the investigation was not to analyze the consequences of COVID-19 directly, during the interviews each actor related tourist activity to the pandemic. The stakeholders all agreed that threats related to the impacts of the tourism activity could compromise the sustainability of the three islands. Thus, alliances have been formed between NGOs and public and private entities to raise funds and assist with environmental conservation to address these concerns.

Analyses from other case studies generally conclude that the stakeholders show significant differences. This is especially true when comparing public and private stakeholders or environment and tourism stakeholders in terms of the balance between tourism development and environmental protection (Santos-Lacueva and Saladié, 2016; Pandy and Rogerson, 2021; Lin *et al.*, 2022; Atay and Saladié, 2022). However, in the Galapagos setting, the interviewees for both the environment and tourism maintain a very similar discourse regarding major problems the archipelago faces related to the use of resources, including waste generation and control of tourist services offerings. As a strategy, it is necessary for tourism and environmental entities from the public and private sectors to have a clearer shared vision and, above all, continue to involve the local community. There is harmony in the different spheres. The only discrepancy is from a single actor who perceived greater concern for the environment than for the population. However, this was an isolated voice.

The aforementioned harmony among the stakeholders can be related to the fact that the Galapagos is a unique destination, not only due to the distinctive flora and fauna found there but also because the islands' economy is sustained by tourism activity based on protected natural resources within a national park. The fragile ecosystem of the archipelago makes it necessary to take measures, many of them strict that govern all economic and productive activities, including tourism. In this sense, the public authorities, supported by the tourism sector, can contribute to avoiding or minimizing the negative impacts of mass tourism detected in other island tourism destinations. Access to the islands is restricted. There is no international airport, unlike other islands, such as those in the Caribbean and the Mediterranean. Due to the influx of tourists arriving in the Galapagos, and considering that 97% of the territory is protected, interviewees see protection of the destination as necessary. The need to protect the destination is based on and supported by tools that are part of the governance of the islands, such as the LOREG (Organic Law for the Special Regime of Galapagos). Another aspect to consider is the entry fee that tourists must pay to enter the archipelago: US$6 for national tourists and US$100 for foreigners. This fee, or tax, is intended for conservation and other needs of the archipelago, such as health, social services, and education.

In conventional sun-and-beach island destinations, the hospitality sector often seeks to increase the number of hotel rooms and attract a greater influx of tourists. For public sphere interviewees, it is not so easy to control the arrival of tourists and as a consequence the management of basic resources and the sustainable management of the destination are compromised. Water scarcity is a recurrent problem in other island tourism destinations, especially in the Mediterranean area (Essex *et al.*, 2004; Gikas and Tchobanoglous, 2009; Tortella and Tirado, 2011). As has been mentioned earlier, in the Galapagos, 97% of the territory is protected. In comparison, many islands have a smaller protected area or a much lower level of protection, such as Cyprus, Mykonos, Malta, Sicily, and the various Caribbean Island nations but receive a greater number of tourists than the Galapagos. The flow of tourists in the archipelago is not currently a serious problem, but stakeholders may perceive it as a potential problem in the future. The protection categories of the Galapagos and the fragility of the ecosystem make the interviewees see the protection of the study area as necessary. Therefore, the need to protect the study area gives rise to convergence between the various stakeholders.

In the future, control over restrictions on activities that are not permitted on the islands, particularly those that conflict with the islands' conservation goals, must be maintained and improved to preserve the Galapagos ecosystem. It is important to emphasize to decision makers that the Galapagos is a protected natural space. Obviously, tourism is important because it generates economic resources, but a balance must be sought to avoid destroying the natural heritage. Because this natural heritage is one of the biggest reasons for the arrival of tourists, measures must be taken to ensure proper conservation. In addition, efforts must continue to limit arrival flows, which must be maintained at a minimum, and it is necessary to put more control in place to avoid the problems detected, which may be incipient in some cases but could worsen.

The interviewees see environmental protection as necessary, and this makes them converge with the other interviewees. To maintain this, coordination between public and private stakeholders, and tourism and environmental policies is vital for the sustainability of the tourism sector on the Galapagos archipelago. Insufficient awareness and lack of coordination between decision makers and the private sector can potentially transform the desired destination into a less desired one (Santos-Lacueva *et al.*, 2019).

Conclusions

The in-depth interviews with the stakeholders have made it possible to determine their perception about tourism activity in the Galapagos Islands, focusing on their concerns about environmental and economic sustainability, but also basic services, tourist attractions, and facilities.

The public administration will have to be the main driver of decision making, providing adequate coordination across sectors and stakeholders, including both those who are directly and indirectly involved with the tourist activity. This case study helps decision makers gain input from actors of different sectors and thus make the best decisions, responding not only to existing or potential problems but also responding to different requirements the tourism sector may have.

This case study opens the way to future research involving tourists and also analyzes other aspects related to sustainability of the Galapagos Islands, such as the application of good environmental practices in tourist establishments and the design and application of sustainable tourism indicators.

Acknowledgements

We would like to thank the ADAPTOUR project (PID2020-112525RB-I00) funded by MCIN/AEI/10.13039/501100011033 (Spain), and the Departament d'Innovació, Universitats i Empresa, Generalitat de Catalunya (2017SGR22).

References

Atay, I. and Saladié, Ò. (2022) Water scarcity and climate change in Mykonos (Greece): The perceptions of the hospitality stakeholders. *Tourism Hospitality* 3, 765–787.

Bernal Escoto, B.E., Montero Delgado, N.I. and Rivera Aguirre, F.A. (2021) Strategic analysis of sustainable tourism in Baja California against COVID-19. *Sustainability* 13(7), 3948. DOI: 10.3390/su13073948.

Dirección del Parque Nacional Galápagos (2014) Plan de Manejo de las Áreas Protegidas de Galápagos para el Buen Vivir. Puerto Ayora, Galápagos, Ecuador. Available at: https://www.galapagos.gob.ec/wp-content/uploads/downloads/2016/07/DPNG_Plan_de_Manejo_2014.pdf (accessed 8 May 2023).

Dirección del Parque Nacional Galápagos (2022) Informe anual de visitantes a las áreas protegidas de Galápagos del año 2022. Galápagos – Ecuador. Available at: https://galapagos.gob.ec/estadistica-de-visitantes/ (accessed 24 April 2023).

Essex, S., Kent, M. and Newnham, R. (2004) Tourism development in mallorca: is water supply a constraint? *Journal of Sustainable Tourism* 12, 4–28.

Gikas, P. and Tchobanoglous, G. (2009) Sustainable use of water in the Aegean Islands. *Journal of Environmental Management* 90, 2601–2611.

González, F. and Cubero, P. (2010) Efecto de actividades turísticas sobre el comportamiento de fauna representativa de las Islas Galápagos. Ecuador. Available at: https://www.scielo.cl/scielo.php?script=sci_arttext&pid=S0718-560X2010000300013 (accessed 19 April 2023).

Grabara, K. and Bajdor, P. (2013) Towards to Sustainable Tourism – Framework, Activities and Dimensions. Available at: https://www.sciencedirect.com/science/article/pii/S2212567113001706 (accessed 24 April 2023).

INEC (2010). Available at: https://www.ecuadorencifras.gob.ec/estadisticas/ (accessed 24 April 2023).

Kumar, J., Hussain, K. and Kannan, S. (2015) Positive vs negative economic impacts of tourism development: A review of economic impact studies. In: *Developments of the New Tourism Paradigm in the Asia Pacific Region*. pp. 405–413. Available at: https://www.researchgate.net/publication/277411573_POSITIVE_VS_NEGATIVE_ECONOMIC_IMPACTS_OF_TOURISM_DEVELOPMENT_A_REVIEW_OF_ECONOMIC_IMPACT_STUDIES (accessed 12 May 2023).

Lew, A.A., Cheer, J.M., Haywood, M., Brouder, P. and Salazar, N.B. (2020) Visions of travel and tourism after the global COVID-19 transformation of 2020. *Tourism Geographies* 22, 455–466. DOI: 10.1080/14616688.2020.1770326.

Lin, C., Wang, W. and Ou, S. (2022) Impact of comparative climate change perceptions on antecedents of tourists' adaptation intentions for a coastal destination in Taiwan. *Journal of Sustainable Tourism* 30, 69–88.

Medina-Chavarría, M.E., Gutiérrez, A. and Saladié, Ò. (2022) The response to increasing numbers of visitors to Catalonia's protected natural areas during the COVID-19 pandemic: A review from publications in digital media. *Boletín de La Asociación de Geógrafos Españoles* 93. DOI: 10.21138/bage.3183.

Pandy, W.R. and Rogerson, C.M. (2021) Coastal tourism and climate change: Risk perception of tourism stakheolders in South Africa's graden route. *GeoJournal of Tourism and Geosites* 37, 730–739.

Parque Nacional Galápagos (2023). Available at: https://galapagos.gob.ec/areas-protegidas-2 (accessed 22 March 2023).

Parra-López, E. and Martínez-González, J.A. (2018) Tourism research on island destinations: A review. *Tourism Review* 73(2), 133–155. DOI: 10.1108/TR-03-2017-0039.

Petrić, L. and Mikulić, D. (2012) Protected Areas and Tourism Development on Croatian Islands: Coexistence or Divergence?. Available at: https://www.researchgate.net/publication/271451690

_Protected_areas_and_tourism_development_on_Croatian_islands_coexistence_or_divergence (accessed 24 April 2023).

Santos-Lacueva, R. and Saladié, O. (2016) Acción pública en materia de turismo y cambio climático: Las percepciones de los stakeholders en la Riviera Maya (México). *PASOS Revista de Turismo y Patrimonio Cultural* 14, 611–630. DOI: 10.25145/j.pasos.2016.14.040.

Santos-Lacueva, R., Ariza, E., Romagosa, F. and Saladie, O. (2019) The vulnerability of destinations to climate change: A comparative analysis of contextual socio-political factors. *Journal of Sustainable Tourism* 27, 1217–1238.

Theobald, W.F. (2012) The Positive and Negative Impacts of Tourism: Brian Archer, Chris Cooper, and Lisa Ruhanen. Available at: https://api.semanticscholar.org/CorpusID:165175437?utm_source=wikipedia (accessed 12 May 2023).

Tortella, B.D. and Tirado, D. (2011) Hotel water consumption at a seasonal mass tourist destination. The case of the island of Mallorca. *Journal of Environmental Management* 92, 2568–2579.

WWF ECUADOR (2022). Available at: https://www.wwf.org.ec/?373650/reservahermandad (accessed 22 March 2023).

10 Tourism Clusters, Nature Research and Visitor Experiences: The Case of Fundación Batis in Barranquilla, Colombian Caribbean

Beatriz-Helena Díaz Solano[1] and Narcís Bassols i Gardella[2]
[1]Universidad Autónoma del Caribe, Barranquilla, Colombia; [2]Division of Quality Science, Uppsala University, Campus Gotland, Visby, Sweden

Summary
Clusters are beneficial to tourist companies in a number of ways as they have been found to promote innovation of both product and market. The case of Fundación Batis, located in the Colombian Caribbean, will be examined as a best practice in terms of adding tourism to a company's supply to achieve diversification and supplemental revenue. After a successful incubation in a cluster, this foundation opened its doors to tourists, and its employees now engage in an open dialogue with visitors about conservation issues. As a result, the region gains an upscale offer in natural tourism, which helps to upgrade and diversify the local tourism product.

Introduction

In many destinations, upgrading the local offer is challenging due to strong inertias from the side of the 'main' stakeholders of the place (Bassols and Leicht, 2021) or the slow-evolving supply of the local enterprises (Bassols i Gardella, 2023). Colombia's destinations are no exception to this rule. Therefore, we take note when cases emerge that successfully improve (or even disrupt) the local supply of products and experiences. This is the case with Fundación Batis, a non-profit organization devoted to researching the natural environment in the Colombian Caribbean. Since this experience has been 'incubated' within the framework of a cluster model, it clearly demonstrates the usefulness of cluster initiatives.

In this case study, we will first offer a brief overview of the concept of 'clusters' and their institutional development framework in Colombia. Subsequently, we delve into the experience of Fundación Batis, underscoring their achievements but also noting the challenges faced. The authors interviewed the founders of the NGO, and also observed visiting groups in a non-participatory fashion. The resulting study gives insight into tourism cluster practices and how companies can evolve within clusters to be more competitive.

A Brief Review of Concepts: Clusters and Competitiveness in Tourism

A 'cluster' refers to a grouping of companies and/or associations which are connected via vertical relations (buyer–seller) or horizontal relations: customers, technology, channels (INNpulsa, Clúster Development and Universidad del Rosario, 2018). The term first emerged in the 1890s with Alfred Marshall's concept of 'industrial districts'. Marshall's research focused on their efficiency and resulting economies of scale. The term continued to evolve until Michael Porter (1990) introduced his famous 'cluster' concept. In his work, Porter presents his 'competitiveness diamond' to illustrate a positive environment for the clustering of industries in any given place.

In tourist destinations, the tourism industry develops thanks to the participation of a significant number of diverse companies. These operate in the destination via a series of activities, some of them in competition, others in cooperation. Therefore, destinations have many complex dynamics in terms of delivering goods and services, as different interests coincide or conflict (Bassols and Leicht, 2021). García-Villaverde *et al.* (2020), Gutiérrez and Mota (2022) and Kachniewska (2013) attempted to articulate the challenges faced within clusters and their respective destinations. According to Nordin (2003), clusters may be a good tool for planning and developing the tourist industry, particularly in rural places (see also Sáez, 2019). However, research on successful cases and how they relate to the cluster's dynamics is still scarce. For this reason, the authors chose to examine the following case study.

A challenge facing any cluster initiative is to achieve greater competitiveness for its members (Jaramillo Panesso, 2003). A cluster only forms when a basic agreement between the different stakeholders emerges. This is necessary to develop common strategies on quality, operations and marketing so that the resulting products bring about superior levels of differentiation. This, in turn, allows the value chain to respond rapidly and efficiently to technological and organizational innovations, as well as market shifts. The final goal is to ensure long-term durability and brand recognition. As companies assimilate new abilities, which may be idiosyncratic or emergent with respect to the market, they will yield more mature products in the long term (Mendoza *et al.*, 2018).

Context: Tourism and Clustering Development in Colombia, South America

Just like in other economies, clusters can be found in Colombia. In the case of Colombia, these tend to be more regionally focused, as they are driven by the regional authorities, with local companies as members. Clusters are an important part of the economic development of Colombia's regions. The Colombian Network of Clusters (Red de Clusters de Colombia), which, in 2017, comprised over 90 cluster initiatives registered throughout the country, governs the initiative. The country's 20+years of experience in cluster development has been instrumental in promoting competitiveness and upgrading different industries. These efforts are still ongoing (INNpulsa, Clúster Development and Universidad del Rosario, 2018). Generally speaking, cluster initiatives in Colombia are highly respected and have become a preferred way for companies to interact in an industrial landscape that traditionally has seen very low levels of cooperation.

Nowadays, Colombia's cluster initiatives are classified according to the economic sectors they belong to. Tourism clusters are among the cluster initiatives in the service industry, where they represent the majority (39%), with a total of 25 clusters nationwide. (INNpulsa, Clúster Development and Universidad del Rosario, 2018). The existence of these clusters has led to substantial advances in the tourism sector, such as improved definition and coupling of the local value chains, and the specialization and addition of more value to tourist products. It has also enabled the showcasing of the local offer to international markets, and the better positioning of certain destinations and products.

Meanwhile, the country has developed several tourist products under the umbrella of a cluster. For example, the 'Safari Llanero' experience in Colombia's Eastern Plains region is a product developed by the tourism cluster in the Casanare region. It consists of an expedition through the region's *sabanas* (low forests). The transportation means used are horse-riding, walking, tractor and 4×4 trucks. The main point of the experience is to observe fauna and flora in a non-invasive way in order to preserve the region's natural quality. At the same time, visitors are shown the distinctive local llanero culture. Currently, about 50 businesses make up this regional cluster and the experience takes place in some of the 30,000 ha of nature preserves in the region (Medina and Vargas, 2020).

Another experience consisting of several products designed and operated by different companies within a cluster is the one developed in the Guajira region, the northernmost region of Colombia's Caribbean coast. 'Guajira Experience: Three Worlds' puts together eco-ethno-tourism experiences focusing on the local indigenous communities. Within this experience, the Wayuu community acts as the 'guardians of the dreams'. The Kogui and Arhuaco communities are the cultural repositories of the nearby Sierra Nevada mountain range, protecting life with their knowledge and traditions. (It is worth mentioning that these communities' cultures were marked as an intangible cultural heritage by UNESCO in 2022). The musical heritage in this product is represented by the *vallenato* music from the region. Finally, the natural component is the wild beauty of the area: desert, sea and mountains with dense woods close to each other. Water plays an important role in this experience, as it can be seen 'being born' in the mountains and 'dying' as it joins the Caribbean Sea. A total of 13 companies are part of this regional cluster, representing the whole value chain needed to produce and operate this product.

A further success story that we want to discuss in detail is the nature tourism cluster in the Atlántico region, officially known as *Cluster de Turismo de Naturaleza del Departamento del Atlántico*. This is a collaborative group of businesses across the whole tourist value chain, whose conceptualization began in 2014 in the city of Barranquilla (the largest on the Caribbean coast of Colombia), thanks to the leadership of the tourist roundtable CRCA (Comisión Regional de Competitividad del Atlántico), the regional committee for competitiveness. A seminal study conducted on the status of the tourism sector in the region concluded that tourist visits were relatively short in terms of numbers of days (Gobernación del Atlántico, 2012). The study also found that the municipalities in the region had, at that time, a very low tourism momentum. These findings raised the need to establish a clear and attractive offer for visitors so that they would stay longer in Barranquilla and its surrounding area. Further research established that the tourism dynamics in the region had not been characterized and the main stakeholders had not been pinpointed (Gutiérrez, 2013). These research efforts resulted in the enumeration of natural and cultural attractions and the businesses that were actually operating within the industry. Another result was the need to support tourism initiatives in the region's municipalities. Based on these findings, leaders decided to create the region's first nature tourism cluster.

In 2016, the CRCA tourism roundtable looked for funding to create the cluster. The regional government stepped up to finance 'Competitive Routes' (the initiative preceding the creation of a cluster in Colombia) as an initial phase. During 2017, the institutional and legal entry conditions to the national programme, 'Competitive Routes', were fulfilled. The cluster was ranked as 'emergent' (i.e. its products were not yet mature enough and had to be further developed) (Puello *et al.*, 2019). In March 2018, the cluster was officially registered with the Colombian Network of Clusters.

Since 2019, the cluster has been coordinated by the Hospitality and Tourism Program at the Universidad Autónoma del Caribe, a tourism research and teaching institution in Barranquilla. The goal of this coordination is to expand the cluster, in terms of numbers of participating enterprises, and to implement actions to reinforce and develop it. A major tool for achieving this is the implementation of the 'macro-strategy', a defined path for growth and development. The macro-strategy of the nature tourism cluster was first defined in 2016 while setting up the aforementioned 'Competitive Routes'. The core of the strategy was to establish 'experiential and sustainable tourism'. Consequently, it was expected that all products generated within the cluster would have a strong experiential component and the member companies would strive towards sustainability.

The macro-strategy has evolved over the past 5 years. There was the need, expressed by many stakeholders in the cluster, to develop products and experiences that would also protect local endangered species of flora and fauna. In fact, in recent years, protecting biodiversity has become an issue of concern across the Colombian Caribbean region. Thereafter, a new strategy was unveiled by the cluster called 'Regenerative tourism'. In recent times, three products have emerged under this new umbrella: the 'Crave Route', an initiative led by the company Asitours in Barranquilla; 'Experiencing the Guaimaro Tree' by the Association of Farmers of the Los Limites municipality; and finally 'Recotono Batis', led by Fundación Batis. More information about the cluster, the member enterprises and their product offer may be found on the following webpage: www.turismoatlantico.com. The remainder of this case study is devoted to analysing and contextualizing Fundación Batis and a particular experience they offer.

Fundación Batis: Studying Local Nature in the Colombian Caribbean Coast

One particular member of the Atlántico tourism cluster is Fundación Batis. This non-profit foundation defines itself as a 'research center with a holistic vision towards environmental sustainability with the goal of undertaking actions for the social well-being and the conservation of nature' (Fundación Batis, 2023). This shows a convergence of interests of the foundation and the cluster. The foundation is made up of a group of biologists working together with practitioners from other disciplines, who are all passionate about applying scientific methodologies to the environmental and socio-cultural domains.

Since the foundation and the cluster first made contact, several topics of mutual interest have been uncovered. Through participatory planning, mutual engagements and responsibilities emerged. This included the foundation's more in-depth approach to the concepts and practices of tourism via an appropriate training plan. It also included participative workshops with other tourist companies along the value chain. Through internships, tourism students helped to create tourist visions and plans. It has been a win-win situation, where the cluster has experienced growth in competitiveness and diversification, and Batis has received a considerable amount of know-how from the cluster to be able to put sustainable and conscientious tourism principles into practice. This was done via webinars, on-the-ground training and common fieldwork. This involved a profound change within the foundation on an organizational level; see further below.

After several months of collaborative work with the cluster, Fundación Batis decided to set up a walking eco-tour that would help visitors reconnect with nature (hence the Spanish name of the tour). The tour takes place in the local mangrove forests and the dry tropical forests on the foundation's premises. It is a chance to educate visitors while extending the foundation's actions on preserving and regenerating these ecosystems.

Re-Connecting Visitors with Nature: The 'Recotono Batis' Tourist Experience

In order to create this tourist product, the methodology proposed by the Chilean National Council for Culture and Arts was adopted (Consejo Nacional de la Cultura y las Artes del Gobierno de Chile, 2008). This outlines four steps in the creation of a tourist product: (1) the conceptualization of the product and its attributes of the experiential moments provided; (2) the product design; (3) the economic analysis of the product and (4) the product's distribution, including sales.

Therefore, concerning the development of our new product, the first step was to diagnose the existing capacities and possibilities of the foundation's premises (consisting of two ecosystems covering an area of 4 ha) and the existing human abilities to develop the product. In order to draw a feasible conclusion, participative planning techniques were used as well as design thinking techniques. Thanks to these methodologies, the product and its attributes were developed. Once the product's attributes

Fig. 10.1. A moment in the 'Recotono Batis' experience: Planting trees.

became clear, each attribute, or moment, was associated with an experience, ensuring the creation of a truly experiential concept. In a further step, a script was designed covering some aspects of environmental teaching, the availability of spaces for interacting with flora and fauna, and the recovery of the local food culture, culminating in an experience activating all the senses called 'Recotono Batis'. The developed product consists of a walking tour through the area, with four distinct attributes/ moments of experiential interest for visitors. The first one is the reception of the visitors. The script for this touchpoint outlines the creation of expectations and the education of visitors about how to behave in the natural surroundings. Visitors are given an 'explorer kit' consisting of a full-body-length net to protect themselves from mosquitoes, a pair of rain boots, a magnifying glass, binoculars, scissors and a souvenir, all provided in a traditional Colombian rucksack. Visitors are informed about when the tools will be used as well as the route's main features (see Figs 10.1–3).

The second moment or experience is the visit to the mangrove forest. This is a completely natural area where visitors learn to distinguish the different species of mangrove trees. They also learn about the different species of animals and flowers upon which the mangrove trees depend for their subsistence. This is done via a walk through the mangrove forest along an elevated pathway. Then, the visitors perform the seeding of a mangrove tree, thereby learning about this tree's physical features and the care it requires to develop and thrive. If a visitor wishes, they may adopt one of the mangrove species.

The third moment on the route is a visit to the tropical dry forest. It is located 1 km away from the mangrove forest, allowing visitors to appreciate the changes in flora and fauna along the way. Once visitors reach the spot, standing in a natural dome formed by the branches of the trees, they are shown a video containing some night recordings taken by the hidden cameras located all over the forest. These show some species using the paths visitors are walking on. This provides an understanding of the nightlife in the forest and the need to preserve it.

As a fourth and final moment on the tour, visitors are directed to the ecological grocery garden. They are given a basket and invited to start walking through the garden, designed as a spiral, where different plants and flowers may be found. Furthermore, visitors are taught about agroecological principles, the possibilities of growing crops on limited land areas, the need for food security and the rescued vegetal species that are no longer produced but were part of the native Amerindians' staple diet. In this area, visitors are also shown different fruits, leaves and flowers and, once the walk is over, visitors

Fig. 10.2. Testing the welcome kit before start (visitors wear nets to protect themselves against mosquitoes).

Fig. 10.3. Another moment in the 'Recotono Batis' experience: Co-creating a food experience by harvesting vegetables and preparing them under supervision.

are led to an open space with tables and chairs where they are invited to a food experience. Here, assisted by their guide, they produce their own meal, invoking all the senses and their newly acquired knowledge.

This product has evolved in terms of quality and level of experience while different test visits have taken place. For example, businesspeople belonging to the cluster were invited to undertake the experience and give their feedback. Once the cluster considered that the product was mature enough according to the criteria set forth by the National Agency for Innovation and Entrepreneurship, INNpulsa, it was included in a nationwide call for projects by said agency. In this call for projects, which went under the name 'Creative Upscaling', the Fundación was awarded funding for further development of the product.

The team of experts in the 'Creative Upscaling' call divided the submitted projects according to their level of maturity in their potential markets. Three categories were established: activities and products directed to the local market (i.e. with a low level of maturity and a potential for improvement); activities and products for the national market (medium-mature, marked-tested products that make up significant experiences and are backed by solid operations: these are adequate for a national visitor); and finally, the third category was made up of those products and experiences that are adequate for the international market, since they show superior experiences for tourists. These products have been consistently tested and have strong operational backing. They are therefore suitable for international nature tourists. 'Recotono Batis' was ranked as a national tourist product, albeit with some recommendations for improvement. As the goal of the call was to promote the creation of advanced tourist products packed with varied experiential offers, every experience included in them had to be validated by experts.

The recommendations for 'Recotono Batis' included switching the communication from an academic language to one better suited to the experiential features of the product, providing more interactive participation in the food experience from the visitors' side, making stronger differentiation between each of the four moments making up the product and undertaking some facilities improvements for greater visitor comfort. Other recommendations touched on procedural matters, such as shortening the timing for a more efficient delivery of the experience, making it more comfortable for visitors and including bilingual guides.

One year after the project was completed, INNpulsa sent out another call for projects: 'Creative Upscaling II'. Fundación Batis saw in this new call the opportunity to consolidate its product design. Having already implemented previous recommendations by experts, this second call resulted in great success for the Fundación. The experience they offer is now classified as a quality product for an international visitor.

There have been some challenges on the way to becoming a supplier of an international tourist product. For example, a deep organizational change was needed for Batis in order to switch from a 'pure' research centre to a 'mixed' institution doing research and welcoming visitors. It required time for scientists to speak about eco-challenges or their research in layman's terms. Another issue was the need for multi-lingual guides (English and Portuguese). This was solved via a collaboration with the Universidad Autónoma, who designed specific language courses aimed at nature guides. Again, this shows the usefulness of cooperating within a cluster framework.

Conclusions

This case study has highlighted some of the challenges and opportunities for entrepreneurship within tourism clusters. We have focused on the issues of organizational change, training needs and the fulfilment of legal requirements for product development, all of which have been facilitated with the cluster's support, demonstrating its effectiveness.

'Recotono Batis' is currently open to the public and is still an evolving product, as each product attribute is being expanded, with the possibility of becoming independent experiences. There is the ever-present effort to make each attribute more structured and experiential.

Although we do not have any data on customer satisfaction levels, our observation and the impressions we gathered from the

Fundación's team show that this product boasts a high level of visitor satisfaction, probably due to the correlation of higher experiential levels with higher levels of satisfaction. These observations, however, are general impressions that need to be researched further.

A final word is one of appreciation for the cluster initiatives in Colombia, which have brought together both public and private stakeholders for the good of the industry. These initiatives have seen the development of high-value products thanks to the cooperation between companies and different upscale drivers, so we can say without hesitation that this is a good practice that may be beneficially replicated in other countries.

References

Bassols i Gardella, N. (2023) Guided tours and tourist attractions in the postpandemic world: Thematic evolutions and spatial changes. *International Journal of Tourism Cities*. DOI: 10.1108/IJTC-08-2023-0149.

Bassols, N. and Leicht, T. (2021) Place marketing, evolutionary theories and the management of multifaceted destinations: Cartagena, colombia, as a study case. *Tourism and Hospitality Research* 21(2), 169–182.

Consejo Nacional de la Cultura y las Artes del Gobierno de Chile (2008) Guía metodológica para proyectos y productos de turismo cultural sustentable. Official document, Santiago de Chile.

Fundación Batis (2023) Nosotros. Fundación Batis. Downloaded 27 March 2023. Available at: https://batisbatis.org/nosotros/ (accessed 24 January 2023).

García-Villaverde, P.M., Elche, D. and Martínez-Pérez, A. (2020) Understanding pioneering orientation in tourism clusters: Market dynamism and social capital. *Tourism Management* 76, 103966.

Gobernación del Atlántico (2012) Plan sectorial de turismo departamento del Atlántico 2013–2015. [Tourism Plan for the Atlántico región]. Editorial de la Gobernación, Barranquilla.

Gutiérrez, F.R. and Mota, F.V.E. (2022) Clúster y Turismo: Modelos de gestión y Competitividad de los Destinos Turísticos. Análisis del Clúster de Turismo de Nuevo León. In: Mora Cantellano, M.P., Serrano Oswald, A., Eréndira, S. and Mota Flores, V.E. (eds) *Reconfigurando territorios a partir de la cultura, el empoderamiento de las mujeres y nuevos turismos*. UNAM-AMECIDER, Mexico, pp. 441–460.

Gutiérrez, P.F. (2013) *Levantamiento del inventario de atractivos turísticos del departamento del Atlántico*. Universidad Autónoma del Caribe, Barranquilla.

INNpulsa, Clúster Development and Universidad del Rosario (2018) Iniciativas clúster en Colombia, instrumentos de desarrollo económico y competitividad. INNpulsa, Bogotá.

Jaramillo Panesso, R. (2003) Apuntes sobre los clústeres turísticos. *Turismo y Sociedad* 2, 35–46.

Kachniewska, M. (2013) Towards the definition of a tourism cluster. *Journal of Entrepreneurship, Management and Innovation* 9(1), 35–56.

Medina, A.M.A. and Vargas, A.M. (2020) Reserva natural El Encanto de Guanapalo, propuesta transmedia. Trabajo para optar por el título de Profesional en Comunicación Social. Pontificia Universidad Javeriana, Facultad de Comunicación y Lenguaje, Bogotá, Colombia.

Mendoza, G.J.M., Mulford, H.M.D. and Barrios, H.K. (2018) Competitividad del clúster turístico del caribe colombiano. In: Mendoza Guerra, J.M., Dorina Mulford, M.B., García Gonzále, J.R., Messino Sosza, A., Blanco Ariza, A.B. *et al.* (Comp) *Turismo Corporativo y TIC: Una Puerta a La Competitividad*. Universidad Simón Bolívar, Barranquilla, pp. 13–60.

Nordin, S. (2003) Tourism clustering and innovation: Paths to economic growth and development. European Tourism Research Institute, U2003: 14, Östersund, Sweden.

Porter, M. (1990) The competitive advantage of nations. *Harvard Business Review* 1990, 73–91.

Puello, P.N., Hugueth, A.A.M., Díaz-Solano, B.H. and Ovallos, G.D.A. (2019) Análisis del nivel de competitividad en clústers. Caso iniciativa turismo de naturaleza en Atlántico, Colombia. Revista Iberica de Sistemas e Tecnologias de Informacao. *RISTI* 24, 234–249.

Sáez, C.A. (2019) Los micro-clusters turísticos ¿una herramienta para el desarrollo turístico en zonas rurales deprimidas. *Semestre Económico* 12(24 (Special Issue)), 57–68.

Section 3: Culture and Heritage in Tourism

11 Cultural Identity and Tourism Awareness of Rímac Residents

Milagros Granda and Brendali Carrillo
School of Hospitality and Tourism Management, Universidad Peruana de Ciencias Aplicadas, Lima, Peru

Summary
Rímac, World Heritage City, houses important monuments from pre-Inca, colonial, and republican times, of great attraction for tourism. However, there are indications that the elements that comprise Rímac's cultural heritage are little recognized and valued by their residents, thus, making them less willing to participate in its conservation and use for tourism purposes. For this reason, this study aimed to explore the cultural identity and tourist awareness of the residents of Rímac through in-depth interviews, where 30 youngsters participated. Participants showed little knowledge about the cultural heritage of Rímac and its tourist potential, but they reported feeling proud to live there because Rímac holds their most cherished childhood memories and family values. Results support the importance of analyzing the memories and emotions that connect residents with their city's cultural heritage and ultimately influence their willingness to preserve it and support the tourism activity in their district. Future research could explore the role of sport and religion as part of the residents' cultural identity and their potential for tourism.

The Value and Interest of the Case Study

This study shows that the residents of Rímac are willing to commit to the conservation and preservation of their cultural heritage. However, they expect the political authorities to take an active role in disseminating the local culture through public cultural events at heritage sites and through social media networks. Results also suggest the need to reinforce the residents' knowledge of local history, traditions, sites, and monuments from early education.

Background

Cultural heritage is composed of the tangible and intangible resources of a community that bring together individual and collective memories (Heersmink, 2021). It reflects a particular place's uniqueness and can serve as a resource for a community's economic and social development through activities such as tourism (Vinuesa and Torralba, 2018). There is evidence that those who live in a place that has unique and distinctive characteristics develop more pride and self-esteem as a result (Amin,

Table 11.1. Relevance of Rímac's main cultural heritage.

Rimac's cultural heritage	Relevance
Alameda de Los Descalzos	The oldest pedestrian walkway in America, built in 1611 as a place for socialization for Lima's citizens. It has been the scenery of love affairs and ghost stories, according to the oral tradition.
Plaza de Acho	Founded in 1766, it is the third oldest bullring in the world, and it is still active today.
Paseo de Aguas	The first urban modernity project in Latin America. Built by the Spaniard Viceroy Amat & Juniet for his lover 'La Perricholi'.
Quinta de Presa	One of the first mansions of a French-style built in Lima in the 18th century, that hosted aristocratic families of the viceregal era, until 1824.
Old Churches	The churches were built between the 16th and 19th centuries to house slaves, devotees, and Franciscans. Their facilities have survived earthquakes for years.
Music and literature	Rímac's history, urban landscapes and citizens' lifestyle and customs have served as inspiration for traditional poems and songs.
Rímac's foundation anniversary	This is celebrated with music and traditional food prepared by residents of the district.
San Cristóbal Hill	It was considered sacred by the families that inhabited the Rímac Valley in pre-Inca times. During the Spanish conquest, in the 16th century, a wooden cross was placed on top of the hill as a symbol of imposition of the Catholic religion.
Señor Crucificado del Rímac festivity	The most representative religious event of the district, with ceremonies dating back to the mid-19th century. In 2020, Peru's Ministry of Culture declared this festivity as National Cultural Heritage, recognizing its contribution to the social cohesion of Lima's population.

Adapted from: Dargent Chamot (2021); Resolución Viceministerial N° 000177–2020-VMPCIC/Resolución Viceministerial N° 000177-2020-VMPCIC/MC (2020); I. Ardiles (personal communication, 12 March 2020).

2018). This can be used to preserve the dynamic cultural features associated with the territory so that they persist from generation to generation and continue to be representative of the individuals' cultural identity (Settimini, 2021).

This case study is centered in Rímac, a district located in Lima, Peru's capital city, which features the city's first signs of urbanization. The district was founded in 1920 under its current name, Rímac. However, this same territory was previously called Barrio San Lázaro, due to the construction of its first church (San Lázaro) in 1586. For this reason, Rímac is considered an old district, with over 400 years of history, housing about half of the 608 monuments of the historical center of Lima. Beyond its architectural value, it also preserves the traditions of Lima's families, especially those related to the celebration of religious festivities, and it has been the stage for historical figures, especially of Republican times, between 1822 and 1923. Table 11.1 summarizes Rímac's main cultural heritage and its relevance. These are the most representative elements of Rímac that attract international tourists and national visitors to this destination, while they serve as recreation places for residents.

In 2015, Rímac was recognized as a World Heritage City by the Organization of World Heritage Cities (2015) due to its 450-year history and the value of its tangible (e.g. monuments) and intangible (e.g. religious festivities) cultural heritage that attracts international tourists. However, there is evidence of a lack of cultural identity among Rímac residents, which manifests in a low level of knowledge about the history of the district and its tourist attractions (Flores-Pérez, 2019). This could also mean that the Rímac residents do not have a strong tourism awareness that influences their commitment to the sustainable use of their cultural resources and positive treatment of tourists (Chng and Narayanan, 2017).

This study tackles the issue of the cultural identity of Rímac residents due to evidence of a lack of knowledge and self-identification of the local population with their history and heritage sites, and the customs of their city (Flores-Pérez, 2019; Ruiz Palacios and Pozo Trigoso, 2020). This is the first study that provides information about the cultural identity of Rímac residents along with their tourist awareness, stating their importance in the conservation of this Latin American city's cultural heritage with high tourist value and potential.

The study applied a qualitative methodology based on online semi-structured interviews with 30 young adult Rímac residents (16 female and 14 male) between 20 and 29 years of age with a minimum of secondary school completion in a local institution. Previous studies on Rímac residents' cultural knowledge had not considered participants of this age range. However, it is important to know if these young adults have learned about their district's cultural heritage and its tourist importance during their school years and, if not, understand how they have formed their cultural identity and tourism awareness, be it either weak or strong.

The residents of Rímac are known for being loyal to their district, meaning that families live there for many generations. Thus, if this is the case, it is important to explore the young adults' perception of their district's cultural value, because if they do not know it, they will not be able to pass it on to the next generation of Rímac residents and the oral tradition might get lost.

The interviews included questions based on the method of spontaneous recall to identify the elements that they could remember and answer without the help of stimuli, such as: 'Which tourist attractions do you know in Rímac?' Additionally, the assisted recall method was used, which consists of giving participants help to enhance their memory (Schiffman and Kanuk, 2005). Thus, participants were shown a picture of Rímac's coat of arms and a brochure with pictures of its tourist attractions to probe their knowledge about those elements.

Table 11.2 shows the general and specific objectives and their relationship to the variables of the study.

Data were analyzed and grouped into categories and subcategories that responded to the study objectives. Findings will be explained in the following sections and supported with statements from participants. It is important to mention that each participant was assigned a code to respect their privacy.

Table 11.2. Study objectives and variables.

Objectives	Variables
General objective	
Explore the cultural identity and tourist awareness of Rímac residents.	Cultural identity and tourist awareness.
Specific objective 1	
Gather the perspectives of Rímac residents about the district's and the population's identity.	Identity
Specific objective 2	
Unveil the residents' perception of Rímac's cultural heritage that granted them the title of World Heritage City.	Culture
Specific objective 3	
Understand why certain tourist attractions of Rímac are more strongly remembered by its residents.	Tourism knowledge
Specific objective 4	
Gather the residents' voice in respect to the tourism activity in Rímac.	Tourism impact

Cultural Identity of Rímac Residents

Cultural identity takes place in the consciousness of individuals who recognize themselves as part of a group of people who share knowledge, traditions, beliefs, symbols, stories, and rules that explain the order of things (Mansilla-Torres, 2006; Jameson, 2007) are manifested through their expressions, attitudes, customs, and lifestyles (Richards, 2018). Starting from this definition, six sub-categories were identified that explain Rímac residents' cultural identity.

Rímac's identity

When participants talked about their district, they referred to it as 'El Rímac' (The Rímac) or 'El rico Rímac' (The savory Rímac), alluding to the district's strong and vibrant character. They explained the identifying features of their district so fervently that at times it seemed as if they were talking about a person and not a place.

Rímac is known for being an old district with a long history because it houses many well-preserved cultural elements from colonial times, such as the old mansions that are now used as flats, and the traditional street food, such as *anticuchos* (cow's heart skewers) and *picarones* (traditional dessert).

Rímac is also famous for being noisy, especially some of the streets that hold many street vendors. Soccer was highlighted as a popular sport that is linked to the residents' and the district's identity. This is evidenced in the many artificial pitches found across the area.

Despite participants' appreciation of their district's identity, they stressed that there are some negative aspects that characterize it, such as the lack of cleanliness and the deficiency of the local government to safeguard the security and preserve the cultural heritage. There are those who consider insecurity to be a strong reason to move out, while others believe that this is only a negative stereotype that has been ascribed to Rímac by the media.

Residents' identity

Most participants come from families that settled in Rímac decades ago. They claimed to be remarkably close to their parents, grandparents, and uncles who taught them to be hospitable and to develop values such as empathy, loyalty, and brotherhood: 'Rímac is my house, my home, my family' (F11). They also expressed affection for their district because it houses memories of their childhood.

Participants pointed out that Rímac residents are generally charismatic, cheerful, and helpful. They also highlighted qualities such as being persistent, hard-working, resourceful, and thriving. On the downside, they emphasized that the Rímac residents 'do not follow the rules' (F4), which makes the participants feel disappointed and worried.

Catholicism is the predominant religion in Rímac, especially among the elderly. Participants in this case study were young, and although they did not identify with a particular religion, they affirmed that they take part in religious activities, such as the Señor Crucificado del Rímac festivity, because of the social dynamics that occur in those events: 'Personally, I am agnostic, but I have participated in religious celebrations because people gather in the street' (M1) (Fig. 11.1).

Residents' knowledge of cultural heritage

The topic of cultural heritage was difficult to navigate because most participants did not know what is considered cultural heritage and what is not. Therefore, their answers were dubious, and they made comments like 'I have not seen what cultural heritage looks like and I do not know what it is' (F6). However, a few of them spontaneously related it to the city's old monuments: 'For me, cultural heritage is the assets of the district that are of significant value, such as churches' (M9). After the concept of cultural heritage was briefly explained to participants, they managed to identify the Alameda de los Descalzos and its statues as the most important examples of the district's tangible cultural heritage and the celebrations of the Catholic religion as the most important intangible cultural heritage.

Looking a little closer at the concept of cultural heritage, participants also pointed at the San Cristóbal hill, the Plaza de Acho (bullfighting arena) and Creole music, as 'many

Fig. 11.1. Religious pilgrimage in San Cristobal Hill. Courtesy of Rímac's Municipality. Since pre-Inca times, Rímac dwellers used to walk up the San Cristobal hill as an act of reverence. Nowadays, international tourists visit the mountain to enjoy a panoramic view of the city.

songs refer to the district' (M1). They also recognized the old mansions and Nuestra Señora del Rosario, known as the smallest church in the world, as tangible cultural heritage.

Historical events, figures, and symbols

Study participants showed limited knowledge of the historical events that took place in the district. Almost half could not remember any of them, and the other half mentioned the historically famous love affair between Viceroy Amat and Perricholi, figures they remember because their parents and grandparents once told them a little of this story, but they have vague memories about it.

Most participants did not know that their district had a coat of arms. They associate the light blue color of the shield with the professional soccer team Sporting Cristal, whose origin and development is tied to the district.

There were even some who confused it with the coat of arms of the city of Lima. However, they did recognize some of its elements, such as the arch of the Paseo de Aguas (although they did not remember its name), the San Cristóbal hill, the Puente de Piedra (stone bridge), and the Rímac River. The element that very few recognized was the amancaes flower, which is a representative icon of the district.

Current state of cultural heritage

Participants indicated that no importance was attached to the care and preservation of their cultural heritage by the local government. They believe that this is due to the lack of order and supervision, the deficiency of public lighting, security and surveillance, and the fact that 'the authorities do not interfere at all' (F6) in finding a solution to those issues. They admit that it is particularly important to them that the cultural heritage is preserved, as this would give Rímac a better image and help eliminate some of the negative prejudices that stigmatize it.

Similarly, they said that only a few people cooperate with heritage conservation while many engage in negative acts, such as throwing rubbish on the street and graffitiing statues, monuments, and mansions. Participants said that during their school years, they were never taught about cultural heritage, but they consider that they should be instructed about heritage conservation at school and home.

Participants also stated that there is a lack of information about the history and cultural heritage of the district and that 'if they had more knowledge about [Rímac's] history and cultural value, the district would be in a better state' (M7) because 'one only cares and loves what one knows' (M13). Following this same line of thought, they assured that if they witnessed their cultural heritage being vandalized, they would defend it themselves or report it on social media so that the authorities could intervene. As one participant pointed out: 'I would defend it because it is part of my life, all my life since I was a child' (M14). Additionally, they claimed that if the local political authorities organized activities that promoted the care and conservation of Rímac's cultural heritage, they would be willing to participate and get involved, and they thought that most of the residents would too.

World Heritage City recognition

Study participants did not know that their district is on the exclusive list of the World Heritage Cities Organization, and they reacted with surprise and admiration: 'I love my district more now that I know what we are' (F10). In explaining this recognition, most agreed that the Rímac district absolutely deserves it. They claim that 'this was to be expected with all the cultural richness it has' (M13) and that it is a positive fact for the district 'that foreign people know that there are valuable things in Rímac' (MI10). Nonetheless, many agreed that 'the problem is the bad investments and the mistakes in the administration' (M8) which 'is not improving comprehensively' (M2) and that 'there is still a lot to do' (F11).

Tourism Awareness Among Rímac Residents

Tourism awareness among stakeholders is one of the most important steps in the successful planning of tourism projects (Cárdenas *et al.*, 2015). Stakeholders include the local population, who often do not understand the impact of poor tourism management or the benefits of well-conducted and inclusive tourism planning. More specifically, local tourism awareness refers to the knowledge that people living in tourist areas have about tourists, tourism, and its impacts (Saarinen, 2010). It occurs as a result of an educational process in which residents learn how to manage resources and treat tourists hospitably (Mihalic, 2016) and often leads to the reinforcement of patriotism and national (regional or local) identity (Frew and White, 2011). Thus, it is necessary to assess people's tourist awareness to identify their specific needs for knowledge (Cárdenas *et al.*, 2015). In this case study, the category of Tourism Awareness yielded two sub-categories that will be discussed hereafter.

Knowledge of tourist attractions

Among the tourist attractions that participants remembered the most is the Alameda de los Descalzos followed by San Cristóbal hill, Paseo de Aguas boardwalk and finally, Plaza de Acho. However, they did not remember the exact names of the places and only talked about their characteristics, such as: 'The spring where the moon is reflected' (F2 referring to Paseo de Aguas). When they were shown a tourism brochure with the pictures of the district's tourist attractions, they recognized the Amancaes hills and the Nuestra Señora del Rosario church.

Most participants had not visited the tourist attractions but had simply passed by them because they were on their way home: 'I drive by it in my car' (M16), but they also acknowledged: 'The truth is that I have never entered the Alameda de los Descalzos' (M9). Others consider these places as popular recreational areas where residents spend time and create memories with their families: 'I used to walk around there [Alameda de Los Descalzos] with my grandfather' (M6).

Emotions associated with the tourist attractions

Many of the interviewees explained that they felt nostalgic when they saw the tourism brochure with the pictures in it, because seeing them 'takes them way back to childhood' (F1). Thus, once again, they associate Rímac district and its iconic sites with family memories: 'It's like home, Rímac gives you peace' (M14). Similarly, they mentioned that it was only during the interview that they became aware of the tourist potential of their area and stressed that they are proud of it: 'Wherever I go, I say that I come from savory Rímac' (M13). Even though 'there is still a lot to improve' (M2), they identified with everything that exists there. Their participation in this study made them curious about what the district has to offer in terms of tourism because they conceded that these places, and tourism in general, had not been relevant to them.

Residents' views and opinions of the tourism activity

Most participants stated that they had hardly ever seen or heard of any tourist activities in Rímac and that the only tourist activity they knew about was the bus ride to Cerro San Cristóbal, a famous lookout point of Lima City. However, they associate it with a tragic memory of an accident that happened on this route 4 years ago, which they think was not handled correctly by the authorities.

Participants believe that tourists mostly explore Rímac on their own because there are very few tourism agencies promoting the district. They also stated that the district 'is not prepared for tourism' (F5) and that they do not trust the authorities to develop it with the necessary responsibility and transparency. Participants then elaborated on their views of the role of the local government in conserving Rímac's tourist attractions, stating that they should ensure constant maintenance, increase security and cleanliness, and promote the district's tourism assets through social networks.

Additionally, most believe that tourism would be beneficial and are willing to support it 'if it is positive and not so mercantilist' (M1), otherwise, it would only be a greater threat to their cultural heritage. Participants also had an optimistic expectation in terms of how tourism could contribute to the creation of direct and indirect jobs for residents which in the end could improve their quality of life, as tourism 'benefits many people, from those who take care of the tourists' cars, to those who sell *anticuchos*' (F15). Similarly, they claimed that tourism would improve the district's prestige and popularity.

Conclusions

This study explored the cultural identity and tourist awareness of a sample of young adult residents of Rímac, a World Heritage City. Cultural identity is built by the transmission of information, such as history, customs, and traditions, and through the interaction of residents with their cultural heritage, which is hardly the case for the participants in this study. Thus, it is important that the cultural identity of Rímac's young residents is strengthened to enhance their affective bonds with their cultural heritage and foster their willingness to preserve it.

The fact that participants were young adults with basic school education revealed the little instruction about the cultural heritage they have received during their school years, but it also showed the important role of parents and grandparents in teaching them about their history and traditions in spontaneous ways. This study also found that most participants have lived in Rímac their entire life and if they continue to live there, they should be aware of their cultural heritage to pass it on to their kids and maintain the oral traditions.

In this sense, it is advisable that kids be instructed about tangible and intangible heritage at school, to reinforce their self-identification with their district's culture from an early age. Local schools could also provide tourism lectures and organize walking tours to the district's iconic sites to increase the students' tourist awareness. Moreover, local promoters of tourism and culture could collaborate in the dissemination of the district's history, cultural features, and tourist attractions through social networks, which are the most preferred source of information for youngsters.

Participants were aware of the negative aspects of their district (e.g. insecurity). However, they expressed positive emotions toward it, such as pride, nostalgia, happiness, and love. Participants also showed signs of place attachment that could be attributed to the happy and nostalgic memories of their parents and grandparents sharing time and stories with them at the district's iconic heritage sites during their childhood. Thus, emotions and memories could be playing a role in fostering place attachment among locals, which could be further explored to engage them in cultural heritage preservation.

Overall, participants expressed good acceptance of the tourism activity, only if it were well conducted by an efficient local government that ensures public safety and promotes interaction among residents and tourists. Thus, it is advisable that the local authorities increase security, especially at heritage sites, and organize cultural activities for people of all ages and nationalities.

An important contribution of this study is that it started from a definition of cultural heritage extracted from the literature, but extended the knowledge of Rímac's cultural heritage from the perception of participants. In particular, the local soccer team is embedded into the district's identity, and it seems fair to state that it has become a part of Rimac's residents' cultural heritage.

Additionally, although religious festivities were already recognized as Rimac's cultural heritage (Table 11.1), this study showed their real significance for participants, who even if non-believers, still attended these events for social purposes. It would be useful to explore how these two elements (soccer and religion) may be embedded into the cultural tourism offer of this World Heritage City.

References

Amin, H.M.T.M. (2018) The impact of heritage decline on urban social life. *Journal of Environmental Psychology* 55, 34–47. DOI: 10.1016/j.jenvp.2017.12.002.

Cárdenas, D.A., Byrd, E.T. and Duffy, L.N. (2015) An exploratory study of community awareness of impacts and agreement to sustainable tourism development principles. *Tourism and Hospitality Research* 15(4), 254–266. DOI: 10.1177/1467358415580359.

Chng, K.S. and Narayanan, S. (2017) Culture and social identity in preserving cultural heritage: An experimental study. *International Journal of Social Economics* 44(8), 1078–1091. DOI: 10.1108/IJSE-10-2015-0271.

Dargent Chamot, E. (2021) Ricardo Palma y el Rímac: el barrio del Rímac en la obra de Palma. Ricardo Palma University, Lima, Peru. Available at: https://www.urp.edu.pe/pdf/id/33145/n/ricardo-palma-y-el-rimac.pdf (accessed 20 May 2021).

Flores-Pérez, S. (2019) Conciencia turística de los pobladores frente al patrimonio cultural del distrito del Rímac 2019. B.S thesis. Cesar Vallejo University, Lima, Peru. Available at: https://hdl.handle.net/20.500.12692/42707 (accessed 5 April 2021).

Frew, E. and White, L. (2011) Research directions for tourism and national identities. *Tourism and National Identities* 24, 215–218.

Heersmink, R. (2021) Materialised identities: Cultural identity, collective memory, and artifacts. *Review of Philosophy and Psychology* 14(1), 1–17. DOI: 10.1007/s13164-021-00570-5.

Jameson, D.A. (2007) Reconceptualizing cultural identity and its role in intercultural business communication. *The Journal of Business Communication (1973)* 44(3), 199–235. DOI: 10.1177/0021943607301346.

Mansilla-Torres, S. (2006) Literatura e identidad cultural. *Estudios Filológicos* 41, 131–143.

Mihalic, T. (2016) Sustainable-responsible tourism discourse–Towards 'responsustable' tourism. *Journal of Cleaner Production* 111, 461–470. DOI: 10.1016/j.jclepro.2014.12.062.

Organization of World Heritage Cities (OWHC) (2015) Distrito del Rímac – Patrimonio Mundial. Available at: https://www.ovpm.org/es (accessed 18 March 2020).

Resolución Viceministerial N° 000177-2020-VMPCIC/MC (2020) Declaran Patrimonio Cultural de la Nación a la Festividad del Señor Crucificado del Rímac (3 November 2020). Available at: https://busquedas.elperuano.pe/normaslegales/declaran-patrimonio-cultural-de-la-nacion-a-la-festividad-de-resolucion-vice-ministerial-n-000177-2020-vmpcicmc-1899772-1/

Richards, G. (2018) Cultural tourism: A review of recent research and trends. *Journal of Hospitality and Tourism Management* 36, 12–21. DOI: 10.1016/j.jhtm.2018.03.005.

Ruiz Palacios, M. and Pozo Trigoso, L. (2020) Conciencia turística de los pobladores en el distrito del Rímac-perú. *Estudios y Perspectivas En Turismo* 29(1), 312–330.

Saarinen, J. (2010) Local tourism awareness: Community views in Katutura and King Nehale conservancy, Namibia. *Development Southern Africa* 27(5), 713–724. DOI: 10.1080/0376835X.2010.522833.

Schiffman, L.G. and Kanuk, L.L. (2005) *Comportamiento del consumidor*. Pearson Prentice Hall, México, DF.

Settimini, E. (2021) Cultural landscapes: Exploring local people's understanding of cultural practices as 'heritage'. *Journal of Cultural Heritage Management and Sustainable Development* 11(2), 185–200. DOI: 10.1108/JCHMSD-03-2020-0042.

Vinuesa, M.Á.T. and Torralba, L.T. (2018) Visión territorial del patrimonio y sostenibilidad del turismo. *Boletín de la Asociación de Geógrafos Españoles* 78, 212–244. DOI: 10.21138/bage.2716.

12 Co-Creation and Developing the Creative Economy: The 'Galo da Madrugada' Case

Larissa Almeida[1] and Fiona Eva Bakas[2]
[1]RECRIA – Brazilian Creative Tourism and Experiences Network, Recife, Brazil;
[2]Tourism Department, ECEO, Lusófona University, Lisbon, Portugal

Summary

This case study explores the potential of connecting the creative economy with tourism through the analysis of the collaborative creative process, to rethink a carnival icon called 'Galo de Madrugada' (Morning Rooster) in Recife, Brazil. The icon already existed but the process by which it used to be created was not collaborative and caused few and temporary impacts. Via the description of the new creative process of the 'Galo de Madrugada', which brings together people from various cities in Pernambuco state, the case study illustrates how awareness for sustainable consumption is increased, which simultaneously acts as a tool to build capacity in local communities, create meaningful connections and boost social cohesion via co-working opportunities.

The Value and Interest of the Case Study

This case study foregrounds the potential for utilizing creative economy processes and activities surrounding the creation of an ephemeral art piece made for carnival time, as a valuable identity-making activity that also creates a year-round inspiration for tourism products. This example would be useful for people interested in better understanding the potential links between sustainability, the creative industry and tourism activities. In this case, study the value of creative economy processes is stressed as being essential in building capacity in artisan techniques and knowledge of reusing and sourcing sustainable material and for creating a local identity that can then be utilized to create various tourism-related resources. The case presents a dialogue with several themes besides tourism, namely: culture, sustainability, creativity, governance, economy of culture, economy of carnival, public management and cultural management, which readers are invited to engage with more deeply after reading the case study.

Background – Creative Economy and Tourism in Recife, Brazil

Recife is a city located in the Northeast of Brazil with a population of 1,488,920 people and a human development index (HDI) in 2010 of 0.77, according to IBGE (2022). Recife's carnival has an important role in the city's tourism activity because it attracts a large number of visitors and generates significant economic impacts. According to data from the Brazilian Institute of Tourism (Embratur, 2022), the Carnival of Recife has been responsible for increasing the tourist flow in the city during

 Tourism Case Studies Latin America Region (eds Thiago Allis and Sandro Carnicelli)

the holiday period. On average, in recent years, the event has received about 500,000 domestic and foreign tourists, and according to the city council in 2023 (Secretaria de Desenvolvimento Econômico, Ciência, Tecnologia e Informação do Recife, 2023), the carnival generated 2 billion reais in the local economy. According to the council, there is a 96.12% hotel occupancy rate during the carnival and the event also increases the movement of the local trade. The Carnival of Recife is considered one of the largest in the country in terms of its mobilization of the public and money. However, all this potential is limited to the carnival period as an event and does not extrapolate to the economy during the rest of the year. Considering that the carnival is a catalyst for the city's cultural manifestations and a tourist attraction, how can we use it as a development platform that generates results all year round?

Tourism is a multisectoral and integrating activity whose products and services are developed based on the specificities of the place, such as nature, business environment and cultural heritage. The creative economy comprises the set of social, cultural, economic and environmental policies of the creative goods and services not only in function of their utilitarian outcome but also their creative and emancipatory purpose (Leitão, 2023). The OECD report (2014) on tourism and the creative economy presents creative tourism as an activity based on the integration of tourism and the creative industries. This encompasses consumers, the producers of experiences, political bodies and educational institutions capable of generating meaningful intangible cultural products, creative content and experiences. The relationship between the creative economy and tourism goes beyond the appreciation of local culture or participation in thematic events. It can also be perceived in the ambience of a destination. A creative, vibrant city that stimulates the development of creative activities creates a feedback process that attracts creative people and by attracting them makes the place more interesting. In turn, people, especially tourists, are attracted by the uniqueness, creativity, vibrancy and innovation of the place. This can be achieved by stimulating the creative economy (Bakas and Duxbury, 2018; Corá and Henriques, 2021).

Tourism requires several interconnections to happen, so a city with a pulsating creative economy can use the creative industry to create tourist products and services. How can we think of an integrating dynamic between tourism and the creative economy that strengthens culture and creativity, strengthens identity and productively integrates carnival cultural traditions throughout the year and not only during the event period?

These potential synergies mean that the relationship between these sectors can act as an important stimulus for growth and innovation in tourism (Duxbury and Richards, 2019) and generate results beyond the growth of tourist demand, such as giving value to and legitimizing cultural practices (Belfiore, 2020). For example the synergy between tourism and the creative economy can generate impacts such as: (a) stimulate the development of small businesses as creative spaces that work as nodes in creative networks, allowing interaction between producers and consumers; (b) boost exports through products that link creative content, places and culture in order to increase general levels of interest in local creativity and encourage tourist visitation; (c) produce knowledge and development of new skills from the interaction between professionals in the creative and tourism industries, between producers and consumers, and between consumers and residents; and (d) encourage innovation, adding technology to tourism development and challenging the creative and tourism industries to find new ways of managing their businesses.

The case study presented here illustrates different aspects of this approach to value creation in tourism, via synergy with the creative economy. The 'Galo de Magrugada' case study is based on empirical data from an interview with the contemporary artisan, Leopoldo Nobrega (who has acceded to the use of his real name in this research) in 2022, and the author's participant observation of the carnival rooster phenomenon in its current format in the years 2019, 2020 and 2023.

The 'Galo de Magrugada' (Morning Rooster)

The Galo de Madrugada (Portuguese for 'Morning Rooster') is one of the most traditional carnival groups in Recife, a city located in Pernambuco state in Brazil and opens the city's carnival parade on carnival's first Saturday. A carnival group is a group of people who gather in a semi-organized way, generally around a theme and go out in the streets singing and dancing. These groups have an organizing core, but several people can join the group along the way during the carnival days. The Recife carnival group (Galo de Madrugada) was founded in 1978 and in 1994 was recognized by the Guinness book of records as 'the largest carnival group on the planet' when it attracted 1.5 million revellers to Recife's streets. In 2018, it attracted 2.3 million revellers (Investe Recife, 2022). In 1995, Recife City Council, as a tribute to the Galo da Madrugada carnival group created a sculpture of the rooster approximately 30m high, called 'Galo da Madrugada' which is now an icon of Recife's carnival. It is placed on the Duarte Coelho bridge in downtown Recife in the pre-carnival week and announces to everyone that the carnival has arrived. The 'bridge rooster', as the locals affectionately call it, causes a real stir among people from Recife who wait anxiously to see the rooster and give their verdict, whether it is successful or not and there is a general public reflection and discussion about it the day after it is hoisted into place on the bridge.

Each year the rooster is produced by a different artist and in 2019 and 2020, it was created by the artist Leopoldo Nóbrega using principles of creative tourism [e.g. authenticity, connection to place and co-creation (Almeida, 2021)]in its elaboration.

The artisan rooster and the circus rooster

To build the allegorical carnival statue, the contemporary artist invited people involved with the carnival from both the community of Bomba do Hemetério, territory that houses traditional carnival associations in Recife, and also the artist's studio crew, which is located in Arruda, a neighbourhood next to Bomba do Hemetério. The artist's choice to bring people together who experience carnival not just as a party, but as a constituent part of their identity and their territory's identity, instead of hiring technicians with experience in large structures, was based on the principle of connecting to place and learning. The carnival makers bring their passion for the carnival and their practical knowledge of making the costumes and the allegorical carnival statues for the carnival parades, imprinting territoriality in the creation of the rooster. The professionals who have already worked in the studio bring their knowledge and skill in scenography, in addition to the pride of being able to be a part of constructing an icon of Pernambuco's carnival. The contemporary artist in charge orchestrates this learning and co-creation process.

In the process of producing the icon, the piece was divided into parts (wing, crest, feet, beak, tail...) and people could choose which part they wanted to work on. Based on this dynamic, groups were formed to work on each part of the rooster. Then, research laboratories were made to select the most suitable materials, textures, shapes and techniques where everyone was invited to bring solutions to the questions raised in a dialogical learning process. This process meant that each person involved in building the rooster could understand the overall plan for what was being done, and the importance of each individual's work in the whole collective process of creating the rooster. This enabled the group to develop, in addition to technical skills, collective co-working skills.

The other innovation of the rooster-building process is related to the creation of the rooster's clothes, since each year the main rooster structure remains practically the same, but the rooster changes clothes. In 2019 (Fig. 12.1), the rooster's clothing was made using waste and elements from other creative territories in Pernambuco, using the rooster as a platform for the connection between municipalities in the Pernambuco state. Waste used to make the 2019 rooster's clothes (Fig. 12.4) came from textile shops at the Toritama clothing hub. The rooster's jewellery was made with clay sourced in Alto do Moura in Caruaru–PE, using a traditional pottery technique from that area. This opens up a range of possibilities and interconnections, since until 2019 the rooster was always seen as a solely Recife icon. This strategy expands territorial connections and cohesion and transforms the rooster into a great platform for raising awareness about environmental and

Fig. 12.1. Artisan rooster 2019 (courtesy of Leopoldo Nóbrega).

cultural causes, alternative products and idea generation.

In 2020, the rooster was presented as the 'Circus rooster with an eye on the future' (Figs 12.2 and 12.3) and merged technology with sustainability. The rooster's clothing was made in a broad co-creation process as the contemporary artist in charge organized dozens of workshops in several cities in Brazil and even in New York. More than 700 people participated in the workshops and painted more than a thousand vinyl records that then became the rooster's 'skin' (Fig. 12.5). In addition to the painted records, the rooster's skin had giant

Fig. 12.2. Circus rooster 2020 (courtesy of Leopoldo Nóbrega).

LED inserts, which was the first time that light fixtures were used in the sculpture.

Discussion

This case study illustrates how sustainability has various facets. The rooster is a sculpture linked to the carnival and just like the carnival parties, is always ephemeral, being hoisted on pre-carnival Thursday and dismantled on Ash Wednesday. In terms of sustainability, the artisan rooster sculpture was created with the intention of not being dismantled on Ash Wednesday but rather to be reused. In the coming years, the basic rooster structure was reused with a few adjustments.

Fig. 12.3. Circus rooster 2020 – Looking to the future (courtesy of Renato Venâncio).

For example the 2020 circus rooster reused 70% of the 2019 artisan rooster structure (Figs. 12.4 and 12.5).

During the workshop held to develop the sculpture, experiments were carried out that made it possible to extrapolate the solutions and innovations found, to other contexts. For example the artisan rooster skin was made with waste from the textile hub. This waste textile was also used in the making of costumes for a local ballet group (Fig. 12.6) that performed in the opening show for Recife's 2019 carnival, as well as in other products. The creative process allows for countless developments, ranging from the initial aim of creating the rooster, to making costumes and tourist souvenirs (e.g. rooster-inspired plates and cups), which illustrates how it is possible to develop a range of products inspired by the carnival rooster's elaboration processes as well as the final result.

Fig. 12.4. Skin of the artisan rooster 2019 (photo Leopoldo Nóbrega).

In order for all these ideas to align, there is an interesting issue pointed out by Leopoldo Nóbrega: making the rooster is not about stimulating creativity for the sake of creativity, but rather to define a purpose for each decision that goes beyond what seems to be the end goal i.e. a sculpture. The sculpture is ephemeral like the carnival, but the process of creating it needs to have a purpose, a meaning to be replicated. The potential of the idea related to the creation of the allegorical carnival statue is infinite. As it connects industry, artisans and locals, it reinforces local identity and inspires new product development and new collaborations among people who did not previously connect. More specifically pondering on the point of taking a person with no experience to do something that would, in some ways, better be done by a professional, Leopoldo Norbega says that giving people this chance to co-create and participate broadens people's world view, connects people from different backgrounds and helps redefine the concept of 'generating the best results', as being very much linked to social cohesion and environmental sustainability goals. The more people's minds open to new possibilities, the more likely it is that they give intelligent and authorial creative responses. A multicultural

Fig. 12.5. Skin of the circus rooster 2020 (Photo: Beatriz Rodrigues).

Fig. 12.6. Costumes for the Recife carnival opening ballet, resembling the rooster's skin (Photo: Maurício Ferry).

and transdisciplinary perspective was essential in creating the rooster in the form that it was presented to the public.

The Galo da Madrugada is an icon of Recife's carnival which creates a strong place-based identity. Still, it is underused as a tourist product and as a product of the creative economy because of its ephemeral nature. The innovation proposed by Leopoldo Nóbrega was to make the rooster a creative territory in itself by facilitating interconnections with other territories each year when the rooster changes its clothes. These territorial interconnections facilitate the use of other creative languages and provide various people with the possibility of working on communication, creation and creativity. All this transforms the making of a temporary carnivalesque structure into an event legacy.

Furthermore, this concept of legacy can be commercialized by creating rooster-inspired designs of products such as gifts and souvenirs, supporting a production chain throughout the year. By doing this, the carnival's creative city identity is strengthened, and more creative processes are activated beyond the time when the rooster is exhibited during the carnival. All these processes create a vibrant creative ecosystem that is reflected in tourism offerings. There needs to be a change in the strategic vision regarding the use of this platform to strengthen the creative economy and creative tourism in the city, as the rooster is more than an icon. Instead, it is a process with the potential to strengthen the ecosystem. For this to become a reality, co-creative laboratories need to be held continuously throughout the year and not just as one-off workshops just before the carnival.

Conclusions

The 'Galo de Madrugada' rooster shows that the creative result is, above all, a technical, conceptual and relational process, rather than an end in itself. The co-creation is a practice that needs to be based on respect, a feeling of belonging and complicity. The fact that they were participating, at any level, in the making of a part of 'Galo de Madrugada', stimulates people to do their best, open up and deeply experience the process, allowing themselves to experience creative freedom. The carnival in this way uses the concepts of sustainability, co-creation, inclusion and innovation to empower people in Recife and the surrounding areas in Pernambuco state.

The exchanges that took place between the knowledge, practices and emotions of the technicians and carnival artists involved in the process, as well as the connections between the different territories based on raw materials and co-creation workshops, demonstrate the richness of bringing together people from different backgrounds. The more diverse the group and its experiences, the richer the process and the more innovative the results. These exchanges boost the creative economy and create a learning flow that improves local artisans' work processes by learning various contemporary art techniques in the process of making the carnival rooster. This simultaneously improves the final result. By stimulating collaboration and building capacity among local people in Pernambuco, Leopoldo's art pieces now include an expanded spectrum of references from the perspective of the people who live there.

Increasingly, it is the intangible that connects people, despite us not often realizing it. The intangible is what enchants, what mobilizes affections. There is an increasingly intense quest to capture the charm of the intangible in a shared tourism and creative industry product (Kim *et al.*, 2021). This connection between creative industries and tourism creates opportunities for the development of future creative tourism activities, cultural experiences and creative organizations. It also means that tourism organizations will have to develop a greater role in the production and distribution of creative content. This, in turn, requires an increasingly deeper interconnection between tourism and creative economy initiatives.

References

Almeida, L.F.L. (2021) *Turismo Criativo: Uma viagem por culturas, encontros e experiências*. Bambual Editora, Rio de Janeiro.

Bakas, F.E. and Duxbury, N. (2018) Development of rural areas and small cities through creative tourism: The CREATOUR project. *Anais Brasileiros de Estudos Turísticos (ABET)* 8(3), 74–84.

Belfiore, E. (2020) Whose cultural value? Representation, power and creative industries. International Journal of Cultural Policy 26(3), 383–397.

Corá, J.M. and Henriques, C. (2021) O turismo criativo como basepara as políticas focadas no desenvolvimento sustentávellocal: o caso de Brasília e do Recife – Brasil. Revista Turismo & Desenvolvimento 36(1), 367–379. DOI: 10.34624/rtd.v1i36.9217..

Duxbury, N. and Richards, G. (2019) *Towards a Research Agenda for Creative Tourism*. Edward Elgar, Cheltenham, UK.

Embratur (2022) Impacto do Carnaval do Recife no Turismo. Available at: www.embratur.gov.br/carnaval-recife-impacto-turismo (accessed 1 August 2023).

IBGE (2022) Cidades e Estados: Recife. Instituto Brasileiro de Geografia e Estatística. Available at: https://www.ibge.gov.br/cidades-e-estados/pe/recife.html (accessed 1 August 2023).

Investe Recife (2022) Carnaval do Recife gera R$ 2 bilhões em negócios e turismo na cidade. Recife Prefeitura. Available at: https://investerecife.recife.pe.gov.br/carnaval-do-recife-gera-r-2-bilhoes-em-negocios-e-turismo-na-cidade (accessed 1 August 2023).

Kim, S., Whitford, M. and Arcodia, C. (2021) Development of intangible cultural heritage as a sustainable tourism resource: The intangible cultural heritage practitioners' perspectives. In: Chhabra, D. (ed.) *Authenticity and Authentication of Heritage*. Routledge, UK, pp. 34–47.

Leitão, C.S. (2023) Criatividade e emancipação nas comunidades-rede – contribuições para uma economia criativa brasileira. Itaú cultural, São Paulo; Editora WMF Martins Fontes.

Organisation for Economic Co-Operation and Development (OECD) (2014) *Tourism and the Creative Economy*. OECD Publishing, Paris. (OECD Studies on Tourism).

Secretaria de Desenvolvimento Econômico, Ciência, Tecnologia e Informação do Recife (2023) Investe Recife. Available at: https://investerecife.recife.pe.gov.br/carnaval-do-recife-gera-r-2-bilhoes-em-negocios-e-turismo-na-cidade (accessed 1 August 2023).

13 Real Argentinian Food: Tensions and Conflicts in Value-Added Processes of Products to Commercialization and Tourism

Sandra Tolosa[1] and Cecilia Pérez Winter[2]
[1]Institute of Anthropological Sciences, Faculty of Philosophy and Literature, University of Buenos Aires, Buenos Aires, Argentina; [2]Institute of Geography, Faculty of Philosophy and Literature, University of Buenos Aires/National Scientific and Technical, Research Council (CONICET), Buenos Aires, Argentina

Summary

Wine and *dulce de leche* (Caramel sauce) were recognized as cultural national food heritage in Argentina due to their importance in daily consumption and gastronomic identity. The formal recognition aligns with global food valorization processes, such as the United Nations Educational, Scientific and Cultural Organization's (UNESCO) immaterial heritage and the importance of gastronomic tourism. Also, public policies in different countries seek to add value to food and beverages in order to foster the development of rural areas, favour tourism commercialization, strengthen local identities and enhance traditional/ancestral knowledge. These show the symbolic and economic value of food.

In this context, several types of formal certifications and commodification processes, which recognize territorial anchorage, of how certain food is produced and consumed may lead to its standardization, homogenization or monopolization, generating various tensions. Hence, we question the relationship between those who promote value-adding initiatives, how farm producer communities participate in them, which sectors or actors really benefit from these processes and what tensions may arise from them. We will address these questions by reconstructing and analysing the heritage and tourism valorization of high-altitude wine in the Calchaquíes Valleys area and *dulce de leche* production in the province of Buenos Aires.

The Value and Interest of the Cases Proposed

High-altitude wine and *dulce de leche* raise a novel and little studied problem in relation to the role played by the processes of adding value when talking about three mechanisms: heritage valorization of food, food certifications and food tourist circuits and how they promote identity legitimization in the producing territories.

Background of the Cases

Wine and *dulce de leche* are foods that are produced, marketed and consumed in different countries, although in each case they may have different connotations and values. In Argentina, both products are highly significant in the daily diet of its citizens. They are also identified as part of Argentina's identity and, as such, have been recognized as a cultural national food heritage, which has given them added value in terms of their commercialization, both for tourism and global market circuits. In this way, these processes (heritage valorization and tourism commercialization) link different aspects, which feed into each other. On the one hand, towards the outside world, the identification of certain foods with a specific territory establishes marks and differences with respect to other producing countries, even those countries that also consider these products to be part of their nationality configurations. This legitimization of discourses of 'ownership' of foodstuffs is currently fed by strategies of tourism and heritage valorization that add value to them, allowing them to enter and compete in new markets (Lacoste *et al.*, 2014). On the other hand, within the territory, these processes aim to strengthen regional economic development through the commercialization of these products and the construction of tourist circuits linked to the productive origins of these foods, as well as to strengthen territorial identities (Barrera and Alvarado, 2008).

However, the complexities involved in the processes of adding value to these foods can end up catalysing tensions (such as promoting the re-appropriation of practices and knowledge, among others). This can increase socio-economic inequalities and pit the productive sectors against those who promote policies of valorization, commercialization and touristification from hegemonic positions. Hence, in this case study, we make a joint and synthetic presentation of these aspects, based on the cases of dulce de leche and high-altitude wine. The joint approach allows us to offer the international public a general overview of the main elements that operate in the heritage processes and tourism marketing of food in Argentina. *Dulce de leche* and wine are very important food products that are well known overseas due to their importance for domestic consumption, and their role in the construction of Argentine food identity and its projection towards a broader market. Along these lines, we have articulated the results of our individual anthropological investigations that show that, despite the specificities of each product, there are common features in the processes in which they are involved that justify a joint approach.

The information presented in this case study was obtained from the identification and analysis of regulations and documentary and statistical sources produced by government organizations. These data are combined with those gathered from anthropological fieldwork carried out in the province of Buenos Aires and in the Calchaquíes Valleys, in the Argentine north-west. This corpus allows us to investigate the valorization and commercialization of certain foods associated with national identity and the effects and tensions that these processes generate in the producing populations.

The case study is structured as follows: The first section provides some quantitative data on the production of wine and *dulce de leche* to show their relevance in the country. Then, we refer to the various mechanisms of heritage processes and certification of these products, which constitute the value-added aspect produced by the state. Next, we consider the forms and circuits that are opened up with regard to their tourist commercialization and consumption. Finally, we propose some lines of debate regarding the tensions that heritage and tourism processes generate in Calchaquies Valleys and Buenos Aires province.

Some Facts About the Production and Consumption of *Dulce De Leche* and Wine

In this section, we will offer some data about the production of *dulce de leche* and wine that allow us to understand not only the importance of its consumption at the national level but also how its growth has expanded the consumer market in which these products are inserted, transcending borders.

Dulce de leche

To begin with, *dulce de leche* (Caramel Sauce) is derived from (cow's) milk. It is a sauce made with milk and sugar as the main ingredients. It is a product that can be found in different countries and in each one we can identify its common points and particularities based on the production system, commensality and the myths of its invention that legitimize this food as each country's own. In Argentina, it is identified as a

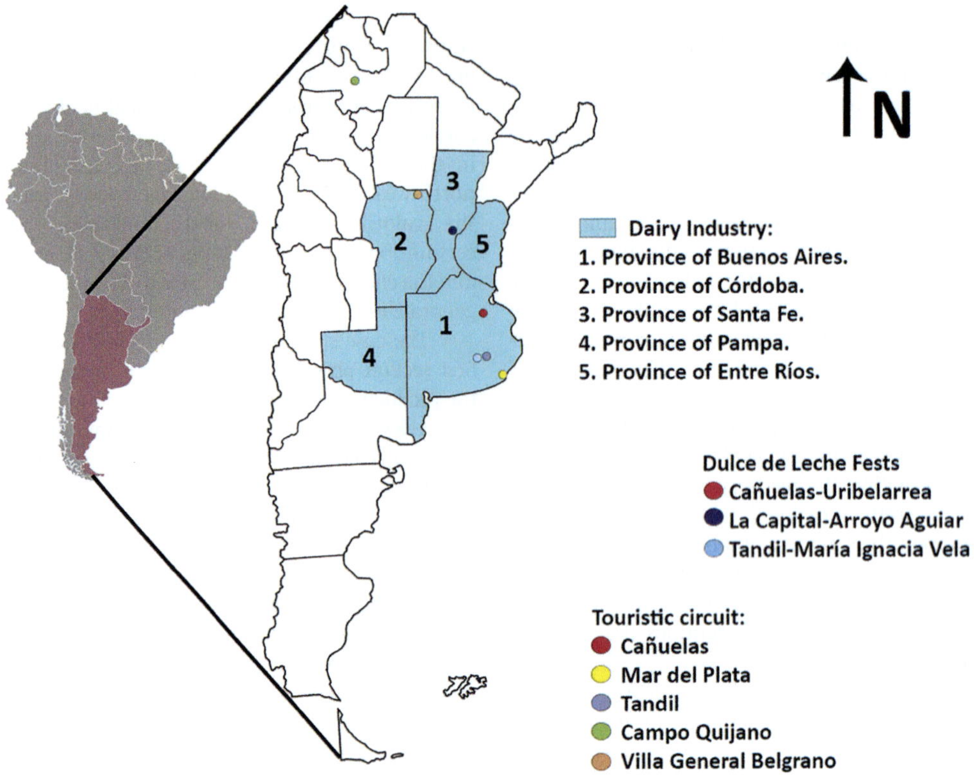

Fig. 13.1. Illustrative *Dulce de Leche* map.

symbol of national identity and gastronomy, but some data can also show its relevance. Although the dairy industry is significant in the country today, it began to develop incipiently from the first dairy farms registered in the mid-19th century. By 1889, one of the first dairy industries in the country, La Martona, was already being set up in the province of Buenos Aires. Others followed, such as Estancia Plomer and La Vascongada (Carman, 2014).

Dairy production in Argentina is currently significant. Indeed, in 2021 more than 11,553 million litres of cow's milk were produced, of which 2820 million were exported and 8647 million were destined for domestic consumption (Minagro, 2016). The provinces with the largest milk production are Córdoba, Santa Fe, Buenos Aires, La Pampa and Entre Ríos (Fig. 13.1). Of the companies located in Argentina, the most important in terms of production and technology are Saputo, Mastellone S. A./La Serenisima, Illolay, Punta del Agua. Some of these are foreign and were able to establish themselves through the acquisition of well-known national companies. For example, Saputo, originally from Canada, in 2003 bought La Paulina, founded in 1921 in Villa María in the province of Córdoba (Saputo, 2023). There are also cases where the opposite has happened, such as La Serenisima, which expanded its market by buying the Brazilian company Leitesol in 1996 (La Serenisima, n.d).

Milk production in Argentina is important, as is the production and consumption of its derivative, *dulce de leche*, of which it is estimated that there is a monthly production of at least 128,000 tonnes. In 2021, 4079 tonnes were exported (Telam, 2022). In this context, it should be noted that in 2006, the Ministry of Agriculture, Livestock and Fisheries designed the protocol for the production of '*dulce de leche* premium' in order to obtain the 'ARGENTINE FOOD A NATURAL CHOICE' quality certification, which was created a year earlier, with the export of the product in mind. In the introduction of the case study, it was highlighted that *dulce de leche* is part of Argentina's gastronomic heritage and that it is important to guarantee the quality production of this food because it is 'typically Argentine' (Minagro, 2006). This is justified by the historical account that points out that *dulce de leche* began to be consumed in 1829 in Argentina as a result of an accident by a maid who inadvertently invented it within the framework of a political pact that we will discuss later.

As for the domestic market, the country currently consumes more than three kilos per capita per year of *dulce de leche* (DeCepas, 2022). This product is highly accessible as it can be found in all supermarkets and grocery stores throughout the country and is present in a variety of food products. These include ice cream, teas, pastries (both traditional and innovative), desserts, and sweets. It can also be easily made in any kitchen with just a few ingredients.

Wine

Wine is a beverage that although it is produced, marketed and consumed almost all over the world, has singularities according to its different producing regions. In Argentina, wine is considered a national drink. There is a consensus that it was introduced to the country around 1550, when the first grape varieties brought from abroad were planted (Ambrosetti, 1895). Since then, its consumption has become commonplace for the Argentinean population. It was first produced in the individual home (a custom that still continues in different areas) and then industrialized and massified.

The latest data published by the National Institute of Viticulture (INV), in 2022 indicates that the area under vine in Argentina is 207,047 ha, distributed across 23,090 registered vineyards. This year, it resulted in a harvest of 19,368,031 quintals. A part of the grape production is marketed as fresh fruit or sultanas. However, most are destined for wine production. A total of 8,276,098 hl was sold to the domestic market and another 2,673,811hl was exported in 2022, for a value of US$839.963.000 (Instituto Nacional de Vitivinicultura, n.d.).

These general data show the importance of the wine sector in Argentina, especially since the 1990s when the so-called 'new viticulture' (Martín, 2009; Neiman, 2017) was introduced as part of the conversion of classic agricultural models into various forms of agrarian extractivism (Manzanal, 2000; Giarracca and Teubal, 2008) as part of the neoliberal policies of the time. This new model, characterized by the investment of foreign capital, the incorporation of new grape varieties, and the technologization of the process of wine production and international commercialization, led to an increase and diversification of production. This even reached regions with no vine-growing tradition, such as Patagonia (Fig. 13.2). The process resulted in the concentration and foreignization of the business of wine and the territories involved.

Different wines are now offered according to the varietals, climate, soils and altitude of the different viticulture areas in Argentina. They have become tourist circuits associated with wine. Within this framework, the Calchaquí Valleys, in the Argentine north-west, constitute a particular geographical point that combines wine production and wine tourism of high purchasing power. The area has reached an exclusive position in the international market, both for the commercialization of premium wines, as well as for being one of the most important destinations in Argentina (Fig. 13.2).

The Calchaquíes Valleys constitute a system of inter-serrano valleys that develop over 500 km between the Quilmes and Aconquija ranges, crossing three provinces (Salta, Tucumán and Catamarca). In the 1990s, the process of external capitalization of the wine-tourism sector led to an increase in vine cultivation, which placed this area in second place in the north-west region of Argentina according to INV registers. Salta, in particular, has concentrated this production, increasing its area of grape production from 2552 to 3608 ha between 2010 and 2021, an increase of 41.4%. There has also been a qualitative difference in terms of the quality of the wines produced, a large proportion of which are destined for the international market. The INV registers a growth from 29,910 hl (US$12,475,000) exported in 2010 to 47,443 hl (US$21,223,000) in 2021 (Instituto Nacional de Vitivinicultura, 2018). These premium signature or boutique wines

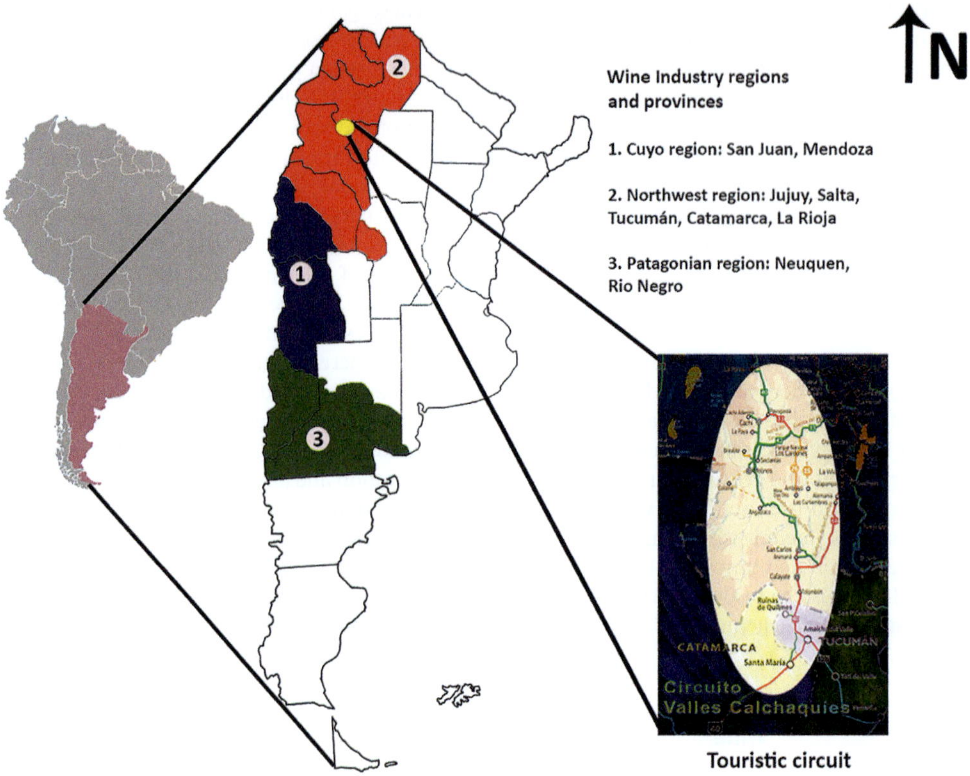

Fig. 13.2. Illustrative wine map.

follow international market standards and/or are aimed at consumers with high purchasing power, which in turn allows the development of high-end wine tourism.

Heritage and Tourism Enhancement

Heritage valorization

The valorization of food – and food with identity – is reflected both in state policies and those of international organizations. The notion of 'food sovereignty' is a concern due to the food crises that organizations such as the Food and Agriculture Organization (FAO) have been warning about since at least 2009 (Manzanal and González, 2010). Related to this, Argentina has created its own policies which tend to protect food sovereignty. What is important to point out is that 'the most vulnerable families should be able to access adequate and sufficient food, in accordance with the particularities and food identity of each region of the country' (Argentina.gob.ar, n.d.). This quote is relevant because it indicates that the protection of and access to food is not only based on guaranteeing nutrition but also takes into account the particularities and identities of each region.

Also, we find another aspect whose relevance lies in valuing certain foods as part of the national identity based on other types of state recognition that reinforce the consumption by two social sectors. These are wealthier sectors or those oriented towards the foreign public and tourists with high purchasing power. Within this context, different mechanisms have appeared, such as heritage valorization of foodstuffs or certificates of origin, aimed at making national foods, including *dulce de leche* and wine, visible and legitimizing them.

In the field of dairy production, heritage valorization of *dulce de leche* was promoted through different actions. In 1997, the Estancia San Martín of Vicente Casares farmhouse, located in the district of Cañuelas, province of Buenos Aires, was declared a 'National Historic Monument'. This is where one of Argentina's pioneering milk-producing companies, La Martona, was located. The brand has recently reappeared on supermarket shelves. The farmhouse was also where two opposing political figures – Juan Manuel de Rosas and Juan Lavalle, who signed the so-called 'pact of Cañuelas' on 24 June 1829 to resolve an important national political conflict – met. Although heritage recognition does not specify it, this property also houses the myth of the creation of the most widespread *dulce de leche* version in the country, which coincides with the date of the aforementioned pact. Hence, locally, 24 June was declared *Dulce de leche* Day to commemorate both milestones, giving rise to the organization of the *Dulce de Leche* festival in the town of Uribelarrea, Cañuelas. This local event became a provincial celebration in 2007. Cañuelas is currently seeking to be recognized as the 'capital of *Dulce de Leche*' (Pérez Winter, 2021). In the midst of these recognitions, which were processed by the municipal government of Cañuelas, in 2003 *Dulce de Leche* was declared under the category of 'Gastronomic and Food Cultural Heritage' by the National Secretariat of Culture, along with four other foods, including Malbec wine.

Malbec heritage, among other actions that will be mentioned in the following section, is linked to the productive and tourist change generated since 1990 across Argentina. This led to the declaration of Argentine Wine as a National Drink by law No. 26.870/2013 and the declaration of Argentine Wine Day (Decree No. 1800/2010). The arguments of this last decree are interesting because they cite the five centuries of history of the beverage in the country, the territorial and climatic extent of production, the qualitative and sustainability standards achieved and its social and economic importance for the producing provinces. Its importance in the external market is discussed, due to its high added value and competitiveness. Its importance to the domestic market, due to its mass consumption and nutritional qualities that place it in the basic food basket, is also noted. In particular, it is pointed out that the fruit of the vine is 'a value of our civilisation and a basic element of Argentine identity' (Decree 1800/2010).

In addition to these national and heritage laws, there are a series of other regulations, policies and associations that support viticulture. These include the creation of the Corporación Vitivinícola Argentina (Law N° 25.849), programmes promoted by the state, and the creation of regional wine tourism circuits. At the international level, Argentinian wine has been the subject of commercial negotiations with the Southern Common Market (Mercosur), the European Union, the International Organization of Vine and Wine (OIV) and with the countries of the World Wine Trade Group, of which Argentina has been a member since its foundation.

Due to the regional diversification of the product, Law 25.163 of 1999 established the General Rules for the Designation and Presentation of Wines and Spirits of Argentine Wine Origin, under the control of the Origin Protection Department of the INV. The system seeks the recognition, protection and registration of geographical names to designate the origin of wines and associated beverages, through a series of categories. One of these is the Indication of Origin (IO), which is reserved for table or regional wines. Another is the Geographical Indication (GI), which identifies a product originating from a recognized region, locality or production area of the national territory, produced and packaged in the same place of cultivation or in a different area. It only applies to quality beverages whose characteristics are due to their geographical origin. In the case of the Calchaquí Valleys, the Salta sector of the Cafayate Valley has a GI (Resolution INV N° C.32/2002; Resolution INV N° C.15/04; Order INV N° 08/2005). Finally, there is the Denomination of Controlled Origin (DCO), which identifies a product originating from a region, locality or differentiated area, whose characteristics are due to the geographical environment, including natural and human factors, where the raw material is produced and the elaboration, ageing and bottling are carried out. It is reserved exclusively for wines and beverages of selected varieties or superior quality. In Argentina, this category is limited to two regions: Luján de Cuyo and San Rafael, both in Mendoza.

The categorization of origin is a right, granted by the State, which includes exclusivity, legal protection, use of acronyms and logos, certifications of genuineness and quality assurance. Registration can be requested individually or collectively by producers, winemakers or wine organizations but is decided by the Authority of Application or, in the case of AOCs, by a Promotion Council. These entities study in detail the climatic, geographical and soil qualities, varieties, cultivation systems, historical background, cultural practices, yield, bottling, presentation standards and control. They ensure the law is complied with, issue certifications and determine sanctions for non-compliance.

As we can see, although *dulce de leche* and wine are foods that have been valued as heritage representing Argentinian national identity, wine is the one that has the greatest trajectory in terms of production, export volume and certifications towards its quality.

Links with tourism

The state mechanisms mentioned in the previous section are closely linked to the processes of building tourism circuits associated with these products. Currently, it is common to find in tourism the promotion of destinations by selling 'experiences', not from the anthropological point of view of an encounter with othernesses (Adame, 2002), but rather of an existential type (Arias Castañeda, 2015). This is expressed in slogans such as 'experience nature' or other references to the place or local cultural-identity aspects transformed into 'tourism products'. These intangible aspects, many of which have been recognized as formal heritage, give the destination an aura of authenticity and legitimization. Likewise, in this framework of experiencing and discovering that tourist proposals sell us, gastronomy and food heritage occupy a special place as constituent elements of identities with their territorial anchorage. In this framework, the addition of value is highly significant, and the valorization of food and cuisine for heritage and tourism are processes that are adapted to the logic of commodification and the expansion of capital.

The processes of food valorization have in common the interest in protecting them, adding value to them and recognizing their territorial anchorage. In fact, it is common to find the term 'territories with identity' in the public policies and guidelines of international organizations that seek the economic 'development' of certain places (Carenzo, 2007). In this way, the focus on food and gastronomy as elements of heritage and tourism has been on the rise in recent decades. UNESCO's recognition of the cuisines of Mexico and France in 2010 was a major step forward, as it considered these cuisines as fundamental intangible heritage of a country, reflecting part of its identity and including popular celebrations around them. However, from the point of view of cultural tourism, and under the modality of gastronomic tourism in particular, the consideration of food from its place of consumption in bars and restaurants dates back to the 1980s. In this process, which has continued to grow, typical foods end up being aestheticized by famous cooks and/or chefs, particularly in urban centres, satisfying the consumption associated with the leisure of the middle and upper classes (González Bracco and Arizaga, 2019; Troncoso and Arzeno, 2019). This poses a problem, as these special circuits produce not only segregation in food consumption but also segregation of the original local producers, who are left out of the benefits that such food consumption produces.

As far as *dulce de leche* is concerned, it is possible to find that, with the rise of tourism, a series of attractions and circuits have been created, particularly between the late 2000s and early 2010s. First, as already mentioned, the *Dulce de Leche* Festival in Cañuelas has gained relevance to the point that it has ceased to be a local celebration and has become a provincial one (Pérez Winter and Enrique, 2020) (Fig. 13.3). This is related to the tourist promotion of the celebration within the framework of the policies promoted by the Secretary of Tourism of the province of Buenos Aires under the Directorate of Popular Festivities, created in 2008, and the creation of a tourist calendar, since at least 2011, based on the promotion of festivals in the province (Pérez Winter, 2019a). Within this framework, since 2007, the Artisan *Dulce*

Fig. 13.3. Entrance to the XIX Fiesta del Dulce de Leche (Sweet Milk Festival), Municipality of Cañuelas.

de Leche Festival has been organized in María Ignacia Vela, in the Buenos Aires municipality of Tandil (ElEco, 2023). We also find that since 2013 the provincial *Dulce de Leche* festival has been held in Arroyo Aguiar, Santa Fe province (Pérez Winter, 2021) (Fig. 13.1).

Since at least 2010, in the city of Buenos Aires, boutique shops such as La Vaca Lechera, Dulce de Leche & Co or La Casa del Dulce de Leche, specializing in the sale of sweet milk, have been set up in the historic and tourist district of San Telmo. There you can find a variety of elaborations of high quality, mainly intended for sale to tourists. Some of these outlets, such as Dulce de Leche & Co, also have a shop in Palermo, another highly touristic neighbourhood in Buenos Aires. Also, we would like to mention the recent creation of a *Dulce de Leche* circuit, which brings together a series of localities located in different provinces of Argentina with the purpose of enjoying the consumption of this product. This circuit (El Cronista, 2021) includes three provinces of Argentina:

1. In Buenos Aires: Cañuelas, through the town of Uribelarrea where the *Dulce de Leche* Festival is held; Mar del Plata, where the traditional Sweet Milk factory, El Chimbote, was created in 1937, and the famous Havanna *alfajores*[1] which have existed since 1947; Tandil, where more recent production of El mundo del dulce de leche – since 2013 – Produlac – since 1991 – and Don Atilio – since 1993 – stand out and which also hosts the Artisanal Sweet Milk Festival mentioned above.
2. In Salta, includes the town of Campo Quijano, with its premium *Dulce de Leche* factory that has functioned since 2004.
3. In Córdoba, from Villa General Belgrano through its factory Ramona delicias del Valle which has been in production since 1980 (Fig. 13.1).

These destinations stand out, not only for the traditional manufacture or the premium quality of *dulce de leche* but also for the manufacture of *alfajores*.

Wine tourism has long been present at the national level, if we take into account that in 1936 the Wine Harvest Festival (*La Vendimia*) began to be held in Mendoza, an event that since then has been increasing in size, in terms of its spectacularity and number of visitors. Later, at the end of 1990, specific actions such as the recognition of the heritage of Mendoza's estates and wineries, the creation of the 'Wine Routes', the design of manuals and regulations to formalize wineries with tourist quality, definitively declared the producing regions as tourist consumption circuits.

In the case of the Calchaquí Valleys, tourism began at the turn of the millennium. Since then, the resulting Wine Route has become the most important one in the northwest region (Fig. 13.2). Certain elements especially sustain the circuit and are advertised along with wine consumption. They include the natural landscapes, the beneficial climate, the tranquillity, the local food, the friendliness of the people and the historical and archaeological attractions that the route passes through. Many of the buildings of the estancias and bodegas, some colonial, are conceptualized (and offered) as 'wine heritage' assets, although in most cases they have not yet been officially declared as such. Most of these buildings have been part of the process of foreign capitalization in the sector and have undergone restoration and aesthetization in order to welcome the public (Fig. 13.4). In many cases, they conserve elements of the old installations, forming small museums that are linked to the Vine and Wine Museum, located in the city of Cafayate, in Salta, the nerve centre of the Route.

The close link between wine and tourism activities is the reason for the interdependence of both activities, which has been called the 'Wine & Tourism Model' (Tolosa, 2023). Its success has led the Salta government to promote and invest in the sector in order to maintain its position as one of the most important destinations in the country (Ministerio de Turismo y Deportes de la provincia de Salta, n.d.). Likewise, different activities are being added, such as the organization of its own Wine Harvest Festival, the organization of national and international congresses and meetings related to wine, and the teaching, at a local level, of oenology courses.

The growth of the Wine & Tourism Model has produced, in addition to foreign investors, an important displacement of the resident population in two ways. There has been the installation of gated communities and country

Fig. 13.4. Old winery restored for tourism, Valles Calchaquíes.

clubs, especially on the outskirts of Cafayate. These are the objects of international amenity migration, generally aimed at sectors with very high purchasing power, especially foreigners (Rainer and Malizia, 2014, 2015). In addition, there has been the migration of labour for services linked to construction, tourism, and gastronomy. Within a framework of territorial appropriation, both for crops, wineries, and elite neighbourhoods, there have been important impacts. Environmentally, this has been related to the expansion of crops and private properties at the expense of the native forest. There have also been serious water consequences due to the development of large-scale vineyards in an arid area, with only seasonal flooding of rivers and historical problems of water supply. Socio-economic impacts relate to associated labour displacements with consequent drops in wages and basic working conditions. There has also been loss of territory to foreign ownership and therefore the impossibility of access to land and housing concentration in poor conditions. Also, certifications and quality controls imply a restriction for small producers, who often sell their harvest to the most important firms. This has all resulted in an increase in inequality (Tolosa, 2023).

Tensions from value-adding processes

In the preceding sections, we have pointed out a significant relation between the processes of food valuation, state mechanisms of heritage valorization and certification, productivity oriented towards segregated consumption between different sectors, and the creation of tourist circuits. The economic dimension of heritage valorization can be deduced from this, as processes which, although discursively centred on questions linked to identity – both national and regional – incorporate the commercial logic of adding value and facilitating commercialization and consumption markets, including tourist circuits and routes.

In this sense, both the vertical perspective implied by governmental actions of valorization, the capitalization of production (be it dairy or wine production) and/or the resulting private and state investment in tourism, raise the need to observe whether these processes, promoted with the explicit intention of generating benefits for local communities, have achieved this objective. Moreover, it is necessary to identify whether, instead, the exponential growth of the associated businesses has generated tensions or concrete problems for these communities.

We are interested to close this case study by pointing out some tensions that we have identified from our research on the subject and that can be applied to both the cases of *dulce de leche* and wine. For space reasons, we mention briefly some of the tensions since delving into each one of them would exceed the space possibilities of this work.

As we have just pointed out, value-addition policies identify communities producing certain foods as the general target group. However, what these policies do not make explicit is which sectors of these communities are the real beneficiaries and which are excluded. Often, these policies end up benefiting those who have more economic resources and not small producers. Furthermore, these policies do not take into account the diversity of territories. We note that in the case of certifications – of origin, quality, organicity, among others – the processes and conditions necessary for their approval are arduous and costly, with the result that only those capitalized companies with the necessary contacts and resources required to obtain such recognition are able to obtain them, leaving small producers excluded. In the case of Calchaquí wine, for example, the possibility of obtaining these certifications falls on the wineries with the greatest economic resources, which are also in charge of the production, bottling and marketing phases. In comparison, family wineries cannot meet the costs. Consequently, small producers are forced to sell the primary production of their farms (the grapes) to the largest companies.

Regarding the commercialization phase of 'valued' food products, although the certifications seek to enable ways for this to occur, and many of them allow the export of recognized foods, exporting is not an accessible alternative for every producer. This is not only because of the difficulty, already mentioned, of accessing certifications but also because not everyone has the necessary contacts and relationships to achieve a market expansion. Consequently, this once again remains in the hands of those ventures with sufficient of all types of capital. This especially affects local or small operators who produce *dulce de leche* in small amounts, allowing only the big enterprises that not only make this sweet but other products related to it (milk, cheese, *alfajores*) to gain. Certified foods have various legal problems associated with being marketed outside their place of production, a situation that requires specific resources (legal representation, time, and money) to correct that are not always accessible to small producers.

In addition, those agro-producing families who want to participate in the certification process face a series of obstacles and challenges, not only in the strict production process but also due to the limitations of their own living conditions. For example, in the Clachaquí area, many of these producer families are restricted in their use of water, a fundamental resource in wine production. Hence, local farmers, who produce wine on a small scale, can use water for irrigation by organizing themselves in shifts. This situation is contrasted with large wine producers, who use drip irrigation systems fed by deep water boreholes. This involves an economic cost that is too high for the local, small-scale producers. This system brings great risks in Calchaqui Valleys where water is a scarce resource, though until now a 'water emergency' has not been declared in the area, in contrast to San Luis, Chaco and Mendoza provinces, to name a few. In these provinces, the access to this vital resource is unequal among the population (Saldi and Petz, 2015). This situation further strengthens unequal power relations in local terms and widens the gap between large-scale companies that have the economic, political and social capital that allows them to carry out their productions and small-scale local producers. Moreover, small producers are often forced to sell their raw materials to these companies, as they do not have access to the necessary structure to produce a manufactured product with their own added value, generating dependency and inequality.

The heritage valorization of foods as elements that form part of local and national identities generates disputes over their appropriation between neighbouring countries – as happened with *dulce de leche* between Argentina and Uruguay (Castro and Pérez Winter, 2018) – and even between localities within the same country that end up competing over which food is more authentic. Such disputes invite differentiation and singularization of each food product according to its local particularities.

We have previously pointed out that the processes of heritage valorization of certain foods – and the disputes that it entails – are closely related to the enable marketing and consumption circuits, especially tourism development, which can also promote new heritage. In this sense, there is a feedback loop that operates not only at a conceptual level, in that both fields speak of 'landscapes', 'circuits' – or similar terms such as paths, itineraries – but also in the commodification of the same producing places. This generates a change in local dynamics and in practices that are identified as significant for the communities, producing re-significations, re-appropriations and even direct spoliations by new actors in the territory, which alters and deepens the economic dimension more than the symbolic-identitarian one.

These dynamics produce, in many cases, processes of territorial concentration around tourism businesses – especially those with high purchasing power – that produce profound changes in work dynamics in rural areas (Barrera, 2006), enhance local conflict (Santana Talavera, 2002) and catalyse existing unequal power relations. Likewise, the gap widens between the possible projects for the local community and large-scale companies that have the economic, political and social capital to allow them to carry out their projects. This has led many authors to argue that, regardless of the more active responses that local communities – rural, indigenous, or small – may produce, large-scale tourism produces negative effects such as business concentration, increased land value, speculative real estate, gentrification (Cañada and Gascón, 2016) and decline in livelihoods and native cultures (Gruenewald, 2003; Comaroff and Comaroff, 2011). This has also begun to have an impact on the consideration of this type of tourism as a particular form of extractivism, which finds a clear example in the case of wine tourism (Torres *et al.*, 2018; Tolosa, 2023).

It is necessary to take into account the complexities that arise from the articulation between the processes of valorization and heritage recognition of products and those of touristification, and the effects that this brings into the territories. Understanding the complexity of this would make it possible to take into account the multiple variables that intervene in the implementation of certain policies that, although they are declaratively oriented to increase the well-being of rural populations, many end up increasing certain capitals (often foreign to the zones) and excluding those for whom these policies were initially designed.

Final Conclusions

In this case study, we have presented, based on two paradigmatic Argentinean foods, wine (focusing on Calchaquí high-altitude wine) and *dulce de leche*, the intertwining that occurs between the forms of valorization of these products with a regional economic sense, the sense of identity that heritage recognition adds and tourism development, which emerges as a function of the commercialization of a differentiated part of these foods and the tourist circuits that involve the producing areas. At the same time, we have noted that, in these processes, there is feedback between the phases of the process that makes it difficult to identify which of them generates or is generated by the other, recognizing instead their intertwined nature.

Likewise, we have pointed out that although heritage valorization, origin protection and tourism policies, among others that generate certain foods, are enunciated in favour of increasing their importance in commercialization and, therefore, to the benefit of their producers, the result is not comprehensive and generates tensions of different kinds, especially in the unequal reproduction of access to these policies, according to the economic, social and political position of the producers. In the same vein, the development of tourism in producer areas, promoted to generate a supposed 'spillover of benefits' to local communities, raises a series of problems that not only cast doubt on these benefits but also call for reflection on the scale of the social, environmental and economic impacts they produce.

To conclude, we believe that value-addition policies, and other forms that end up accompanying them, such as the valorization of heritage and tourism, must have a territorial anchorage that considers and incorporates local communities as protagonists. This should include not only those sectors with political and economic power and resources, but also – and above all – those in positions of social, economic and political inequality, who should be the first beneficiaries of these policies aimed at regional development, but from which they are often excluded. Therefore, it is necessary to seek strategies to ensure that value-adding processes provide democratic and horizontal access to them. At the same time, it is necessary to generate consumption circuits that are not restricted to a certain economic power, such as community tourism projects, and direct sales circuits that provide concrete opportunities for the sectors of local communities that need them most. We think that participatory and collaborative strategies are the main way to help ensure farm producers have voice and vote in value-adding processes. These strategies can be promoted by the local community, by creating cooperatives or associations or by public policies (Guastavino *et al*., 2010; Pérez Winter, 2019b) that can help farm producers create networks with public and private sectors (Chiodo *et al*., 2019; Lacko, 2019).

Note

[1] Alfajores are traditional Argentine pastries made with two cookies filled with dulce de leche. There exists a variety of flavours and fillers.

References

Adame, M.A. (2002) Exotismo posmoderno: la antropología de los encuentros turísticos interculturales. *Cuicuilco* 9(24), 333–364.

Ambrosetti, J.B. (1895) La región vinícola de la provincia de salta en los valles calchaquíes. *Boletín Del Departamento Nacional de Agricultura* XIX(XX), 565–579.

Argentina.gob.ar (n.d.). Available at: https://www.argentina.gob.ar/argentina-contra-el-hambre (accessed 10 May 2023).

Arias Castañeda, E. (2015) La experiencia existencial como modo de interpretación del turismo. *Revista Hospitalidade* 2, 586–599.

Barrera, E. (2006) Turismo Rural. Nueva ruralidad y empleo rural no agrícola. OIT-CINTERFOR. Available at: https://www.oitcinterfor.org/sites/default/files/file_publicacion/barrera.pdf (accessed 14 October 2023).

Barrera, E. and Alvarado, O. (2008) Food trails: tourist architectures built on food identity gastronomic sciences. *Food for Thought* 3(8), 38–43.

Cañada, E. and Gascón, J. (2016) Urbanizar el paisaje: turismo residencial, descampinización y gentrificación rural: una introducción. In: Cañada, E. and Gascón, J. (eds) *(Coords) Turismo Residencial y Gentrificación Rural*, Vol. 16. Colección PASOS Edita, pp. 5–36.

Carenzo, S. (2007) Territorio, identidades y consumo: reflexiones en torno a la construcción de nuevos paradigmas en el desarrollo. *Cuadernos de Antropología* 26, 125–143.

Carman, R.L. (2014) *La lechería argentina, sus comienzos*. Fundación de Historia Natural Félix de Azara, Buenos Aires.

Castro, H. and Pérez Winter, C. (2018) ¿Alimentos con identidad?: el dulce de leche como producto argentino. In: de Oliveira, MASA, Vanzella, E and Brambilla, A (eds) *Processos Sociais: Sistemas Culinários Em Contexto de Deslocamentos, Construções de Identidades, Memórias e Patrimônios 235–260, Serie Alimentação e Processos Sociais*, Vol. III.

Chiodo, E., Adriani, H.L., Navarro, F.P. and Salvatores, R. (2019) Collaborative processes and collective impact in tourist rural villages – insights from a comparative analysis between Argentinian and Italian cases. *Sustainability* 11, 432. DOI: 10.3390/su11020432.

Comaroff, J. and Comaroff, J. (2011) Etnicidad S.A. buenos aires, katz. p. 251.

DeCepas (2022) Dulce de leche argentino, crece su consumo en el mundo. Available at: https://decepas.com/economia/dulce-de-leche-argentino-crece-consumo-mundo-20221010-3464 (accessed 12 May 2023).

El Cronista (2021) VERANO 2021 ruta del dulce de leche clásico, repostero artesanal: cómo es, dónde está para vacaciones o escapada. Available at: https://www.cronista.com/clase/break/VERANO-2021-ruta-del-dulce-de-leche-clasico-repostero-artesanalcomo-es-donde-esta-para-vacaciones-o-escapada-20200907-0003.html (accessed 19 May 2023).

ElEco (2023) Fiesta del Dulce de Leche. Available at: https://www.eleco.com.ar/etiqueta/fiesta-del-dulce-de-leche (accessed 19 May 2023).

Giarracca, N. and Teubal, M. (2008) Del desarrollo agroindustrial a la expansión del 'agronegocio': el caso argentino. In: Mançano Fernández, B. (ed.) *Campesinado y Agronegocios En América Latina*. San Pablo, Clacso-ASDI, pp. 139–164.

González Bracco, M. and Arizaga, C. (2019) Retóricas de lo auténtico: estética y sensibilidad en los negocios gastronómicos de la ciudad de Buenos Aires. In: Cortés, A. and Licona, E. (eds) *Narrativas Culinarias. Miradas Etnográficas*. Mauritius, Editorial Académicas Española, pp. 102–118.

Gruenewald, R. (2003) Turismo e etnicidade. *Horizontes Antropológicos* 9(20), 141–159.

Guastavino, M., Rozemblum, C. and Trímboli, G. (2010) El turismo rural en el INTA. In: *Proceedings of the Primer Encuentro de Economía Agraria y Extensión Rural*. Available at: http://www.aader.org.ar/XV_Jornada/trabajos/espanol/Estrategias_y_experiencias/ensayos/Trabajo%2075%20Completo.pdf (accessed 27 July 2023).

Instituto Nacional de Vitivinicultura (2018) Regiones Vitivinícolas Argentinas Noroeste (Provincias: La Rioja, Salta, Catamarca, Tucumán y Jujuy). Available at: https://www.argentina.gob.ar/sites/default/files/region_noroeste_18.pdf (accessed 12 June 2023).

Instituto Nacional de Vitivinicultura (n.d.) Principales datos vitivinícolas. Available at: https://www.argentina.gob.ar/inv/vinos/principales-datos-vitivinicolas#:~:text=Superficie%20cultivada%3A%20211.099,U%24S%201.077.292.00 (accessed 10 July 2023).

Lacko, E. (2019) Reflexiones acerca del contexto de surgimiento de políticas turísticas ¿indigenistas? El proyecto de Red Argentina de Turismo Rural Comunitario (RATURC) del Ministerio de Turismo de la Nación (Argentina, 2008–2019).

Lacoste, P., Briones, F., Jiménez, D. and Rendón, B. (2014) La denominación de Origen Pisco en Chile: algunos problemas nacionales e internacionales. *IDESIA* 32(2), 47–56.

La Serenisima (n.d). Available at: https://www.laserenisima.com.ar/empresa-actualidad.php (accessed 10 May 2023).

Manzanal, M. (2000) Neoliberalismo y territorio en argentina de fin de siglo. *Economía, Sociedad y Territorio* II 7, 433–458.

Manzanal, M. and González, F. (2010) Soberanía alimentaria y agricultura familiar. Oportunidades y desafíos del caso argentino. *Realidad Económica* 255, 12–42.

Martín, F. (2009) Las transformaciones recientes en la agricultura de oasis en Mendoza, Argentina. Una aproximación al caso de la reestructuración vitivinícola desde la economía política de la agricultura. Master thesis, Facultad Latinoamericana de Ciencias Sociales/FLACSO, Buenos Aires, Argentina.

Minagro (2006) Ministerio de Agricultura, Ganadería y Pesca. Dulce de leche protocol. Available at: https://www.argentina.gob.ar/normativa/nacional/resoluci%C3%B3n-12-2018-306790/texto (accessed 10 May 2023).

Minagro (2016) Ministerio de Agricultura, Ganadería y Pesca. Available at: https://www.magyp.gob.ar/sitio/areas/ss_lecheria/estadisticas/balance/_archivos/BL004.php (accessed 10 May 2023).

Ministerio de Turismo y Deportes de la provincia de Salta (n.d.) Ruta del Vino. Available at: http://turismosalta.gov.ar/contenido/71/ruta-del-vino (accessed 15 May 2023).

Neiman, G. (2017) La 'nueva vitivinicultura' en la provincia de mendoza. un análisis cuantitativo de su crecimiento durante la última década del siglo XX. *Revista de Estudios Sociales Contemporáneos* 16, 40–60.

Pérez Winter, C. (2019a) La diversificación y promoción turística en tiempos de 'nuevas ruralidades': de la costa atlántica al campo pampeano. *ROTUR* 13(2), 68–85.

Pérez Winter, C. (2019b) El 'retorno al campo' desde la mirada local. Redescubriendo los atractivos patrimoniales pampeano-bonaerenses (Argentina). *PASOS* 17(4), 779–793.

Pérez Winter, C. (2021) Los procesos de valorización patrimonial y turística en la creación y legitimación de Fiesta provincial del Dulce de Leche en la provincia de Buenos Aires (Argentina). *RIVAR* 8(23), 153–170.

Pérez Winter, C. and Enrique, L.A. (2020) Entre lo tradicional y lo estratégico: disputas por la autenticidad como valor en las fiestas gastronómicas para la revitalización turística de los poblados bonaerenses (argentina). En I Seminario Internacional de Fiesta y Agricultura XVIII Seminario Iberoamericano Viticultura y Ciencias Sociales. Santiago de Chile, Chile.

Rainer, G. and Malizia, M. (2014) Los countries en el country. migración de amenidad, vino de altura y urbanizaciones cerradas en Cafayate (Salta, Argentina). *Journal of Latin American Geography* 13(1), 39–66.

Rainer, G. and Malizia, M. (2015) En búsqueda de lo rural: Migración de amenidades en los Valles Calchaquíes, Argentina. *Journal of Latin American Geography* 14(1), 57–78.

Saldi, L. and Petz, I. (2015) Aguas ajenas, tierras extrañas. desigualdad hídrica al sur de la cordillera de los Andes en Mendoza (Argentina) a principios del siglo XXI. *Cuadernos de Desarrollo Rural* 12(75), 123–144. DOI: 10.11144/Javeriana.cdr12-75.aate.

Santana Talavera, A. (2002) Desarrollo y conflictos en torno al turismo rural: claves ydilemas desde la antropología social. In: Riedl, M., Anécio Almeida, J. and Lima Barbosa, A. (eds) *Turismo Rural: Tendências e Sustentabilidade*. Universidade de Santa Cruz do Sul, Santa Cruz do Sul.

Saputo (2023) Dulce de leche argentino, crece su consumo en el mundo. Available at: https://saputo.com/es/nuestros-productos/sector-internacional/division-de-productos-lacteos-argentina (accessed 11 May 2023).

Telam (2022) Con las exportaciones en alza, se celebra el Día Internacional del Dulce de Leche. Available at: https://www.telam.com.ar/notas/202210/607187-exportaciones-alza-dia-internacional-dulce-de-leche.html (accessed 11 May 2023).

Tolosa, S. (2023) Un extractivismo cool. apuntes sobre el avance del modelo Vino & Turismo en los Valles Calchaquíes tucumanos, República Argentina. *Punto Sur. Revista de Geografía de La UBA* 9, 220–240.

Torres, L., Grosso, M. and Scoones, A. (2018) Turismo de lujo y extractivismo: la ruralidad como presa del capital. Reflexiones a propósito del Valle de Uco (Mendoza, Argentina). Scripta Nova. *Revista Electrónica de Geografía y Ciencias Sociales* XXII 585, 1–32.

Troncoso, C. and Arzeno, M. (2019) Turismo, gastronomía y producción agraria en la provincia de Jujuy (Argentina): actores, dinámicas y transformaciones asociadas a la valorización de productos tradicionales. *Revista Investigaciones Turísticas* 18, 169–192.

14 Looking Behind the Stage in a Beach Paradise: Taganga, Colombia

Freddy Giovanny Quiroga Nieto[1], Luis Leonardo Garavito González[1] and Narcís Bassols i Gardella[2]

[1]Faculty of Social and Human Sciences, Universidad Externado de Colombia, Bogota, Colombia; [2]Division of Quality Science, Uppsala University, Campus Gotland, Visby, Sweden

Introduction and Theoretical Framework

The present case study diachronically presents the case of the beach area of Taganga, in the Colombian Caribbean. Taganga is a good example as it has gone through several stages of development, crises and recoveries that significantly reflect national events and international tourism trends. Through qualitative fieldwork based on observation and interviews with community members (García, 2018), we get an in-depth look into matters such as environmental challenges and the (sometimes uneasy) coexistence of two revenue sources – fishing and tourism – over the course of six decades. The value of this research consists in uncovering the seldom examined behind the scenes happenings of a beach paradise destination.

Given that this case is a chronological narrative, evolutionary frameworks are utilized to contextualize our efforts. The classical theory of the destination life cycle proposed by Butler (1980) defining various evolutionary stages is an ideal framework. Butler – and subsequent literature refining his original work – posits a temporary deterministic evolution from a starting phase to a rejuvenation (or final decay) phase via growth and maturity phases. More recently, Richards (2018) proposes a more nuanced theory based on the evolution of attractions as he states that destinations tend to move away from basic tourist attractors towards more complex ones, specifically from basic attractions to cultural attractions to intangible attractions.

Methodology and Context

The Taganga fishing community is examined through participant and non-participant observation, as well as with semi-structured interviews, social and historical mappings, and primary and secondary documentation. The fieldwork was carried out from October 2020 to January 2021. The interviews included 13 fishermen, 4 community leaders involved in cultural areas and one diving entrepreneur. Some of these are quoted further below. Further, two social maps were made to locate fishing or diving sites, as well as the conflicts around them. Additionally, a historical map was used to represent land changes in the study area from 1960 to 2020. Finally, the literature review included primary documents, located in the local museum, such as: photos and local historical press articles, written communications with local, regional, and national authorities, and local censuses. Other primary sources were located online, such as institutional documents on territorial or tourist planning. It is also worth noting that external experts were called upon to verify different points of view regarding historical events as conflicts and negotiations with local authorities were uncovered. The actions and effects by several local stakeholders across time were examined, and changes in tourist profiles were ascertained via the analysis and triangulation of primary and secondary sources – the latter being mainly scientific papers and press articles.

It is noteworthy that, over the course of the last 60 years, Taganga has become an important regional, national, and even world destination. Its development has gone through stages, transitions, and conflicts which have made this place a complex and challenging one. In order to adequately report these shifts, this study is divided chronologically into three main phases: a first phase from the mid-1950s to the mid-1970s, a second phase from the mid-1970s to the turn of the century, and finally a third phase comprising the current century.

Taganga is located on the Colombian Caribbean coast and is a district belonging to the nearby city of Santa Marta, Colombia's third largest city on the Caribbean shore with around half a million residents (Fig. 14.1). Historically, Taganga has been inhabited by an indigenous population called the 'Tagangas'. Their homeland is a bay on the Caribbean Sea at the foothills of the Sierra Nevada mountain range (the world's highest coastal mountain range). As evident in Fig. 14.1, Taganga occupies one of the natural bays forming the coastline north of Santa Marta. The rest of the bays in the area are protected as they belong to the Tayrona National Park. Until the mid-20th century, the Tagangas were quite isolated due to road access difficulties (still apparent today as the place is connected to Santa Marta via a strenuous mountain road of about 6 km). Until then, and for generations, the most important revenue for the residents had been catching fish and selling it in Santa Marta. Traditional Caribbean fishing techniques such as the 'chinchorro' nets were put to use on a day-to-day basis. Like many sea-fishing communities, catching fish was not just a source of revenue but also a practice that strongly influenced the local culture and social relations, so a good fisherman was also considered an outstanding member of the community (Dussan de Reichel, 1954).

The Slow Birth of Tourism in Taganga: From the Mid-1950s to the Mid-1970s

In 1956, a road connecting Taganga to Santa Marta was built. Thereafter, the dynamics of this small fishing village began to slowly change.

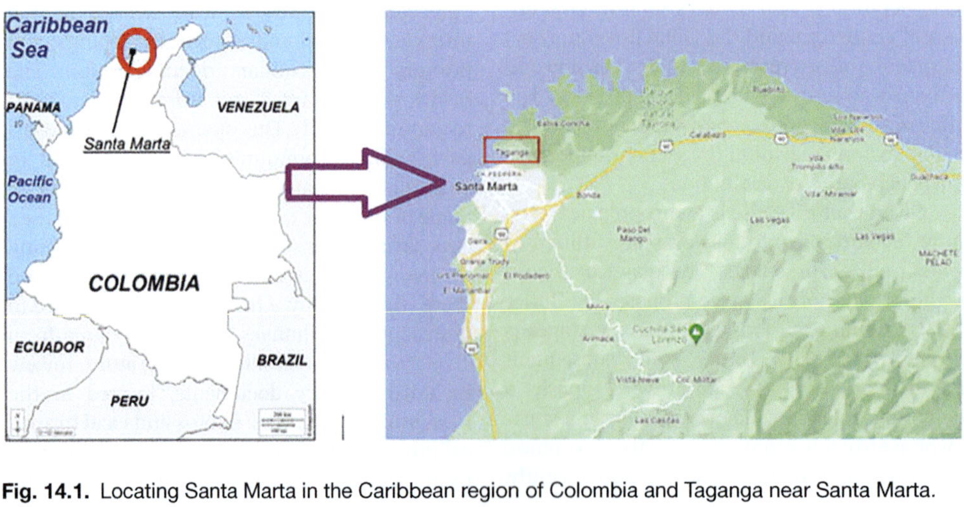

Fig. 14.1. Locating Santa Marta in the Caribbean region of Colombia and Taganga near Santa Marta. Sources: Authors on free maps by D-Maps and Google Earth.

Among notable shifts, the most noteworthy was the arrival of the first visitors. These first visitors were wanderers and 'hippies', who began descending on Taganga. They were part of the socio-cultural movement that expanded throughout the United States and Europe in the 1960s, but also influenced other parts of the world (Ramírez Carreño, 2009) like Colombia. 'Hippies' sought natural and uncrowded areas and so Taganga, with its pristine beaches, its welcoming community, and its guaranteed tranquillity, became a haven to them. In those years, there was no infrastructure providing tourists with services, so everything was homemade and artisanal. The local community began offering their homes as lodgings, sold traditional food and drinks to the visitors, and even invited them to join in the fishing tasks. Generally speaking, 'both groups lived side by side without any trouble' (R. Mattos, 2020, personal communication). By the early 1970s, the first tourist families arrived in Taganga. In this era, 'there was the opinion in the community that tourism was a "good thing", as it did not interfere much with the community's life and some among them earned extra income from the visitors' (R. Mattos, 2020, personal communication). The first (part-time) jobs related to tourism services emerged (Bienestar Estudiantil de Taganga, 1979). Fishing was by far the main source of revenue, and most men would work catching the fish. The women would sell it in the city while also raising the children and caring for the household.

Tourism in Taganga developed this way until the late 1970s as, due to the lack of hotels and restaurants, and also any marketing actions by the authorities, the community welcomed few people. This is in stark contrast to the beach zones south of Santa Marta, 'officially designated' by the national authorities, which supported the building of beach resorts from scratch such as El Rodadero in Gaira (see Fig. 14.1) to satisfy the need for sun-and-sand tourism of a growing middle-class from Colombia's inland cities. While we can only point to this contrast without going into it in-depth due to space limitations, it is evident that the dynamics created by an organic development vs a planned one have resulted in very different paths for both areas from the outset (Mertins, 1972).

Infrastructures and public utilities arrived at the village in this initial phase. Taganga and Santa Marta became better connected to the rest of the country thanks to the 'Troncal del Caribe', the most important regional road, built from 1956 to 1972,, which connects the major cities and towns along Colombia's Caribbean coast. At a national level, Colombia's connections to the world improved thanks to the opening of the El Dorado International Airport in Bogota, the country's capital city, in 1959. Step by step, Santa Marta started its conversion towards a major tourism destination (Ayala, 2000) with the refurbishing of its public and natural spaces. Public utilities came to Taganga in this era, including piped water in the 1960s and electricity in 1974, though they did not cover the whole village and would prove quite unreliable over the first two decades (Rodríguez and Gutiérrez, 1995).

To conclude this subsection, we underscore how the findings reported here relate to Butler's (1980) initial tourism development phase, as little infrastructure means a limited number of arrivals, with the 'hippies' acting as the place discoverers. Tourism does not conflict with anything, as it is a quite marginal activity. In terms of Richards (2018) work, the basic attractions of the place (beaches, sun, and a peaceful community) become the place's main pullers. Commonalities with similar cases across the world may be seen here.

Tourism Grows: From the Mid-1970s to the Turn of the Century

In 1978, Taganga made an important step in its tourism development with the opening of its first hotel, 'La Ballena Azul', located next to the main beach, which is still in operation today. This project, carried out with the community's acquiescence, was led by a female French resident who had lived in the village for some time (Herrera, 2020, personal communication). From then on, tourist stays became longer, which ushered in a new era in the development of local tourism.

Over the course of the 1980s, a significant number of locals became involved with tourism. In 1986, the first public project related to tourism was carried out, as several kiosks were built in Playa Grande (a beach adjacent to Taganga's main bay) to sell food and beverages to the visitors. This stepwise improvement in the quality of the tourist services benefited both visitors and the community. It is worth noting that most of the services geared towards the visitors were offered by the community's women (Atehortua Mejía, 2007), as their husbands' main activity was still fishing. In 1987, the National Tourism Corporation built the seafront along the main beach, giving Taganga's beachfront the appearance it still displays today (Bell Lemus *et al.*, 1992).

Meanwhile, tourism in Colombia was becoming an important source of national revenue and it kept growing due to an ever-expanding supply of visitors. In fact, the country's main cities, such as Bogota, Cartagena, and Medellin, greatly benefited from this growth (Santoyo López and Silva Barrera, 1988). However, the nation increasingly experienced public security issues and civil unrest due to the emerging war between the public forces and illegally armed groups (guerrillas and drug cartels), which became quite intense during this era. This matter would further deteriorate, becoming one of the darkest chapters in Colombia's history. Consequently, travel demand decreased sharply both in foreign and domestic markets (Cotelco, 2019). Therefore, from the 1980s into the 2000s, the decades in which violence peaked in the country, the Colombian tourism industry became a sort of roller coaster, with many ups and downs (Bassols, 2016). Despite all this, the impact on tourism in Taganga and the rest of the Colombian Caribbean region was quite negligible, as it was mostly excluded from the conflict. Tanganga's reputation as an exotic place for foreigners in the Colombian Caribbean continued to rise. Two new hotels opened their doors in 1988. Several restaurants, government-sponsored kiosks, and shops also opened their doors around that year (Bell Lemus *et al.*, 1992).

In the early 1990s, tourism in Taganga became more diversified and complex with the arrival of recreational diving companies, which added to the local supply of activities. Their arrival was mainly due to the discovery of some beautiful underwater grounds in the adjacent Tayrona National Park. These diving sites 'were apt to all levels of divers, and economically competitive in comparison with other diving venues in Colombia or Latin America' (Peláez, 2021, personal communication). This was the first segment-targeted supply in the destination, attracting niche international tourists who 'avoided the security risks on the national roads by travelling by plane to Santa Marta' (Peláez, 2021, personal communication).

A positive result of this expansion of the tourist industry in the 1990s was the creation of more jobs for local women and youth, who traditionally had had limited job opportunities. Women began to work in restaurants or to set up their own fast-food stores and stalls. Tourism was still the women's third job, after fish selling and caring for their households (Cortina and Herrera, 1994). For their part, the young people took over jobs such as dive guide assistants, hotel and hostel employees, and vendors (Cortina and Herrera, 1994). A number of fishermen swapped jobs and went into the maritime transportation sector, which was much better paid. Fishing in the 1990s declined for several reasons, including the setup of large coal ports on the Colombian Caribbean and industrial overfishing across the country (Urieles, 2018). In Taganga, tourism became a new and attractive way for local economic development, both for the native community and the foreigners who had established their businesses there.

While tourism generated new opportunities for a growing number of locals, it started affecting their daily life too. Tourism seasonality had a perceived effect on the community whereas previously it had not, and seasonal beach overcrowding became an issue. As most visitors were unaware of local fishing practices, nets were damaged and the fishermen's work was disrupted. Such is the case reported by Mr Luis González (2021, personal communication) who settled in the village in the 1980s to work as a fisherman and would have 'heavy arguments with bathers' almost every day as he cast his 'chinchorro' nets on the beach. The first public security issues also emerged, as well as problems related to the lack of land planning and the illegal occupation of lands (Peláez, 2021, personal communication), resulting in environmental damage in the surrounding foothills, see Figs 14.2 and 14.3.

By the end of the 1990s, tourism had gained a strong foothold and was an important part of

Fig. 14.2. A picture of the village in the 1940s. Source: Authors in cooperation with Román Mattos, 2021.

Fig. 14.3. The expansion of Taganga in the last decades showing the expansion from 1970 (green perimeter) to nowadays, doubling the built surface. Source: Authors in cooperation with Román Mattos, 2021.

the local economy. Taganga had become an ideal vacation spot, with sun and beach, water sports, and nice food. None the less, local tourism was still far from being a planned system. There was not much control over new buildings, nor the division and functional zoning of the land. The village could not arrive at a consensus for a future vision. Police control was almost non-existent and community cooperation came to an all-time low (Luis González, 2021, personal communication). Drug consumption among youth in the village increased notably, and some fishermen were involved in the drug transportation routes towards North America. However, we underscore again that, throughout the decades of violence in Colombia, tourism kept growing in Taganga and also in the rest of the Colombian Caribbean, as the Caribbean region was mostly spared from this conflict, which mainly concerned the large inland cities.

This subsection has revealed the first conflicts as tourism developed in an organic and unplanned, or rather 'laissez faire', way. Taganga escaped the years of violence in the country for the most part, and, following national and international trends, tourism grew in the village. In terms of Butler's framework (1980), this is clearly the 'growing' phase. As for Richards (2018), we underscore that the diving companies represent a first push towards more value and sophistication in the place based on Taganga's natural resources.

Two Decades of Challenges: Taganga in the 21st Century

The new century brought both positive and negative events to Taganga's tourism. In 2002, strong security policies began to be implemented in Colombia at a national level to counter-terrorism. Security was heightened on national roads, which allowed Colombians to travel safely throughout much of the country. Consequently, domestic tourism rebounded (MINCIT, 2011; Bassols, 2016).

At the local level, in the 2000s conflicts between Taganga's two local main sources of revenue (fishing and tourism) heightened. More than ever before, growing numbers of tourists bathing, diving, rowing or jet-skiing invaded fishing sites, damaged fishing equipment and severely disrupted the fishermen's work due to their ignorance of local traditional practices. This came to a head with a dispute around Isla Aguja, a nearby islet good for both fisheries and diving (Guerra, 2020, personal communication). As tourism grew into a serious hindrance to fishing, it threatened a central revenue source and sociocultural symbol of the Tagangas.

By the end of the 2000s, the village started to experience significant troubles. A group of foreigners, mainly from Israel, established themselves in the area, promoting drug abuse, prostitution and even child exploitation. These practices went on in Taganga for several years, and the place suffered rapid image decay (Gallo, 2018) as these offenders 'became business owners, and community members found themselves entangled in criminal action' (Herrera, 2020, personal communication). The community 'followed a denial process rather than a resolutive one' (R. Mattos, 2021, personal communication). The Tagangas were even accused of placing their children into exploitation. Yahaira Mattos, a fisherwoman and local leader (2021, personal communication), puts it in the following terms:

> They were saying that we were exploiting our own children, but that was simply not true ... but this was of course very hurting for everyone's feelings in the community: our children were singled out, and this was very painful, it was a very bad situation, the community's morale went completely down the floor.

Tourism quickly declined as visitors chose other destinations in the Colombian Caribbean. The issue would be resolved only in 2017, as the suspected head of the gang was expelled from the country and his assets seized (El Tiempo , 2017).

At this time, because of a low demand meeting an excess of supply, Taganga entered a tourism crisis that expanded to the social sphere: 'many businesses, hotels, restaurants, and diving centers were closed or sold, as profitability was no longer the same' (R. Mattos, 2020, personal communication). Despite this decline, the community continued to operate tourism activities, and by 2010, it became the main local source of revenue. By then in Taganga there were over 100 tourism-related businesses, including 22 hotels/hostels, 34 kiosks, 22 bars and restaurants, 8 diving centres and four tour operators (Botero and Zielinski, 2010). In these years, public spaces were upgraded and connectivity was improved. The quality of tourism products generally improved thanks to the diversified supply of accommodation, food outlets, and supplied activities. Natural attractions garnered a high rate of satisfaction, with more than 75% of visitors evaluating them well. Accommodation quality was rated at around 60% (Botero and Zielinski, 2011). Taganga thus completed its transformation from a fishing community into a tourist destination.

Figueroa Cabas (2011) gives a critical vision of Taganga's economy during this time. While locals made up 76% of the tourist workforce, they owned less than half of the businesses, and generally owned the smaller businesses. So, despite the increase in jobs and entrepreneurial opportunities, the Tagangas do not reap the most benefit from their tourist industry. In most cases, they hold jobs that are seasonal and low-paid. However, many locals think of tourism development as a positive force. However, they also see that public and private mismanagement and poorly enforced regulations have caused a series of negative effects (Tejeda, 2020, personal communication).

Therefore, future tourism projects need to be centred on the community, their experience, and their ancestral knowledge. Some inputs have appeared from the academia since 2010 (cf. cited works in this subsection). In the past 15 years or so, the community has become aware that tourism is a double-edged sword, with positive and negative effects to be considered.

The most recent developments have to do with the community regaining their culture and practices. The Colombian government acknowledged the Taganga ancestral identity in 2020. This has led to an increased interest in the local intangible heritage, something that has never happened before since tourism started.

Conclusions

Amidst crises and unexpected events, Taganga has moved towards 'maturity' (Butler, 1980) in its tourism development process. With awareness and appreciation of intangible cultural assets emerging in recent times, Richards' (2018) notion is also confirmed, with a trend from sun and sand via upscale nature tourism (diving tourism built on the basis of the area's rich sea grounds) to the current interest in developing intangible products related to the culture and the identity of the Tagangas.

This case has reviewed the most important moments in the history of a Caribbean destination and has shed some light on the impacts of tourism on the resident community. Under the guidance of two well-established frameworks (Butler, 1980; Richards, 2018), the obvious conclusion is that the destination displays unique traits (which are harder to explain using general theories) such as the rebounds after violent periods in its history. However, it also displays common traits with other destinations since similarities with other organically-grown coastal destinations are clearly visible in the Taganga case. Such is the case of tourism replacing the local traditional sources of income or the social and environmental shifts and changes it produces.

References

Atehortua Mejía, M. (2007) Resistencia cultural y adaptación en Taganga, pueblo de pescadores del Caribe Colombiano. (Tesis de maestría). Universidad Nacional de Colombia– Sede Caribe. Colombia.

Ayala, C. (2000) Gustavo Rojas Pinilla, 100 años. Red Cultural del Banco de la República. Credencial Historia No. Available at: https://www.banrepcultural.org/bibliotecavirtual/credencial-historia/numero-123/gustavo-rojas-pinilla-100-anos-1900-1975 (accessed 5 May 2023).

Bassols, N. (2016) Branding and promoting a country amidst a long-term conflict: The case of Colombia. *Journal of Destination Marketing and Management* 5(4), 314–324.

Bell Lemus, C.A., González, R., Hernández, C. and Pabón, M. del S. (1992) Elementos metodológicos para una planeación urbana participativa en el corregimiento de Taganga. *Investigación & Desarrollo* 2, 110–127. DOI: 10.14482/id.v0i2.3008.

Bienestar Estudiantil de Taganga (1979) Censo de población del centro poblado de Taganga (27 de octubre de 1979). Santa Marta.

Botero, C. and Zielinski, S. (2010) Evaluación del potencial para el desarrollo del turismo sostenible en el corregimiento de Taganga, Distrito de Santa Marta (Colombia). *Anuario Turismo y Sociedad* XI, 10–34. Available at: https://revistas.uexternado.edu.co/index.php/tursoc/article/view/2825/2470 (accessed 21 April 2023).

Botero, C. and Zielinski, S. (2011) Percepción de la comunidad como indicador del ciclo de vida de un destino.*Estudios y perspectivas en turismo* 20, 1366–1388.

Butler, R. (1980) The concept of a tourist area cycle of evolution: Implications for management resources. *Canadian Geographer* 24(1), 5–12.

Cortina, M. and Herrera, J. (1994) Contribución de la Pesca Artesanal al Desarrollo Socio Económico del Corregimiento de Taganga. Santa Marta: Universidad del Magdalena. Available at: https://core.ac.uk/download/pdf/270124939.pdf (accessed 11 April 2023).

Cotelco (2019) 65 años trabajando por el desarrollo de la hotelería y el turismo de Colombia. Uricoechea Publicidad.

Dussan de Reichel, A. (1954) Características de la personalidad masculina y femenina en Taganga. *Revista Colombiana de Antropología* 2, 89–113.

El Tiempo(2017) Quién es el israelí expulsado del país por seguridad nacional. Available at: https://www.eltiempo.com/justicia/delitos/quien-es-el-israeli-expulsado-decolombia-por-seguridad-nacional-155634 (accessed 25 February 2023).

Figueroa Cabas, Y. (2011) Caracterización del turismo en el corregimiento de Taganga, Santa Marta D.T.C.H: Un análisis desde la perspectiva de la sostenibilidad. Master Thesis. Universidad Nacional de Colombia. Available at: https://repositorio.unal.edu.co/bitstream/handle/unal/54203/caracterizaciondelturismoenelcorregimientotaganga%2csantamartad.t.c.h.un%20an%c3%a1lisisdesdelaperspectivadelasostenibilidad.pdf?sequence=1&isAllowed=y (accessed 11 April 2023).

Gallo, I. (2018) A la caza de Moosh, el israelí de las orgías en Taganga. Las 2 Orillas. Available at: https://www.las2orillas.co/el-bunker-hotel-benjamin-y-la-toma-israelita-de-taganga/ (accessed 11 April 2023).

García, J. (2018) Conceptualización Etnográfica. En Aproximación interdisciplinaria al proceso organizativo de la Comunidad Indígena Taganga del pueblo de Taganga (Corregimiento de Santa Marta). Ministerio del Interior: Dirección de Asuntos Indígenas, Rom, y Minorías Grupo de Investigación y Registro, pp. 57–109.

Mertins, G. (1972) El Rodadero (Santa Marta), centro turístico o colonia de vacaciones? Instituto Colombo – Alemán Invest. Cient. Available at: https://aquadocs.org/bitstream/handle/1834/3251/Rodadero.pdf?sequence=1&isAllowed=y

MINCIT (Ministerio de Comercio, Industria y Turismo) (2011) Turismo en Colombia: un sector clave para la prosperidad. Available at: https://www.mincit.gov.co/prensa/presentaciones-de-laministra/doc-presentaciones/turismo-en-colombia-un-sector-clave-para-la-prospe.aspx (accessed 21 April 2023).

Ramírez Carreño, S. (2009) Hippismo Criollo 40 años después. Tesis de grado. Pontificia Universidad Javeriana, Colombia. Available at: https://repository.javeriana.edu.co/bitstream/handle/10554/5362/tesis343.pdf?sequence=3 (accessed 5 May 2023).

Richards, G. (2018) Cultural tourism: A review of recent research trends. *Journal of Hospitality and Tourism Management* 36, 12–21.

Rodríguez, S. and Gutiérrez, F. (1995) Evaluación socio económica de los asociados de la Cooperativa de Pescadores de Taganga 'Coopestaganga' [BA thesis]. Universidad del Magdalena. Available at: https://core.ac.uk/download/pdf/198274845.pdf (accessed 21 April 2023).

Santoyo López, N. and Silva Barrera, C. (1988) Aproximación a la cuantificación de las divisas generadas por turismo en las ciudades de Cartagena, Santa Marta y San Andrés (Periodo 1984–primer semestre de 1987) [BA thesis]. Universidad Externado de Colombia, Bogotá, Colombia.

Urieles, R. (2018) El ocaso de la pesca en el corregimiento de Taganga. Available at: https://www.eltiempo.com/colombia/otras-ciudades/el-ocaso-de-la-pesca-en-elcorregimiento-de-taganga-278764 (accessed 5 May 2023).

15 Qhapaq Ñan as a World Heritage Site: New Tourist Destinations and Inherited Tensions in Argentina

Claudia A. Troncoso
Consejo Nacional de Investigaciones Científicas y Técnicas (CONICET), Universidad de Buenos Aires, Buenos Aires, Argentina

Summary
Qhapaq Ñan, or the Andean Road System, was inscribed on UNESCO's World Heritage List in 2014. This inclusion was the result of a process initiated by six South American countries (Peru, Bolivia, Ecuador, Colombia, Chile, and Argentina). The process identified the selection of features that distinguished the road and justified the inscription. How was this selection made? Was tourism involved in it? This case study analyzes the details of the Qhapaq Ñan nomination process in Argentina. It especially focuses on the northwestern region of the country, an area with a long (and controversial) history of tourist valorization of Andean heritage. It is argued that tourism was involved in the heritage selection processes in Argentina by two means. Firstly, tourism interests have oriented the choosing of some archeological sites that could be tourist attractions. Secondly, the definition of heritage sections and sites has sought to avoid previous heritage areas where the presence of tourism has been highly conflictive. Avoiding potential conflicts is also at the core of recent tourist projects designed for Qhapaq Ñan in Northwestern Argentina that aim to encourage 'off the beaten track' (and supposedly non-controversial) forms of tourism.

The Value and Interest of the Case Study

This case study presents some of the aspects that characterize the current relationship between tourism and heritage. It highlights the relevance that heritage distinctions have as a tourist attraction, especially highly visible ones such as UNESCO. The case study also shows the tensions and conflicts that may be involved in heritage nomination processes in relation to the tourist use that is sought to be given to the properties.

Qhapaq Ñan: Heritage for Global Tourism

In recent decades, we have witnessed a process of increasing interest in heritage recognition. Monuments, historical buildings, archaeological sites, and cultural expressions are formally distinguished as heritage by national or subnational institutions and also by international organizations. Among the latter is the UNESCO World Heritage Committee that created the World Heritage List in 1972. The Committee includes sites and expressions of intangible heritage to the list that, according to established criteria, are considered of outstanding universal value.

These heritage distinctions are a result of social processes through which identities are defined and strengthened. Technical and scientific knowledge (and competent institutions) involved in this process legitimize the heritage condition of goods and places. However, the heritage-making process is also related to political interests, economic initiatives, and leisure/tourism practices that characterize contemporary societies (Prats, 2009).

Several authors have discussed heritage-making processes in relation to tourism dynamics and trends. They have pointed out that heritage has a special role as a tourist attraction in a context that is defined by a strong interest in culture in general and in the past, in particular. This is related to the importance that tourism has achieved as a leisure practice in contemporary societies (Urry and Larsen, 2011). In addition, the incorporation of places into heritage lists has been interpreted as part of global strategies to multiply tourist attractions and consumption events based on the development of economic activities (Harvey, 2002). Considering these strategies, some authors have argued that tourism is highly involved in defining which goods and places are officially distinguished as heritage (Prats, 2009; Hernández Ramírez, 2011; Santamarina Campos, 2013). This interpretation includes the inscriptions in UNESCO's World Heritage List.

In this context, in 2014, some fragments of the Inca's road system that stretches over Western South America were inscribed as a Cultural Route in the UNESCO World Heritage List (Fig. 15.1). This road network is known as Qhapaq Ñan, which means main Andean road in Quechua.

This case study analyzes the Qhapaq Ñan nomination process in North-western Argentina. Personal interviews with governmental actors who held public positions in provincial agencies were conducted to provide details for this case study. These interviews

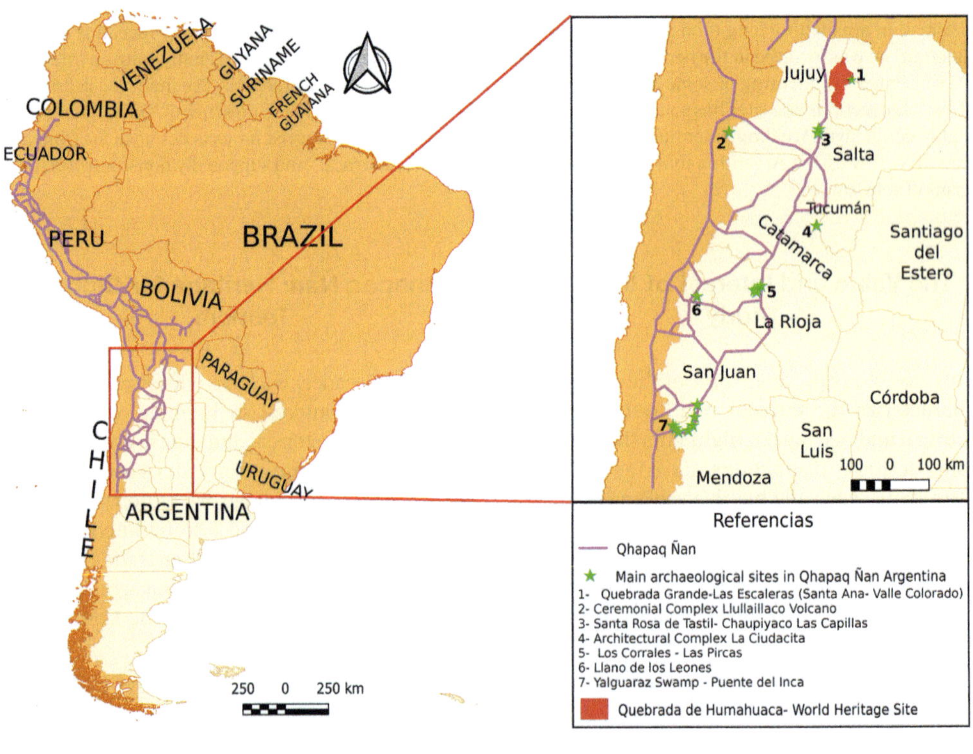

Fig. 15.1. Qhapaq Ñan in South America and Argentina. Source: Luca Rosasco.

took place between 2009 and 2017 and were designed to identify the existing positions related to the Qhapaq Ñan nomination process in these agencies.

Main Andean Road: The Inscription on the World Heritage List

In 2003, Peru invited other South American countries to undertake the nomination process aiming to recognize the Andean road system as World Heritage. The system consists of a pedestrian network that stretches over the Andes. It was built by pre-Incan cultures but enlarged and improved by the Incan Empire. Six countries (Peru, Argentina, Bolivia, Chile, Colombia, and Ecuador) agreed to foster the project, aided by technical assistance from the Inter-American Development Bank (IDB) and UNESCO itself ((BID (IDB), 2006).

In order to inscribe something on the World Heritage List, it is necessary to justify its outstanding universal value. The official nomination file prepared for Qhapaq Ñan inscription includes a justification based on the extension of this infrastructure network (used for the movement of people, armies, goods, and resources) that connects cities, villages, and rural areas located in different environmental regions of the empire (high mountains, coasts, forests, deserts, and valleys). The outstanding value of this road system was also based on the fact that some of its fragments are currently in use by the people living in the area.

In June 2014, the World Heritage Committee decided to inscribe Qhapaq Ñan on the World Heritage List. UNESCO recognized its outstanding value based on the following selection criteria established by the organization (World Heritage Committee, 2014):

- to exhibit an important interchange of human values, over a span of time or within a cultural area of the world, on developments in architecture or technology, monumental arts, town-planning, or landscape design;
- to bear a unique or at least exceptional testimony to a cultural tradition or to a civilization which is living or which has disappeared;
- to be an outstanding example of a type of building, architectural or technological ensemble, or landscape which illustrates (a) significant stage(s) in human history; and
- to be directly or tangibly associated with events or living traditions, with ideas, or with beliefs, with artistic and literary works of outstanding universal significance.

The nominated property did not include the entire road network that comprises thousands of kilometers and archaeological sites. Rather, only 720 kilometers and 291 sites across the six countries were selected for inscription (Argentina, Bolivia, Chile, Colombia, Ecuador and Peru, 2013). Thereby, World Heritage Qhapaq Ñan is a group of chosen road fragments and sites that make sense through the abstraction exercise of linking them into a broader historical network spread across Western South America. Scientific knowledge helps to join and organize the spatially scattered sites through a narrative that invokes uses and functions of the historical road network.

Technical interests and established criteria are not the only elements that guide heritage nomination processes. They are also permeated by political projects. In the case of Qhapaq Ñan, there is a geopolitical dimension that cannot be underestimated. The nomination was part of a strategy to reinforce bonds among the South American countries interested in defining a common identity based on their pre-Hispanic past. This position challenges the historical dependency of South American states on the Western world. It was also a way to strengthen the political cooperation among some South American nations led by progressive governments (Sosa, 2019).

What about tourism? What was its role in this process? There is practically no mention of tourism in the file sent to UNESCO. The activity is only referred to as a potential threat to the property because of its lack of control and the inadequate infrastructure at Qhapaq Ñan heritage places to support it. It is also recognized that tourism could be a way for local people to access an income, but it would require special training on the subject. Finally, it is proposed to develop 'culturally appropriate' forms of tourism that 'involved the active participation of the communities' (Argentina, Bolivia, Chile, Colombia, Ecuador and Peru, 2013, p. 1173). This is a point highly valued by UNESCO for nominations (Salin, 2021). Although tourism has a very limited presence in the document, the fact is that different countries have sought to develop public use plans for the sites because as an interviewee pointed out: 'tourism arrival is inevitable at heritage sites' (Qhapaq Ñan Salta Programme, 2016, personal interview).

Qhapaq Ñan Argentina: Heritage and Geographical Selections

In Argentina, there are around 100 sections of the Andean road network. A total of 13 of them were selected as part of World Heritage nomination together with 32 archaeological sites associated with the pedestrian system. These chosen places (Fig. 15.1) are located in the North and West of the country across seven provinces (Jujuy, Salta, Tucumán, Catamarca, La Rioja, San Juan, and Mendoza) (Sosa, 2019). In general, most of them are located far from tourist destinations.

In organizational terms, the National Ministry of Culture is responsible for national coordination tasks related to the Qhapaq Ñan nomination and management. In turn, each province appointed a working team to define the road sections and archaeological sites selected for the nomination. It was also in the provincial context that the management strategies for the sites were designed. For the decision making process, participatory methodologies were used (Sosa, 2015). These methodologies sought the involvement of the 18 local communities near Qhapaq Ñan sites. They have participated in workshops created to inform about the property and decide about its management (Sosa, 2015, 2019).

In relation to tourism, the intention was to develop sustainable tourism alternatives and formulate a public use plan for Qhapaq Ñan (Ministerio de Turismo de la Nación, 2013). This plan, prepared in 2013, proposes some touristic options for the heritage sites, including themed tourist experiences related to the Incas, the multiple uses of the road, and the ongoing daily life of the people that inhabit heritage places associated with Qhapaq Ñan. Trekking, mountaineering, and rural or community-based tourism were among the proposed tourist experiences (Ministerio de Turismo de la Nación, 2013). Implicit in these options is the decision to avoid forms of mass tourism in these heritage areas.

Heritage, Tensions, and Conflicts in Qhapaq Ñan, Argentina

The influence of previous heritage experiences

In each Argentine province, it was decided which sections of the road and archaeological sites would form part of the Qhapaq Ñan World Heritage Site. What were the criteria involved in the selection of them? The relevance of the sections, the degree of conservation of the road, and the functions of the archaeological sites were considered. Tourism-related criteria also played a role, particularly in the provinces of Jujuy and Salta (at the North of the country).

Tourism has influenced the heritage selections because the relationship between World Heritage recognition and tourism has had a conflictive history in this region. In 2003, the Andean valley Quebrada de Humahuaca (province of Jujuy) was included in the World Heritage List as a Cultural Landscape. This valley, which contains small urban localities and extensive rural areas inhabited by peasants and indigenous people, received the UNESCO distinction and experienced rapid growth in tourism. This generated conflicts and unrest among local residents. Quebrada de Humahuaca includes a low-income population living in precarious conditions. During the nomination process, the provincial government argued that this distinction would bring tourism and generate development for the region. The local population, with little access to information on the potential negative aspects of this kind of processes, supported the nomination in all the participatory instances where the issue was discussed. The growth of tourism, however, has benefited mainly tourism entrepreneurs who came from other Argentine cities to set up their businesses in the emergent destination. In addition, the growth of tourism went hand in hand with the acquisition of land occupied by the local population in an informal manner. Many residents were expelled from the land because they did not have ownership titles. As a result, conflicts and criticism arose. They included the questioning of the heritage and tourism policy deployed by the government that did not guarantee the conditions for the local population to access the benefits brought by tourism. In addition, there was objection to outsiders who received benefits from tourism but are not culturally related to Andean heritage. The local population pointed out that they only had access to poor quality jobs (temporary, informal, and low paid jobs) and were reduced to a mere 'attraction' as representatives of the 'exotic' Andean culture in the eyes of the tourists (Troncoso, 2012).

This background was a key factor in the heritagization process of Qhapaq Ñan in Argentina. The authorities involved, both at the national and provincial level, did not want to be exposed to a situation similar to the one that had arisen in Quebrada de Humahuaca. This case became nationally known as 'the model to avoid' (Sosa, 2015) and the 'example of what not to do' (Programa Qhapaq Ñan Salta, 2013, personal interview).

So, how did this background play in the heritage selection processes associated with Qhapaq Ñan? In the province of Jujuy, part of the Inca trail crosses the Quebrada de Humahuaca. However, the provincial authorities tried to avoid the selection of these sections due to the conflicts mentioned above. Instead, they selected fragments of the road in another relatively isolated area of the province with no tourism presence (Jujuy Culture Department, 2013, personal interview). Similarly, in the neighboring province of Salta, the Quebrada de Humahuaca precedent also influenced the heritage selection process. There, heritage sites away from tourist destinations were chosen to avoid problematic situations related to the potential increase in tourist visits (Programa Qhapaq Ñan Salta, 2013, personal interview).

Objections from the local population

Tensions around Qhapaq Ñan in Northern Argentina involved other issues. Some indigenous communities refused to be included in the heritage area (Copa and Herrera Salinas, 2020). This occurred in Queta (Jujuy) where the local population objected to the heritage process on the grounds that they had not been consulted on the issue as required by current legislation (Jujuy Culture Department, 2013, personal interview). This community was clearly aware of Quebrada de Humahuaca's background when making this decision. For them, it was no longer a matter of lack of information as in the previous case.

As mentioned above, in Jujuy, the selected areas of Qhapaq Ñan do not coincide with Quebrada de Humahuaca. They are located in the eastern end of the province, near the towns of Santa Ana and Valle Colorado. Here, too, conflicts arose. The section of the Inca trail selected is part of the road linking the two localities, and it is still in use by local inhabitants. When the provincial government began the survey work for the nomination to UNESCO, they realized that fragments of the Inca road were being destroyed to build a modern route that would improve connectivity conditions between these towns. Jujuy's Highway Department, which carried out this task, was unaware of the historical character of this road and replaced it with a new paved route. Archaeological evidence of the Inca road partially disappeared. As a result, there was a dispute within the provincial government as to what to prioritize, the route or the heritage (Jujuy Tourism Office, 2009, personal interview).

This dilemma was transferred to the local population, some supported the heritage protection but many preferred the construction of the route that would improve their mobility. In addition, among the local population, the historic road did not have a heritage or special character. It was simply the road that they habitually used and that their parents had used. Provincial and national governments carried out 'awareness-raising' activities to inform the population about the heritage significance of the road (Jujuy Culture Department, 2013, personal interview) because they needed the support of the inhabitants to legitimize the project. But, beyond the relevance of preserving the road to fulfil the nomination, the experts and officials involved recognized that, for some locals, the paving was more relevant and necessary than the protection of the heritage property. Some of these experts had also participated in the heritagization process of Quebrada de Humahuaca. One of them expressed her dilemma: 'after Quebrada [de Humahuaca] I promised myself to respect what the population wanted and in Santa Ana I found that they were not interested in the heritage but in the road. So, what do we do in these situations?' (Jujuy Culture Department, 2013, personal interview).

The issue was raised: heritage is destroyed in pursuit of the will of those who considered themselves heirs and users of that heritage or heritage awareness is raised among local people in order to move forward an international project to achieve the highest heritage distinction. Eventually, the route was built elsewhere and the two plans could be carried out: the construction of a paved route and the protection of the Qhapaq Ñan. However, the dilemma between protection and destruction raised by the users of heritage properties remains.

In the province of Salta, the UNESCO nomination process also generated tensions. One of the selected places is the archaeological site of Tastil. There, participatory workshops were carried out with the local population to inform them about the heritage relevance of the site. The tourist importance of the nomination was also pointed out since the construction of an interpretation center was planned. This center would generate some jobs for the local population.

In this case, the opinions were also divided. There were those who supported the project and bet on the advantages that it could bring and those who opposed it, knowing the possible negative effects of these initiatives. In this context, some sectors of the community demanded details of the nomination file and the tourism management plan before moving forward with the project. They also requested that the land access issues be resolved. The local population who make their living from agricultural activities do not have land ownership titles, so they are highly vulnerable to the dynamics of the real estate market associated with tourism. The knowledge that these communities had about the events in the Quebrada de Humahuaca also played a central role in this case (Salta Heritage Department, 2014, personal interview). Despite the opposition of part of the population, the project went ahead with the participation of only those locals who support the initiative (Salta Heritage Department, 2014, personal interview). This situation once again exposes the question of whether the opinions of local people on heritage issues are taken into account or, on the contrary, consultative events are mere formalities to be fulfilled.

All these issues highlight heritage dissonance (Graham *et al.*, 2000), with respect to the meanings and uses proposed by the different actors in relation to heritage. What happens when ideas about heritage do not coincide, when locals and experts and competent institutions do not share the same opinions about what is proposed, when heritage is not heritage for local people? Is it possible to challenge the authorized heritage discourse (Smith, 2006) held by experts and institutions and to go back on a heritage project of the magnitude of Qhapaq Ñan? Is this what explains attempts to avoid conflictive sites and look for isolated places where tourism is not yet present to add them to the project?

Tourism and Qhapaq Ñan: 'Off the Beaten Track' Destinations in Argentina

In Argentina, the inclusion of Qhapaq Ñan in the World Heritage List was accompanied by tourism promotion of sections and sites as new attractions. In this respect, the international distinction was a way to launch some unvisited places as tourist destinations. The Qhapaq Ñan's status as an 'off the beaten track' destination has led to the creation of tourism initiatives (from the state and private sector) aimed at attracting tourists to isolated, remote destinations that were not yet touched by tourism. At the same time, the tourism proposed for these places is introduced as a conflict-free one. Here, we are dealing with tourism modalities assumed as innocuous, low impact, respectful of environments, and people but at the same time able to contribute to the improvement of the living conditions of the local population. Community-based tourism and adventure tourism were the modalities chosen to fulfil these functions.

Following these premises, in Jujuy, the areas of Santa Ana and Valle Colorado were subject to a project developed between 2018 and 2020 and financed by the Corporación Andina de Fomento ('Socio-productive empowerment of vulnerable communities in Jujuy') that encouraged community-based tourism. The invocation to this tourism modality, which involves the distribution of incomes obtained by tourism, seeks to avoid the processes of unequal appropriation of benefits that occurred in Quebrada de Humahuaca.

In the province of Salta, the local government has also drawn upon the promotion of community-based forms of tourism that include the local population. This was especially encouraged in Quebrada del Toro, near Tastil archaeological site. Likewise, adventure tourism was promoted on the grounds that this is a modality that involves few people and, therefore, would not generate major negative impacts. This is the proposal to develop in the Llullaillaco volcano, which is usually climbed only by skilled mountaineers (Salta Tourism Office, 2017, personal interview).

These initiatives are recent, and there are some questions that arise: will the creation of these new global heritage places serve only to multiply leisure and consumption options? Or will they also constitute an alternative to the conflicting and unequal appropriation of benefits that historically marked the relationship between heritage and tourism in the region?

Conclusions

Tourism and heritage have a very close relationship. As many scholars claim, we should think about tourism demands when properties are nominated to be included in the World Heritage List. Creating formal heritage is a way of generating tourism and public interest in cultural and historical diversity. At the same time, tourism is present in other moments of the heritagization process. Conflictive backgrounds involving heritage and tourism were essential to understand the Qhapaq Ñan heritage process, at least in Argentina. On the one hand, the inherited tensions defined the trail selections made either because of prudence on the part of the authorities or because of objections on the part of the communities. On the other hand, this background generated new knowledge in the local societies that they brought into play to evaluate their participation in the project. Finally, the intention to avoid conflicts and negative consequences for local people has defined the tourist modalities for these places. Modalities assumed as innocuous and beneficial for the locals have been chosen to inaugurate new destinations and contribute to the process of opening 'off the beaten track' spaces for global tourism.

References

Argentina, Bolivia, Chile, Colombia, Ecuador and Peru (2013) *Qhapaq Ñan, Andean Road System Qhapaq Ñan, Système des Routes Andines World Heritage Nomination, Executive Summary*. UNESCO, Paris, p. 2765.

BID (IDB) (2006) *El Qhapaq Ñan- Camino Principal Andino.Plan de Acción Regional para un proceso de integración y cooperación*. Peru, p. 292.

Copa, M. and Herrera Salinas, C. (2020) El Qhapaq Ñan como memoria ancestral de los pueblos andinos, In Ministerio de Cultura de la Nación *Camino ancestral Qhapaq Ñan Una vía de integración de los Andes en Argentina*. Ministerio de Cultura de la Nación, Buenos Aires, pp. 53–66.

Graham, B., Ashworth, G. and Tunbridge, J. (2000) *A Geography of Heritage. Power, Culture and Economy*. Arnold Publishers, London.

Harvey, D. (2002) The art of rent: Globalization, monopoly and the commodification of culture. *Socialist Register* 38, 93–110.

Hernández Ramírez, J. (2011) Los caminos del patrimonio. Rutas turísticas e itinerarios culturales, *Pasos Revista de Turismo y Patrimonio Cultural* 9(2), 225–236.

Ministerio de Turismo de la Nación (2013) *Plan de uso público del Qhapaq Ñan-Sistema vial andino. Argentina*. Buenos Aires, p. 217.

Prats, L. (2009) *Antropología y patrimonio*. Ariel, Barcelona.

Salin, E. (2021) Routes patrimoniales et narrations plurielles: le Chemin de l'Inca (Qhapaq Ñan, *Camino principal andino/Andean road system*) le cas particulier d'une route patrimoniale non touristique? *Via*, 20.

Santamarina Campos, B. (2013) Los mapas geopolíticos de la UNESCO: entre la distinción y la diferencia están las asimetrías. El éxito (exótico) del patrimonio inmaterial, *Revista de Antropología Social* 22, 263–286.

Smith, L. (2006) *The Uses of Heritage*. Routledge, New York.

Sosa, V. (2015) 'Gestión participativa del Qhapaq Ñan Argentina'. Jornada Protección, conservación y gestión del Qhapaq Ñan – Sistema Vial Andino, Patrimonio Mundial. Cátedra UNESCO de Turismo Cultural Untref-Aamnba.

Sosa, V. (2019) La lógica de la aporía. Comentarios sobre la patrimonialización del Qhapaq Ñan de un sujeto detrás del señuelo. *Diálogos Patrimonio cultural* 2, 87–104.

Troncoso, C. (2012) *Turismo y patrimonio en la Quebrada de Humahuaca. Lugares, actores y conflictos en la definición de un destino turístico argentino*, *Pasos Edita*, La Laguna.

Urry, J. and Larsen, J. (2011) *The Tourist Gaze 3.0*. Sage, London.

World Heritage Committee (2014) *Decisions Adopted by the World Heritage Committee at its 38th Session (Doha 2014)*. UNESCO, Doha, p. 274.

16 Tourists Perceptions of Cultural Experiences in Heritage Hotels in Queretaro City

Driselda P. Sánchez-Aguirre[1] and Domingo Martínez[2]
[1]Escuela Nacional de Ciencias de la Tierra UNAM, Ciudad de México, Mexico; [2]LIIGH, UNAM Juriquilla, Querétaro, Mexico

Summary

In this case study, we identify themes related to cultural tourism experience in boutique and non-boutique heritage hotels in Queretaro City, Mexico to determine whether there are differentiating elements that could be useful as a competitive strategy. We did a content analysis using web scraping and supervised text mining techniques, allowing us to statistically analyze 1040 guest reviews collected on TripAdvisor from January 2016 to December 2019. We made and validated a codebook to identify and characterize cultural tourism experiences in Mexican heritage hotels. We found that the experience of escapism, the perceived quality, and authenticity are significant differentiating elements between boutique heritage hotels and non-boutique heritage hotels.

Background (and/or Context)

The city of Queretaro, Mexico is located in the Mexican Bajio region and has significant cultural attractions. It is part of the Colonial Treasures tourist route and has been recognized as one of Mexico's World Heritage Cities. Since 2010, the city has had a constant investment in internal security and tourist promotion, causing important economic and demographic growth (Nicolas and González-Gómez, 2014). As in many other tourist cities, gentrification has turned several of the historic buildings located in the city center into organizations that provide temporary lodging services (Novo Malvárez, 2019; Haro-Aragú and García Mestanza, 2021), mixing in historic city centers an authentic landscape with a commercialized one (González-Alcantud, 2018).

Most of the buildings that have been converted into hotels in downtown Querétaro, date back to the 18th century and can be classified as 'new' heritage hotels, defined as spaces 'that have been converted into lodging establishments with retention of parts of historic character' (Xie and Shi, 2020, p. 670). The utilization of heritage houses or buildings as hotels not only aids in conserving their intrinsic value but also brings about various advantages, encompassing financial, social, and cultural aspects, for both tourist destinations and hotel enterprises (Elshaer *et al.*, 2022). Consequently, due to these multifaceted benefits and their unique charm, some of these establishments have garnered the boutique designation given by the Queretaro Association of boutique hoteliers. This designation typically signifies an emphasis on providing intimate, personalized experiences within a distinctive and often historic setting, thus aligning well with the characteristics of repurposed heritage properties. Acquiring the boutique distinction can incur additional expenses for organizations, such as investment in unique decor, personalized services, and maintaining the historical integrity of the property. It also entails higher operating costs due to the emphasis on quality and exclusivity. These elements often necessitate charging a premium price compared to competitors. Tourists are often willing to pay the premium if they perceive a significant difference (Liu *et al.*, 2020) in the experience offered by boutique establishments, based on bespoke services, curated experiences, cleanness and a sense of individuality that sets them apart from standard accommodation providers (Sánchez-Aguirre *et al.*, 2017).

In the context of heritage hotels located in downtown Querétaro City, it is important to distinguish between two specific categories: heritage hotels and heritage hotels with the boutique distinction. Heritage hotels usually refer to accommodations set within historic buildings, often restored to preserve their original architectural and cultural significance. These establishments offer guests a glimpse into the region's rich history and cultural heritage. Heritage hotels with the boutique distinction emphasize a unique and intimate guest experience and also boast historical significance. Boutique heritage hotels regularly offer individualized services, bespoke amenities, and carefully selected aesthetics that appeal to discerning travelers looking for a fusion of luxury and authenticity. Even though both categories have similar characteristics, such as historical locations and architectural charm in Querétaro City, the difference may be in the overall guest experience.

Although the guest perspective has been utilized in numerous studies to draw comparisons between boutique and lifestyle category hotels (Quadri-Felitti *et al.*, 2022) or to assess memorable tourism experiences in boutique hotels (Sánchez-Aguirre *et al.*, 2017; Parolin and Boeing, 2019), there is a lack of evidence regarding cultural tourism experiences in boutique hotels and the possible differences between heritage and heritage boutique hotels. Rectifying this is important as identifying distinguishing elements is of utmost importance in promoting and enhancing the competitiveness of organizations.

Something that matters, especially for differentiating tourism services, is customer perception (Tribe, 2010). The tourist's perspective is a triggering factor in the competitiveness of tourism companies because the opinion of others helps the decision making process of both, the buyer and the service providers. For the buyer, customer perception is an element that provides confidence before making the purchase. For the service provider, it is a reference to improve the design and execution of their services (Chatterjee, 2020). Therefore, offering memorable experiences to tourists can be used as an effective differentiation strategy in the hotel industry (Seyfi *et al.*, 2020).

Experiences have a temporal structure through stages. One of the later stages is memory, a stage in which communication with others about the learning obtained is encouraged. A satisfied tourist has created a memory from his/her trip that, when communicated, may be able to convince other tourists to take a similar trip, or even convince him/herself to travel again to the previously visited destination and choose the same services.

Gilmore and Pine (2002) are the main reference points about the importance of designing memorable experiences as a differentiation strategy in organizations. They proposed a typology of four realms to analyse the experiences that take place in organizations. These are educational experiences, aesthetic experiences, escape experiences, and entertainment experiences. These types of experiences are mutually compatible and often merge, presenting multi-modal experiences in organizations. The principal aim of each of the four realms of memorable experience is to motivate people to talk about them.

Broadly defined, an educational experience is one in which the tourist attains some learning or his/her curiosity to learn is awakened. In the aesthetic experience, the tourist feels captivated by the harmony or beauty of the hotel or its surroundings. The entertainment experience considers the non-participant observation of situations or activities in the hotel or the destination. The experience of escape from reality occurs when tourists get the feeling of forgetting their daily routines or reality, or even get the feeling of living in another time (Oh *et al.*, 2007). In other hotel studies, the aesthetic experience is the major factor determining tourist satisfaction in hotels, followed by the experience of escapism (Gongalves *et al.*, 2013).

Although the memorable experience theory from Gilmore and Pine (2002) has had multiple applications in tourism organizations (Elshaer *et al.*, 2022), it has not been related before to cultural tourism experiences in hotels.

Cultural Tourism Experiences

The academic literature on memorable cultural tourism experiences is limited regarding the experiences in hotels. It gives more emphasis to the study of cultural experiences in destinations (Cetin and Bilgihan, 2016; Chen and Rahman, 2018). Several authors have attempted to identify and analyse the elements that define cultural tourism experiences in destinations. They have identified the following elements: expectations, quality of service, authenticity, social interactions and emotions. These are briefly described below.

Expectations

Expectations can be defined as a series of minimum requirements that the tourist expects to be met during his/her trip (Horvath, 2013). These requirements have been previously formed, in the tourist's mind, due to information he/she was exposed to. This information generates an image of the destination and the services to be acquired in the mind of the tourist. This image is the result of the reputation that the destination or the organization has forged (Seyfi *et al.*, 2020). Reputation can be defined as 'management behaviors intended to identify and affect how external audiences perceive an organization, including their perceptions of what the organization is, what it strives to be, and the organization's positive impact on society more broadly' (Østergaard Pedersen and Houlberg Salomonsen, 2023, p. 40). The main determining dimensions of reputation are variety of product and/or service offerings, brand image, business leadership, ethics, culture and social responsibility, financial position and innovation (Martínez and Olmedo, 2013)

Quality of service

Quality of service plays a key role in heritage hotel stays (See and Goh, 2019). The concept of quality is primarily based on the tourist's perspective, and a substantial portion of its worth is attributed to the employees. Having human resources that are patient, empathetic, helpful, and responsive to customer needs contributes to customers' perception of better quality service (Su and Hsu, 2013; Cetin and Bilgihan, 2016; Sánchez-Aguirre *et al.*, 2017). Perceived service quality contributes to the fact that the traveller successfully or unsuccessfully copes with various challenges, such as noise, language, security, and parking (Cetin and Bilgihan, 2016; Seyfi *et al.*, 2020)

Authenticity

Authenticity can be defined as an intangible value attached to certain products, services, or places from the perspective and judgment of the observer (Seyfi *et al.*, 2020). It is, for tourists, the original, local and unique cues well differentiated from what they can find in their place of origin (Cetin and Bilgihan, 2016; Wickens, 2017). Tourists staying in heritage hotels are looking for an experience based on historical decoration, old architecture, and traditional services (Chittiprolu *et al.*, 2021).

Authenticity *per se* is a complex construct that is constantly changing. However, several authors (Ramkissoon and Uysal, 2011; Cetin and Bilgihan, 2016; Wickens, 2017; Seyfi *et al.*, 2020) have made a significant effort to characterize authenticity to understand the tourism experience, finding four main typologies: (i) objective authenticity, which refers to inherent characteristics of objects or the projection tourists make on objects to consider them authentic; (ii) existential authenticity, defined as the connection between the tourist and the destination through participation in activities such as events and festivals; (iii) dynamic authenticity, which considers society as a pillar of authenticity and therefore is constructed and reconstructed through narratives, this concept aligns with perceived authenticity where it is included in the lifestyles of locals such as food preparation; (iv) static authenticity, which is based on the roots of the past, where history takes on real importance for the tourist experience.

Social interaction

Interaction with other tourists, with the staff involved in tourist services and with the locals of the destination contributes greatly to the experience of cultural tourists as it allows for the growth of social development (Horvath, 2013). Recent research has highlighted the characteristics that contribute to defining the social interaction of tourists with locals and/or service providers. They incorporate adjectives such as helpful, kind, friendly, polite, calm, hospitable, interesting, sharing, service-minded, goodwill, and generous (Cetin and Bilgihan, 2016; Seyfi *et al.*, 2020).

Emotions

Culture-related attractions can be visualized as tourism products capable of generating emotions (McIntosh, 1999). The emotional factor consists of the perception of pleasure or displeasure during the experience (Titz, 2008). Hotel room, food, staff behaviour, hotel amenities, and traditional services are the main elements associated with the perception of pleasure in heritage hotels, while issues with services, reservation, stay, and pricing are the dimensions related to guest displeasure (Chittiprolu *et al.*, 2021).

Although several of these elements have been studied independently as factors of competitiveness in cultural tourism destinations, there is no integrative study of these elements as a comparative factor in boutique and non-boutique heritage hotels in Mexico at the time of the research. Gilmore and Pine's theory of experience economy and the work of the previously mentioned authors can help fill this gap in the literature by studying these characteristics with an organizational focus.

Method

Sample

TripAdvisor is one of the most popular websites related to user-generated content on the specific topic of travel (Calheiros *et al.*, 2017). It is through this type of website that tourist satisfaction can be identified, accompanied by opinions that allow us to visualize the emotions lived during the trip in the specific use of certain services, such as accommodation, food, and tours (Li *et al.*, 2015; Xu, 2020).

TripAdvisor's comments were compiled from the following facilities: All heritage boutique hotels, members of the Association of Boutique Hotels of Queretaro at the time of the investigation, located in downtown Queretaro; all heritage hotels that were located 2 km or less from said boutique hotels (i.e. in the center of the city). The final sample consisted of 16 hotels in total: nine boutique heritage hotels and seven heritage non-boutique hotels.

With the support of the *rvest* package on Rstudio, we used a webscrapping technique to extract 1040 comments in Spanish from the TripAdvisor travel platform, over a period from January 2016 to December 2019. We chose this temporality to avoid possible bias due to the COVID-19 pandemic that began in 2020 which limited travel and thus the use and supply of temporary lodging services.

From the 1040 reviews collected, 56% were from boutique heritage hotels and 44% from non-boutique heritage hotels.

Operationalization

Based on the proposal of a Spanish Dictionary of Affect in Language (Gravano and Dell'Amerlina, 2014), we classify the words used in the reviews considering a continuous distribution from 1 to 3, rounded to a single decimal value. A word with a value equal to or greater than two belonged to a positive category (pleasure), while any word with a value less than two was determined as negative (displeasure).

Then we proceeded to identify the presence of words associated with pleasure and displeasure in each review.

We made two codebooks, the first of which was made to analyse the cultural memorable experiences in hotels. It was made by adapting the literature review about cultural tourism experiences at destinations. We determined five categories (authenticity, emotions, quality of service, expectations, and social interaction) and identified a minimum of 20 keywords for each category. The second codebook was made to analyse whether the identification of cultural elements in hotels coincide or differ from the experience economy theory. We considered the four experience types (aesthetic, educational, escapism, entertainment) and identified 20 keywords for each type of experience.

Data analysis

To validate the codebooks, we followed Saldaña's (2016) technique for team coding and did a pilot test with 104 additional reviews where two coders were asked to work independently and to identify in each pilot-review the presence or absence of words associated with memorable cultural experiences and with experience economy theory. An inter-coding percentage of 72% allowed us to validate both codebooks.

Once the codebooks were validated, we proceeded to identify in each of the 1040 reviews the presence or absence of words associated with memorable cultural experiences and with experience economy theory. Agreements from these coders were calculated with a dichotomous value, indicating whether the code is present or absent. As an adjustment technique, Krippendorff's (2004) alpha was used, which yields values between 0 and 1, an *alpha* of 0.652 was obtained. By obtaining an alpha greater than 0.6, it is possible to reach acceptable conclusions (Krippendorff, 2004)

With R version 4.2.1 we calculated the frequency of words associated with memorable experiences and cultural experiences in each review. To identify possible differences between groups, we ran permutation tests and the Mann–Whitney–Wilcoxon test.

We assume that the greater the frequency of word identification in a review the greater the likelihood that guest experienced pleasure or displeasure, had a memorable experience, or had cultural tourism experiences in the hotel. With this assumption, we can identify patterns and understand guests' perceptions, as has been done in other studies (Alrawadieh and Law, 2019; Gunasekar and Sudhakar, 2019; Zhao *et al.*, 2019). However, this premise should be approached with caution, as it does not account for the full complexity of human experience. Therefore, complementary studies will be needed to reinforce these findings.

Results

Memorable experiences based on the experience economy

Although the identification of a higher frequency of words associated with the aesthetic experience (present in 75% of the reviews) is consistent with previous results from other researchers (Gongalves *et al.*, 2013), no significant statistical difference was found between boutique heritage hotels and heritage hotels in this aspect. So, it can be inferred that hotels *per se* have sufficient resources to provide aesthetic experiences beyond the 'label' they hold. The same situation was found with educational and entertainment experiences, which coincided with previous studies (Sánchez-Aguirre *et al.*, 2017).

We found, with a statistically significant difference (p-value = 3.051e−07), that the escape experience was mentioned more frequently by guests of boutique heritage hotels. This may mean that the boutique organizations are doing good exposure and promotion of the history and culture of their facilities, a characteristic of boutique hotels that has been previously documented as having the ability to evoke in the guests a feeling of living in another time (Jones *et al.*, 2013; Sánchez-Aguirre *et al.*, 2017). Although several non-boutique heritage hotels obviously also have a history to tell, guests do not seem to remember it or do not mention it when sharing their experience on TripAdvisor.

An example of a TripAdvisor review coded as an escape experience is:

> Excellent place to get away as a couple. I highly recommend you to book a table on the terrace, have a good wine and enjoy dinner. Excellent service from all the staff and a very nice view. We recommend requesting a room with a balcony facing the street. (Text 671. Boutique Heritage Hotel)

Memorable cultural experiences

Most identified elements were about the expectation theme (present in 62% of the reviews), which leads us to infer that it is of high importance for guests to corroborate that they acquired the service they expected, since they make continuous mention of this in their public opinions after the stay. This confirms what Seyfi *et al.* (2020) mentioned about how cultural heritage experiences are modelled by visitors' previous information and expectations. This is also true for any lodging experience, regardless of category.

The second theme most identified was authenticity (present in 57% of the reviews). Significant differences were identified (*p value* 3.722e−07) regarding more reference to authenticity, specifically as a cognitive factor of memorable tourism experience (Pulido and Navarro, 2014), in boutique heritage hotels compared to non-boutique heritage hotels.

The authenticity of heritage hotels is not just based on the buildings (Al Riyami *et al.*, 2022). According to our results, most of the reviews alluded to a static authenticity where history takes on real importance for the tourist experience, but we identified a touch of dynamic authenticity as well. In several reviews typical food of the region, served in the hotel's restaurants, was mentioned. This aligns with the concept of authenticity as those local, original, and unique cues are well differentiated from what tourists can find in their place of origin (Cetin and Bilgihan, 2016; Wickens, 2017).

The TripAdvisor review identified with the highest number of words related to authenticity was:

> What a beauty of a hotel, the house is a handful of stories as it was inhabited by many characters from the history of Mexico, and I'm sure they spent a lot of money on all the details of the decoration, we were all my family for a wedding, and it turned out that the main the groom is his room. The air conditioning did not work, and they solved it with 3 floor fans the rooms are sealed windows and there is no way to ventilate it. All the staff are very friendly and cordial, and the breakfast is good. (Text 54. Heritage Boutique Hotel)

In service quality as an element of the cultural tourism experience, we found a statistically significant difference (*p value 0.003755*). Tourists staying in heritage hotels without boutique distinction wrote more about deficient service quality.

The TripAdvisor review identified as the one with the highest number of elements related to quality services shows the opinion of an unsatisfied guest:

> It is 1145pm, and I arrive after reporting the air conditioning in my room, and it does not work, instead of receiving a solution the person in charge on duty, Mr. Ricardo Garcia lying about the hotel is full (which I verified on hotel.com, and there are rooms available) does not give me a solution. Irresponsible and liars, far from giving 5-star service! (Text 38. Heritage hotel)

Emotions

Although the element of emotions did not present significant differences, when we made a more detailed analysis and considered texts associated with pleasure or displeasure, we found a greater presence of positive-valence-words in the heritage hotels with the boutique label (*p value* of 0.022 in the permutations test)

Preliminary evidence suggests that staying in heritage hotels has a very different emotional value compared to staying in commercial hotels (Chittiprolu *et al.*, 2021). Having a boutique label provides an added value in the perception of pleasure at the time of lodging.

Examples of the TripAdvisor reviews coding for this category are:

> **Pleasure**
>
> We had the opportunity to stay in this beautiful hotel with only 19 rooms, I loved the concept because besides being a hotel, it is an art gallery that has a spa with the best treatments, being a gallery, all the spaces in the hotel are very well done, clean, well maintained, the rooms resemble the old haciendas, with beautiful decor, high ceilings, and with all the amenities of a modern hotel. The breakfast that is included in the rate is delicious and with ingredients of the best quality, we went to spend my birthday, and it was our first visit to Queretaro, it also has a Spa with many treatments and options to relax

either individually or as a couple, and the staff is very well qualified, it is a must if you stay at the Hotel to take a treatment or just to know it is beautiful. The prices are very affordable for all that the hotel offers and for the quality that they give to the client, without a doubt, we will return delighted. (Text 84. Boutique Heritage Hotel)

Displeasure

To begin with, when we arrived we were attended by a lady, supposedly the 'manager' and she told us with a bad attitude and bad manners that our reservation was not in the system. Despite the bad time we had with the lady, we decided to stay, and it was the worst thing we could do. In general, the whole hotel is very dirty and super neglected, the rooms are disgusting and to top it off, there was no water! The breakfast is horrible and miserable. The hotel smells disgustingly of cigarettes. The worst of the case is that we tried to talk to the same lady who attended us when we arrived and never gave us the face!!!! We will not go there again. (Text 306. Heritage hotel)

Conclusions

The gentrification of Queretaro's historic center has allowed tourism to bring guests of converted heritage hotels closer to the history of the city, changing the meaning of historic buildings into tourist experiences. Although it sounds obvious that cultural elements are interwoven into these experiences, little had been explored previously in small-medium-sized cities in Mexico.

Based on the content analysis conducted, we identified in temporary lodging organizations the presence of all the elements that have been documented as characterizations of cultural tourism experiences in destinations (See and Goh, 2019; Seyfi *et al.*, 2020). This could help hoteliers to provide a more accurate description of the possible experiences to be offered in heritage hotels.

According to previous research, the market segment targeted by boutique hotels seeks clear differentiation between the lodging services they purchase (Parolin and Boeing, 2019). However, there are no explicit differences between heritage hotels that have the boutique distinction and those that do not. This makes the traveller's planning process to find differentiated lodging more difficult. Our findings show that guests do perceive important differences between heritage boutique hotels and heritage hotels, which is an important factor for boutique hoteliers to consider when differentiating themselves between similar lodging styles.

This case of study suggests that the heritage hotel companies (boutique and non-boutique) should do more:

i. To promote the attributes and benefits of heritage boutique hotels in terms of quality, authenticity, and escape experience. For example, authenticity was perceived not only by the old buildings and their furniture but also by gastronomic elements, a situation that could be useful in the promotion of these organizations.
ii. In addition to the previous point, it is crucial to begin to differentiate the narrative between heritage hotels and heritage boutique hotels in such a way as to guarantee an image that is consistent with the service offered, where the trinomial expectation/perceived quality/emotions becomes a building block of a differentiated brand identity. This could increase the motivation of travelers to discover the ideal type of accommodation for them based on the coupling of their desires/needs and the differentiated experiences offered. At the same time, precise segmentation and a clear value proposition could attract investors to maintain or improve the profitability of this type of organization.
iii. To develop, in all those involved in providing the heritage lodging service, skills and competencies necessary for the design and management of experiences (Al Riyami *et al.*, 2022) to ensure a service that can be remembered and recommended by those who lived the experience. Training for the development of these skills should be based on identifying and highlighting the components of static and dynamic authenticity inherent to each boutique and non-boutique heritage hotel in the city of Querétaro.

It is important to note that although an educational experience is inherent in cultural tourism experiences, its presence has not been identified as a statistically significant differentiator in the temporary lodging organizations of the study. Like the entertainment experience, it remains a latent opportunity for research, development, and management of experiences for heritage hotels.

Besides the managerial implications mentioned above, this research also has social and cultural implications. Societally, the study highlights the importance of preserving and promoting cultural heritage through the hospitality industry. By fostering an authentic and immersive cultural experience in heritage hotels, communities can benefit from increased tourism revenue and improved appreciation of local customs and traditions. Culturally, the study emphasizes the value of showcasing heritage in contemporary settings, allowing for a dynamic interchange between tradition and modernity. By leveraging the boutique distinction to present cultural heritage in innovative ways, the hospitality industry can contribute to the preservation and revitalization of cultural practices, enriching the cultural identity of the destination.

There are some research limitations that should be acknowledged to maintain a balanced perspective on the findings. First, the study was confined to Queretaro City, which might limit the generalizability of the results to other cultural destinations and heritage hotels worldwide. Expanding the research to different locations and cultural contexts could yield a more comprehensive understanding of the impact of the boutique distinction on cultural tourism experiences.

Second, some factors, which were not considered in this study, such as personal, social, cultural and geographic characteristics of the tourist profile may influence what they share publicly about their perception of cultural tourism experiences in a hotel. Future research could analyse those data and identify if there is any correlation between them.

Another limitation lies in the narrow focus on tourists' perceptions, neglecting the perspective of hotel management and staff. Incorporating the viewpoints of hotel managers and staff could offer a more holistic view of the challenges and opportunities in providing cultural tourism experiences in boutique heritage hotels.

Funding

Consejo Nacional de Humanidades, Ciencia y Tecnología, (Grant/Award Number: 'Posdoctoral Fellowship 2023–2025 Domingo Martínez')

References

Alrawadieh, Z. and Law, R. (2019) Determinants of hotel guests' satisfaction from the perspective of online hotel reviewers. *International Journal of Culture, Tourism and Hospitality Research* 13(1), 84–97.

Al Riyami, H., Almuhrzi, H. and Scott, N. (2022) Heritage hotel experiences in Nizwa, Oman. *International Journal of Tourism Research* 24, 169–180. DOI: 10.1002/jtr.2492.

Calheiros, A.C., Moro, S. and Rita, P. (2017) Sentiment classification of consumer generated online reviews using topic modeling. *Journal of Hospitalilty Marketing and Management* 26, 675–693.

Cetin, G. and Bilgihan, A. (2016) Components of cultural tourists' experiences in destinations. *Current Issues in Tourism* 19(2), 137–154. DOI: 10.1080/13683500.2014.994595.

Chatterjee, S. (2020) Drivers of helpfulness of online hotel reviews: A sentiment and emotion mining approach. *International Journal of Hospitality Management* 85, 102356. DOI: 10.1016/j.ijhm.2019.102356.

Chen, H. and Rahman, I. (2018) Cultural tourism: An analysis of engagement, cultural contact, memorable tourism experience and destination loyalty. *Tourism Management Perspectives* 26, 153–163. DOI: 10.1016/j.tmp.2017.10.006.

Chittiprolu, V., Samala, N. and Bellamkonda, R.S. (2021) Heritage hotels and customer experience: a text mining analysis of online reviews. *International Journal of Culture, Tourism and Hospitality Research* 15(2), 131–156. DOI: 10.1108/IJCTHR-02-2020-0050.

Elshaer, I.A., Fayyad, S., Ammar, S., Abdulaziz, T.A. and Mahmoud, S.W. (2022) Adaptive reuse of heritage houses and hotel conative loyalty: Digital technology as a moderator and memorable tourism and hospitality experience as a mediator. *Sustainability* 14(6), 1–13. DOI: 10.3390/su14063580.

Gilmore, J.H. and Pine, B.J. (2002) Differentiating hospitality operations via experiences: Why selling services is not enough. *The Cornell Hotel and Restaurant Administration Quarterly* 43(3), 87–96. DOI: 10.1016/S0010-8804(02)80022-2.

Gongalves, J., Fraiz, J. and Manosso, F. (2013) Calidad de la experiencia en los hoteles termales de Galicia, España. Un análisis a través de la reputación online. *Estudios y Perspectivas en Turismo* 22, 492–525.

González-Alcantud, J.A. (2018) Turismo y patrimonio inmaterial, una alianza obscena. *EREBEA Revista de Humanidades y Ciencias Sociales* 8, 89–111.

Gravano, A. and Dell'Amerlina Ríos, M. (2014) *Spanish DAL: A Spanish Dictionary of Affect in Language*. Buenos Aires: Facultad de Ciencias Exactas y Naturales. Universidad de Buenos Aires, 1–24. Available at: https://bibliotecadigital.exactas.uba.ar/download/technicalreport/technicalreport_n00001.pdf (accessed 21 January 2020).

Gunasekar, S. and Sudhakar, S. (2019) Does hotel attributes impact customer satisfaction: A sentiment analysis of online reviews. *Journal of Global Scholars of Marketing Science* 29(2), 180–195.

Haro-Aragú, M. and García Mestanza, J. (2021) Metamorfosis del Turismo y Transformaciones Urbanas: Hacia el Desarrollo de un Modelo de Turismo Sostenible. DOI: 10.6035/Kult-ur.2021.8.15.1.

Horvath, Z. (2013) *Cultural Value Perception in the Memorable Tourism Experience*. In: Smith, M. and Richards, G. (eds). Taylor and Francis (Routledge), Abingdon, UK, pp. 375–382. DOI: 10.4324/9780203120958.ch47.

Jones, D.L., Day, J. and Quadri-Felitti, D. (2013) Emerging Definitions of Boutique and Lifestyle Hotels: A Delphi Study. *Journal of Travel and Tourism Marketing* 30(7), 715–731. DOI: 10.1080/10548408.2013.827549.

Krippendorff, K. (2004) *Content Analysis: An Introduction to its Methodology*. SAGE Publications, Segunda, Reino Unido, 215–243.

Li, S., Scott, N. and Walters, G. (2015) Current and potential methods for measuring emotion in tourism experiences: A review. *Current Issues in Tourism* 18(9), 805–827. DOI: 10.1080/13683500.2014.975679.

Liu, C.-R., Chiu, T.-H., Wang, Y.-C. and Huang, W.-S. (2020) Generation Y's revisit intention and price premium for lifestyle hotels: Brand love as the mediator. *International Journal of Hospitality and Tourism Administration* 21(3), 242–264. DOI: 10.1080/15256480.2018.1464421.

Martínez, I. and Olmedo, I. (2013) La reputación corporativa como fuente de competitividad en las PyMES de servicios. *Economía Industrial* 389, 125–134.

McIntosh, A.J. (1999) Into the Tourist's Mind: Understanding the Value of the Heritage Experience. *Journal of Travel and Tourism Marketing* 8(1), 41–64. DOI: 10.1300/J073v08n01_03.

Nicolas, D.H. and González-Gómez, C.I. (2014) Gentrificación, simbólica y poder en los centros históricos: querétaro, México. *Scripta Nova. Revista Electrónica de Geografía y Ciencias Sociales* 18. Available at: https://revistes.ub.edu/index.php/ScriptaNova/article/view/15001 (accessed 8 August 2022).

Novo Malvárez, M. (2019) Los nuevos usos del patrimonio: La expansión de hoteles boutique en Palma (Mallorca). *ESTOA: Revista de La Facultad de Arquitectura y Urbanismo de La Universidad de Cuenca* 8(16), 83–95. DOI: 10.18537/est.v008.n016.a07.

Oh, H., Fiore, A. and Jeoung, M. (2007) Measuring Experience Economy Concepts: Tourism Applications. *Journal of Travel Research* 46, 119–132.

Østergaard Pedersen, M. and Houlberg Salomonsen, H. (2023) Conceptualizing and Measuring (Public) Reputation Management. *Perspectives on Public Management and Governance* 6(1), 40–53. DOI: 10.1093/ppmgov/gvac023.

Parolin, C.F., Boeing, R. and University of Wisconsin - Stevens Point (2019) Consumption of experiences in boutique hotels in the context of e-WOM. *Tourism and Management Studies* 15(2), 33–45. DOI: 10.18089/tms.2019.150203.

Pulido, J.I. and Navarro, U. (2014) Identificación de items para medir experiencias del turista en el destino. *Culture* 8(1), 1–31.

Quadri-Felitti, D., Su, N. and Day, J. (2022) Consumer perspectives of boutique and lifestyle hotels: Is there a difference. *Sage Journals* 257–383. DOI: 10.1177/14673584211054605.

Ramkissoon, H. and Uysal, M.S. (2011) The effects of perceived authenticity, information search behaviour, motivation and destination imagery on cultural behavioural intentions of tourists. *Current Issues in Tourism* 14(6), 537–562. DOI: 10.1080/13683500.2010.493607.

Saldaña, J. (2016) *The Coding Manual for Qualitative Researchers*. SAGE Publications, Inc, London. Available at: https://www.amazon.com/Coding-Manual-Qualitative-Researchers-Third/dp/1473902495 (accessed 9 August 2022).

Sánchez-Aguirre, D.P. *et al*. (2017) Experiencia de usuarios en hoteles boutique. *Revista Global de Negocios* 5(1), 79–90.

See, G.-T. and Goh, Y.-N. (2019) Tourists intention to visit heritage hotels at George Town World Heritage Site. *Journal of Heritage Tourism* 14(1), 33–48. DOI: 10.1080/1743873X.2018.1458853.

Seyfi, S., Hall, C.M. and Rasoolimanesh, S.M. (2020) Exploring memorable cultural tourism experiences. *Journal of Heritage Tourism* 15(3), 341–357. DOI: 10.1080/1743873X.2019.1639717.

Su, L. and Hsu, M.K. (2013) Service fairness, consumption emotions, satisfaction, and behavioral intentions: The experience of Chinese heritage tourists. *Journal of Travel and Tourism Marketing* 30(8), 786–805. DOI: 10.1080/10548408.2013.835228.

Titz, K. (2008) Experiential consumption: Affect—emotions—hedonism. In: *Handbook of Hospitality Marketing Management*. Butterworth-Heinemann, Inglaterra, 324–352.

Tribe, J. (2010) *Strategy for Tourism*. Goodfellow Publishers Limited, Londres, 2–37.

Wickens, E. (2017) The consumption of cultural experiences in city tourism. *Tourism and Hospitality Research* 17(3), 264–271. DOI: 10.1177/1467358416646606.

Xie, P.F. and Shi, W.L. (2020) Elucidating the characteristics of heritage hotels. *Anatolia* 31(4), 670–673. DOI: 10.1080/13032917.2020.1744456.

Xu, X. (2020) Examining consumer emotion and behavior in online reviews of hotels when expecting managerial response. *International Journal of Hospitality Management* 89, 102559. DOI: 10.1016/j.ijhm.2020.102559.

Zhao, Y., Xu, X. and Wang, M. (2019) Predicting overall customer satisfaction: big data evidence from hotel online textual reviews. *International Journal of Hospitality Management* 76, 111–121. DOI: 10.1016/j.ijhm.2018.03.017.

Section 4: Destination Management and Tourism Marketing

17 Nation Branding in Latin America: Repositioning Brazil to Reset Its Brand Image

Marco Ocke[1] and Debora Cordeiro Braga[2]
[1]Business Management Department, Federal University of Santa Catarina, Florianopolis, Brazil; [2]Public Relations, Propaganda and Tourism Department, University of Sao Paulo, Sao Paulo, Brazil

Summary
By performing country brand initiatives and hosting mega sporting events as international promotional tools, Brazil, the largest economy in Latin America, has tried to leverage its brand awareness as a promising hub for investment for a great variety of sectors and add new attributes to the country's image, such as a strong, innovative and sustainable economy. However, the country has accumulated instability and turbulence over the years, which has discredited reputation-building efforts. The lack of public financial management, economic crisis associated with widespread corruption, inappropriate responses to the COVID-19 pandemic and lack of environmental management affected the country's image negatively. Currently, the relaunch of 'Marca Brasil' represents an effort to reset its brand image based on a repositioning strategy oriented to stand the country as a protagonist on the global stage, adding value to how the country is perceived.

The Value and Interest of the Case Study

The contribution of this case study is based on the discussion of the country brand image concept built from the brand identity and, to some extent, the achievement of place reputation. Its originality provides Latin America, especially Brazil, a unique case to develop such understanding.

Background

Away from severe geopolitical conflicts and the competition among superpowers, Latin America is a multicultural region rich in food, ore and commodity production. Despite being a diverse region, full of resources and human capital, many Latin American countries routinely face a broad array of challenges to attract stakeholders by improved reputation. In this context, some countries are managing and positioning their brands as a reputation-building effort. However, the activity of branding Latin American countries requires a consciousness that many of the contextual assumptions of the business environment in developed economies such as relatively free markets, comparatively minimal corruption, extant infrastructure and rare direct government interventions are not necessarily applied or applicable.

With the world moving forward post-pandemic reopening and the lasting war in Ukraine, the pitfalls of lasting supply chain disruptions, governance crises, high inflation rates and fiscal and debt outlook are still on radar. Latin America and the Caribbean are facing a complex economic scenario, with projections of deceleration in 2023 and growth of only 1.3% (ECLAC, 2023). The inflationary pressures that had been occurring as a result of the supply shocks generated by the COVID-19 pandemic were followed by slower economic and trade growth, weak job creation, falling investment and growing social demands. Also, the rise of the American dollar and the tightening of global financial conditions are expected to negatively affect countries in the region. Added to domestic turbulence is an international crisis in which the war between Russia and Ukraine has led to increasing geopolitical tensions, less dynamic global growth, reduced food availability and increases in energy prices.

All these events have emphasized the need to understand the role of country brand management in times of crisis (Kavaratzis and Florek, 2021). Quick and effective government responses had a significant impact on nation brands. On the contrary, countries that reacted slowly and ineffectively have been negatively impacted. In this context, some Latin American countries have successfully managed the worst of the pandemic's humanitarian, economic and reputational toll. Those nations that attempted to benchmark best practices and adopted European-style lockdowns – a group that includes Chile, Peru and Colombia – find themselves in a better position than Brazil, which downplayed the crisis and suffered a steep decline in brand value (Brand Finance, 2021).

The continent's largest economy has lost 12% of its brand value and dropped out of the top 20 in the Brand Finance Nation Brands for 2021. Famous for its vibrant culture, lifestyle and sports, Brazil has previously been ranked the highest South American nation, but the combination of high COVID-19 cases and damage to the agriculture sector from severe droughts has caused substantial harm to the economy in recent years. Nevertheless, the reasons Brazil has been negatively affected in terms of country image and reputation among domestic and foreign markets lie beyond current events. Ocke and Platt (2022) describe cumulative mistaken actions damaging the country's reputation over the years, resulting in brand value loss. First, the intent to raise awareness and brand the country as a promising hub for investment by holding mega sports events – the FIFA World Cup in 2014 and the Olympic Games in Rio de Janeiro in 2016 – was undone by poor public management accompanied by strong economic recession and spreading corruption within the left-sided federal government. Later, the weak response to the COVID-19 pandemic, lack of environmental management efforts and constant threats to democracy performed by the right-wing president Jair Bolsonaro harmed the country's image and reputation worldwide.

The following sections blend literature review with the description of current management and positioning features of country brands among some Latin nations. Examined country brands were the ones where information could be found on official country brand platforms, nation brand rankings and the media. Later, some thoughts on Brazil 's reputation and the decline of its brand value over the years are presented, followed by a socio-economic and political analysis of current events in Brazil. In this context, the main objective of the case is to analyse and discuss the repositioning of brand Brazil considering the challenges and opportunities of a nation rebranding in a competitive global scenario.

This discussion highlights how with the recently elected and former President Lula, an international recognized statesman, and the relaunch of 'Marca Brasil', the country could take the lead in a comprehensive agenda and consensus building on issues of common interest on the global stage. However, the intent to reset the country brand image and contribute to Latin American growth is accompanied by domestic challenges to drive economic recovery with policies aimed at controlling inflation and making public finances sustainable within a polarized society. Eventually, strong leadership and efficient planning for socio-economic development should prevail over political interest, otherwise the country's image and reputation are at risk of being compromised by multiple political players, periodic elections, corruption, ideologies and populism.

Branding Countries in Latin America

Due to Latin America's natural attractions as a strong competitive asset of its brand identity, nation branding in the region, especially in Central America and the Caribbean, has tended to be focused on destination branding aimed at international touristic markets. In recent years, as a reputation-building effort, some Latin American countries have been managing and positioning their brands based on market-driven planning by diversifying the country's competitive assets and developing mechanisms to flexibly adapt to conditions, fostering and cultivating entrepreneurial characteristics. While nation branding research and practice are widespread in Europe, North America and Asia, Latin American countries have only recently gained popularity and investment growth on official country brand programs to leverage competition for exports, resources, investors and visitors.

Most definitions of nation branding are attributed to the economic perspective which discusses it as a strategic effort to support public diplomacy by promoting a country to its markets and make it appear more attractive on the international stage (Olins, 2002; Kaneva, 2011; Aronczyk, 2013). In this sense, nation branding might be considered a metaphor (Anholt, 2020) for how effectively countries compete for a positive overall image on the international stage, covering political, economic and cultural dimensions. The existing literature suggests that nation branding relies on feature mapping and diagnoses to reach a deep understanding of how these assets can benefit the country image. This perspective suggests that nation branding relies on a continuum between the identity and the image of a nation where its competitive identity is an emitting concept, which refers to a dynamic phenomenon between the political dimension, the territorial tangibles and the results of socio-economic and technological interactions of people in space, whereas the image is a receiving concept because it reflects the way it is perceived by the public (Dinnie, 2015). As a result, nation branding becomes a reputation-building initiative centered on impressions built around numerous country images and understood as the aggregate of these images over time within a long-term cumulative effort.

Favourable performance among the worldwide competitive nation branding rankings may leverage the number of country brand programs performed by governors with the support of consultancy firms and advisory personnel (CityNationPlace, 2023). All countries in Latin America are captured in brand rankings but not all of them perform official nation branding programs. However, even when a country does not officially manage its brand, stakeholders build an image of it that results from several characteristics which influence attitudes and behaviour in relation to the country and that can be brought up by the simple mention of its name. The Good Country Index, for example, periodically measures over 160 countries using a wide range of data from the United Nations and other international organizations. Currently in Latin America, Chile comes first, followed by Uruguay and Costa Rica (Good Country Index, 2022). Measurements are made in seven major categories: science and technology, culture, international peace and security, world order, planet and climate change, prosperity and equality, and health and wellbeing.

Costa Rica highlights a consolidated status as a top country brand performer in different rankings (Future Brand Country Index, 2020; Anholt-Ipsos Nation Brand, 2022; Bloom Consulting, 2022). Its positioning *'Essential Costa Rica'* is based on a qualitative international study of perception within 79 countries, which resulted in an understanding that this is a green-oriented country and its rainforested wildlife and human traits are considered a differentiator that enhances the country competitiveness, improving its image worldwide (Essential Costa Rica, 2022). By betting on biodiversity, preservation and sustainable development as the engine of its economy, Costa Rica became a benchmark for other countries opting for such a positioning strategy.

Despite great performance in tourism and hospitality among regional competitors, Mexico seems to have developed a better understanding on nation branding effectiveness instead of concentrating efforts on destination branding acting isolated on touristic markets (UNWTO, 2022). Despite cases of drug-related insecurity and the controversial debate on immigration that is driving certain sectors in the USA, Mexico is reaping the benefits of years of investment in promotion to be firmly recognized for attributes and features linked to experiences based on its heritage and culture, tourism and 'made in' labels (Mexico Tourism Board, 2022). Mexico's business potential, which includes advanced technology and infrastructure, led the country to shift its geography from formerly pertaining to Central America and now to North America. Also, its proximity to the USA, the unrivalled and indisputable regional leader in the Americas, has become a competitive advantage for its country brand in terms of attracting visitors, investors and foreign trades. The essence of this network is seen in the FIFA World Cup 2026, jointly hosted by three nations for the first time: Mexico, Canada and the USA.

Countries in Central America such as Panama, El Salvador and Nicaragua have been trying to diversify their competitive assets but are currently relying on boosting the traditional tourism economy. The same is found in the Caribbean, which has experienced economic growth despite being shaken by natural disasters in the last few years. The region, as well as most Latin countries, has suffered outsized economic and humanitarian hardship as a result of the COVID-19 pandemic. The region might see the tourism economy increase after reopening borders to international visitors, owing to the concerted effort by governments to kickstart economic recovery by getting local businesses up and running (UNWTO, 2022).

Peru performs the most traditional country brand platform among South American countries, and it has been well ranked over the years while Colombia and Bolivia continue to leap forward and reinforce the trend towards growth by positioning '*the culture experience*' as the main asset explored in branding strategies in the region (Future Brand Bolivia, 2022; Marca Peru, 2022; Procolombia, 2022). Defining culture as a strategic asset of their competitive identity aimed at tourism, trades and investment attraction, these countries leverage specific features such as arts, gastronomy, heritage and people by developing branding programs in terms of corporate licensing, brand ambassadors, promotional campaigns and business events.

Further south, a consistent umbrella branding strategy is observed in countries such as Argentina, Uruguay and Chile (Marca Chile, 2022; Marca País Argentina, 2022; Uruguai XXI, 2022). The intent of these is to develop sectorial brands and integrate stakeholders' co-creation in the decision making process. This strengthens the sense of place, so sectors are marketed and branded in accordance with their competitive assets under one brand identity. This brand architecture is essential to leverage the country images across distinct areas by performing a coherent narrative across entities. Notably, Chile's positioning as '*creating future*' is supported by local talent and infrastructure underpinning a unique ecosystem of science and innovation to stand out in fields like marine biodiversity, forestry and astronomy (Marca Chile, 2022). By developing outstanding human capital and state-of-the-art infrastructure, the country becomes an attractive destination for students, professors, researchers and scientists from around the globe who seek to address challenges regarding sustainable development.

Brazil's Reputation

Since the 1990s in Brazil, efforts to support regional tourism development have achieved only moderate success. A lack of a coordinated approach between the various levels of government and a shortage of tourism data led Brazil to miss several opportunities to become a leading destination in the global market (Lohmann *et al.*, 2022). By performing country brand initiatives and hosting mega sporting events as international promotional tools – 2014 FIFA World Cup and the 2016 Olympic Games in Rio de Janeiro – Brazil sought to consolidate itself as a nation featuring high organizational capacity, global economic strength, a thriving democracy and a well-developing economy, as well as being one that is keen on international cooperation (Ocke, 2013). The intention was to highlight awareness of Brazil as a promising hub for investment for a great variety of sectors and add new attributes to the country's image, such as a strong, innovative and sustainable economy.

These initiatives were part of an international marketing plan entitled 'Plano Aquarela' aimed at touristic markets. It was managed by Embratur – Brazilian Tourism Institute, the agency of the Ministry of Tourism focused on the promotion, marketing and support of

Fig. 17.1. Marca Brasil. Courtesy of visitbrasil.com.

the commercialization of Brazilian products, services and touristic destinations abroad. This marketing plan launched 'Marca Brasil' in 2005 (see Fig. 17.1), the first attempt to create a country brand to leverage international marketing campaigns aimed at the touristic trade.

Two years of studies carried out by Embratur indicated that awareness of the country brand grew from 11 to 20% between 2006 and 2009. This research also pointed to a 25% increase in tourist arrivals in Brazil, which helped to an increase in the offer of tourist products by more than 120% between 2003 and 2010 (Visit Brasil, 2023).

Despite strengthening the positive characteristics for which the country was already known (its friendly and hospitable people alongside its privileged vast natural beauty), Brazil ended up creating a mood of uncertainty by demonstrating a lack of financial management which led to political and economic crises associated with wide-spread corruption, especially between 2012 and 2016. This demonstrates that the intended positive country image efforts designed earlier were instead dominated by problems and unhappiness within inward markets and became a drawback for international investors, as it negatively affected the country's reputation.

Later, the COVID-19 pandemic and the government's weak response, which was widely covered by international media (*The Economist*, 2021; *The Guardian*, 2021; *The New York Times*, 2021) characterized the Brazilian government and President Jair Bolsonaro as one of the worst leaderships dealing with the crisis. Besides, the president's constant rhetorical attacks on democracy and lack of environmental management turned the country into an international one. This impacted the country's global brand image, its attractiveness and economic development.

When managing a country brand, the role of a president is a critical element that directly affects the image of a place, especially in Latin American countries where the reputation of a political leader can improve or deteriorate a country's image. It is associated with their acting and thinking, and how well they collaborate and communicate. This impacts perceptions of governance, exports, tourism, investment, immigration, culture, heritage and people. In terms of discrediting the reputation-building efforts of nation branding initiatives, mistaken political measures can weaken place brand equity when they fail to align with a place's distinctive characteristics that are attractive to potential markets.

A political perspective

Despite a history of democratic instability and ideological flux at the highest levels of power, Latin American countries have always been united in their will for international recognition. At the same time, the region has a dynamic historic proclivity towards political swings from left to right and back. Since 2018, leaders on the left of the political spectrum have reached the presidency of Mexico, Bolivia, Peru, Honduras, Chile and Colombia. Although other countries in the region have chosen governments from different sides of the political spectrum in recent years, such as Uruguay and Argentina, the comeback of former president Lula in Brazil, one of Latin America's most influential and enduring politicians, drives Latin America's seven most populous nations and its six largest economies in the hands of the left.

Between 2003 and 2010, then President Lula used a windfall from a commodities boom to help reduce poverty and became a respected international statesman. Brazil turned into a significant global player, being nominated to lead the World Trade Organization and the Food and Agriculture Organization of the United Nations. Also, by helping create the G20 and establishing international relations with the BRICS group – Brazil, Russia, India, China and South Africa – his government directly engaged with a foreign public in pursuit of reputation management among external audiences. This was achieved by establishing a dialogue designed to inform and influence how the nation conducted its foreign policy. In this sense, nation branding and public diplomacy are areas of convergence and the desire for international recognition sets an important channel of information through diplomacy and influence on every part of the country image.

Repositioning Brand Brazil

After discontinuing the usage of its country brand 'Marca Brasil' for 4 years, the government elected in 2022 decided to relaunch it under the management of Embratur and the Ministry of Tourism. Once more, it is focused on repositioning the country image and resetting Brazil as a prime international touristic destination. It counts on the traditional and well-known competitive assets of its identity, such as its vibrant culture, lifestyle and natural environment. Personnel in charge consider this relaunch a consequence of the country's brand legacy – a consolidation and revived relevance of the country brand that is still recognized worldwide among tourism markets due to investments made by the agency, federal and state governments, and the tourism industry over 15 years (Visit Brasil, 2023).

It is important to highlight that positioning may be considered the cornerstone of brand management as a tool of competitive strategy by differentiating a brand from its competitors (Ries and Trout, 1981). In this context, brand positioning provides stakeholders with an understanding of differences among competing brands by creating a distinctive brand image or brand value in their minds. However, the exclusive usage of a country brand aimed at international touristic markets seems to be limited in comparison to nation branding programs being practiced among Brazil's neighbors in Latin America as well as developed countries worldwide (Place Brand Observer, 2022).

In other words, country brand management must be understood as a reputation-building effort based on a positioning strategy that enhances the country's competitiveness. The brand reflects its identity of distinguishing features, not only impacting the nation itself but also companies, products and services. As a result, it makes diversified economic sectors more attractive among consumers by adding value to the country's overall perception

(Anholt, 2020). In this sense, official country brand management in Brazil should be set under a brand architecture program where sectorial brands are marketed in accordance with their competitive assets with the support of modern marketing and communication techniques. This way, Brazil rebranding must set a strategic business unit to coordinate programs and actions to benefit different industries and sectors of the economy, not only tourism, within a consistent umbrella branding strategy, integrating stakeholders' co-creation in the decision making process.

The global reality draws fierce competition among nations to convince investors, talent and visitors to trust their public management and consider them worthy of their time and money. The globe moving towards post-pandemic reopening, the lasting war in Ukraine, the dangers of enduring supply chain disruptions, governance crises, high inflation rates, and the fiscal and debt outlook are shaping the short-term environment for nation branding. This scenario invites Brazil to work on a repositioning exercise, considering contemporary global challenges where strong leadership and efficient planning for socio-economic development surpass political ideologies and populism. Otherwise, the country's image and reputation are at risk of being continuously compromised.

Rebranding Brazil might strengthen its reputation. This path can also include the return of the country as a protagonist into the Community of Latin American and Caribbean States (CELAC) and the Southern Common Market (Mercosul) in order to promote regional integration within a comprehensive agenda in international forums to encourage Latin American and Caribbean countries to develop a vision as a community of nations, capable of dialogue and consensus building on issues of common interest. With the new government in charge, Brazil could undertake joint work across diverse sectors in a conscious and orchestrated repositioning of the nation branding strategy to reset its brand value and contribute to Latin American growth. By leading the continent towards contemporary challenges, a positive overall image of Brazil contributing to the world outside its own borders might reset its attractiveness and generate economic development in the long run.

The context of increasing recognition and awareness of environmental expectations and growing concerns about degradation, especially climate change, is driving countries to demonstrate environmental leadership on the global stage. In this sense, a nation rebranding oriented toward sustainable leadership could be a marketing opportunity for Brazil since its natural environment and potential clean energy matrix are strong assets of its competitive identity. Where environmental values constitute the nation branding essence, short-term planning to boost economic activity does not hamper social and environmental sustainability at the expense of fast economic recovery. This may represent a particular challenge for Brazil regarding inward markets, since the domestic situation represents major challenges for macroeconomic agents, which must drive economic recovery with policies aimed at controlling inflation and making public finances sustainable within a political and social polarizating context.

Conclusions

Countries in Latin America are adopting contemporary nation branding practices with regard to a favourable perception and improved reputation. On the other hand, Brazil, the largest economy in the region, adopted political measures that harmed the country's image, reputation and socio-economic development. After discrediting reputation-building efforts and weakening place brand equity, Brazil can reset its brand image and value within international markets and contribute to the continent's growth by taking the lead in the regional and global agenda. However, this intent is accompanied by domestic and global challenges. In this context, this case draws attention to a nation rebranding in the Brazilian context with the support of 'Marca Brasil' to enhance the country's competitiveness regarding exports, resources, investment and tourism. If properly managed, this country brand could strengthen the sense of nation by adding value to the country's overall perception and highlight a brand identity able to reflect distinguished positioning assets in a competitive global scenario.

References

Anholt, S. (2020) *The Good Country Equation: How We Can Repair the World in One Generation*. Berrett-Koehler Publishers, Oakland, CA.

Anholt-Ipsos Nation Brand (2022). Available at: https://www.ipsos.com/en/nation-brands-index-2022 (accessed 20 June 2022).

Aronczyk, M. (2013) *Branding the Nation: The Global Business of National Identity*. Oxford University Press, Oxford, UK.

Bloom Consulting (2022). Available at: https://www.bloom-consulting.com/index_pt.php (accessed 10 May 2022).

Brand Finance (2021). Available at: https://brandfinance.com/offices/brazil (accessed 20 March 2022).

CityNationPlace (2023) The Ultimate Guide to Place Brand Rankings. Available at: https://www.citynationplace.com/the-ultimate-guide-to-place-brand-rankings (accessed 20 January 2023).

Dinnie, K. (2015) *Nation Branding: Concepts, Issues, Practice*, 2 edn. Routledge, Abingdon, Oxfordshire, UK.

ECLAC (2023) Preliminary Overview of the Economies of Latin America and the Caribbean. Available at: https://www.cepal.org/en/publications/47670-preliminary-overview-economies-latin-america-and-caribbean-2021 (accessed 10 January 2022).

Essential Costa Rica (2022). Available at: https://www.esencialcostarica.com/eng/ (accessed 30 August 2022).

Future Brand Bolivia (2022) Boosting the Image of Bolivia. Available at: https://www.futurebrand.com/our-work/bolivia (accessed 10 July 2022).

Future Brand Country Index (2020). Available at: https://www.futurebrand.com/country-brand-report-latin-america-2019-2020 (accessed 20 June 2022).

Good Country Index (2022). Available at: https://index.goodcountry.org/ (accessed 20 September 2022).

Kaneva, N. (2011) Nation branding: Toward an agenda of critical research. *International Journal of Communication* 5, 117–141.

Kavaratzis, M. and Florek, M. (2021) Special section: The future of place branding. *Place Branding and Public Diplomacy* 17, 63–64.

Lohmann, G., Lobo, H.A.S., Trigo, L.G.G., Valduga, V., Castro, R. *et al*. (2022) Tourism in Brazil: From politics, social inequality, corruption and violence towards the 2030 Brazilian tourism agenda. *Tourism Review* 77(1), 72–96.

Marca Chile (2022). Available at: https://marcachile.cl/creating-future/ (accessed 20 October 2022).

Marca País Argentina (2022). Available at: https://www.argentina.gob.ar/turismoydeportes/marca-pais (accessed 5 October 2022).

Marca Peru (2022). Available at: https://peru.info/es-pe/marca-peru (accessed 30 October 2022).

Mexico Tourism Board (2022). Available at: https://visitmexico.com/ (accessed 10 November 2022).

Ocke, M. (2013) Marca e Imagem de um País: Considerações sobre a Copa do Mundo de 2014 e a Promoção do Brasil. *PODIUM Sport, Leisure and Tourism Review* 2(2), 95–120.

Ocke, M. and Platt, A.A. (2022) Place branding Brazil: Research agenda and advisory panel for the 21st century. *Brazilian Journal of Marketing* 21(Special Issue), 29–63.

Olins, W. (2002) Branding the nation – The historical context. *Journal of Brand Management* 9(4/5), 241–269.

Place Brand Observer (2022). Available at: https://placebrandobserver.com/tag/panel/ (accessed 5 February 2022).

Procolombia (2022). Available at: https://www.colombia.co/marca-pais/marca-colombia/ (accessed 20 July 2022).

Ries, A. and Trout, J. (1981) *Positioning: The Battle for Your Mind*. McGraw Hill, New York.

The Economist (2021) Brazil's Dire Decade. Jair Bolsonaro is not the Only Reason His Country is in a Ditch. The Political System that Helped Him Win Office Needs Deep Reform. Available at: https://www.economist.com/leaders/2021/06/05/jair-bolsonaro-is-not-the-only-reason-his-country-is-in-a-ditch (accessed 3 June 2021).

The Guardian (2021) Tens of Thousands of Brazilians March to Demand Bolsonaro's impeachment. Available at: https://www.theguardian.com/world/2021/may/29/tens-of-thousands-of-brazilians-march-to-demand-bolsonaros-impeachment (accessed 5 June 2021).

The New York Times (2021) Complete Insanity': Virus-Stricken Brazil Agrees to Host Soccer Tournament. Available at: https://www.nytimes.com/2021/06/02/world/americas/brazil-copa-america-soccer- tournament.html (accessed 8 June 2021).
UNWTO (2022). Available at: https://www.unwto.org/americas (accessed 10 September 2022).
Uruguai XXI (2022). Available at: https://marcapaisuruguay.gub.uy/ (accessed 10 September 2022).
Visit Brasil (2023). Available at: https://www.visitbrasil.com/pt/marcabrasil/ (accessed 15 April 2023).

18 Traversing the Andes on the Most Beautiful Train in the World

Carla Guerrón Montero
Department of Anthropology, University of Delaware, Newark, Delaware, USA

Summary
In this case study, I take the reader on a voyage through Ecuador's railway system. It was inaugurated as a unifying component of the nation's transportation infrastructure at the beginning of the 20th century, then reconstructed and transformed into a heritage tourism attraction in the early years of the 21st century. The reconstruction project, initiated in 2008 by controversial President Rafael Correa Delgado (2007–2017), restored a large portion of the original railway system for heritage and tourism purposes. Even though subsequent administrations have abandoned the project, the revitalization of the railroad system is considered to this day as one of the most emblematic successes of Correa's tourism policy and a model for railway tourism. Based on ethnographic and bibliographic research conducted between 2014 and 2023, I discuss the successes and limitations of using railway heritage tourism to enforce political ideologies and nation-building.

To Matt, my favorite eternal wanderer

Introduction

> Imagine that Parisians [...] said about the Eiffel Tower, a monument built a century ago and only useful as a lookout point: 'Since this is not worth anything, let us demolish the Eiffel Tower.' Never! It is a symbol of France. Similarly, the train symbolized Ecuador; it was a symbol of unity, a symbol of Alfaro, the most crucial Ecuadorian figure of all times, and they left it die. Now, once again, this railway engine brings happiness to our lives with its sound and rattle. Alfaro's train is working again; it is not the most difficult train in the world, but be assured, it is the most beautiful train in the world. (Rafael Correa Delgado, 11 November 2014)

With these words, former Ecuadorian President Rafael Correa Delgado (2007–2017) opened the inaugural ceremony for the Guamote-Socavón railroad route in Central Ecuador in November 2014. Six years earlier, in 2008, he had initiated an aggressive program to restore a large portion of the Ecuadorian railway system for heritage and tourism purposes. Even today, when the railway is once more suffering from neglect, the restoration project is emblematic of Correa's tourism policy.

This case study examines how the Correa regime used the railway system reconstruction project as an attempt to build a unified national identity around their *Revolución Ciudadana* (Citizen's Revolution) and the concept of *buen vivir*, an approach to socialism that embraces the ancestral and communitarian knowledge of the Quichua indigenous peoples. I propose that, just as the original railroad system had been built to unify the country in the early 20th century, its reconstruction 100 years later as a tourism attraction had similar political and cultural objectives under the banner of '21st-century socialism'.

This case study is based on ethnographic research carried out between 2014 and 2023.

I conducted participant observation on board the *Tren Crucero* between 30 September and 4 October 2014, and on four of the five main thematic routes available between 2014 and 2018. I visited some of the unused stations on these routes in 2023.

I documented the stories and themes that emerged through extensive fieldnotes. I also conducted open-ended, semi-structured interviews with key officials of the railway system, tour guides, and domestic and international tourists between 2014 and 2023.[1]

The Railroad that Unified Ecuador

The Ecuadorian railway is unlike any other railroad system in Latin America, both because of Ecuador's rugged topography, and the degree of national integration that this incredible feat of railway engineering generated in a previously fragmented nation (Fig. 18.1). Four of the five largest cities in the country were connected for the first time by the railroad when it was finally inaugurated. But this boon to the country took a long time to develop fully: nearly 70 years of start-and-stop work before the system was complete.

Initial congressional authorization for building the original railway system, the *Ferrocarril Trasandino*, was obtained by conservative president Gabriel García Moreno in 1861. However, by the time of Moreno's assassination in 1875, only 45 km of track had been laid. There was little additional progress until the late 1890s when José Eloy Alfaro Delgado, the leader of the Liberal Revolution, took up the project. To move the undertaking to completion, US entrepreneur Archer Harman incorporated the Guayaquil & Quito Railway Company (G&Q) in New Jersey on 7 September 1897, to build and run the railway system.[2] The finished railroad was inaugurated on Alfaro's birthday, 25 June 1908. Seventeen years later, due to numerous disagreements between the company and the Ecuadorian government, partly because of the unwanted intervention of the US government in Ecuadorian affairs, President Gonzalo Córdova authorized the purchase of 75% of the railway's stock and Ecuador, 'for all practical purposes, controlled the enterprise' (Loewenfeld, 1946, p. 93). Thirty-five years later, in 1970, President José María Velasco Ibarra created the state-owned National State Railway Company (*Empresa Nacional de Ferrocarriles del Estado*, ENFE), which managed limited train routes for passenger and freight transportation (Carrera, 2012).

The Liberal Revolution (1895–1925) gave rise to one of the most important political, economic, and cultural transformations in the history of Ecuador. According to Enrique Ayala Mora (2002, p. 71), the Liberal Revolution produced, 'on the one hand, a change in the political direction of the Ecuadorian state, which went [from oligarchic] to bourgeois hands; on the other hand, it was a moment of unprecedented popular mobilization in our history, where social actors such as peasants and urban populations had an essential participation'. In just 15 years, the Revolution secularized and guaranteed public education, declared freedom of religion, promulgated laws about civil marriage and divorce, established the civil registry of births, marriages, and deaths, secularized cemeteries, and nationalized the properties of religious orders (Clark, 2012, p. 68). The railroad exemplifies the Liberal Revolution's economic and social achievements under Alfaro.

However, in the latter part of the 20th century, Ecuador was changing in ways that diminished the railroad's value. The country's economy up to the 1970s had been based primarily on exporting bananas, coffee, and cacao. But when oil production and exportation became Ecuador's central economic engine, the state turned its attention to constructing roads – particularly to and from the Amazonian region – to facilitate oil transportation, enhance commerce, and accommodate population growth. This transformation resulted in the neglect of the railway system, which was left to deteriorate under challenging climate conditions. By the 1990s, the Ecuadorian railway system served the transportation and commerce needs of just a few rural areas. Only two routes in the Andes were maintained for tourism purposes, the Cotopaxi National Park and the *Nariz del Diablo* (Devil's Nose) routes. These routes directly benefited only a few local populations involved in tourism (Barreno Lalama and Brito-Henriques, 2015, p. 1924).

Fig. 18.1. Map of Ecuador.

Although it had once been an essential component of Ecuador's commercial infrastructure, the railroad was sinking into irrelevance, and its roadbeds and equipment were succumbing to decay. President Correa gave it a new lease on life through his vision of marketing it as a heritage tourist attraction.

Railway Tourism as Heritage Tourism

Heritage tourism is travel to experience cultural, historical, and natural resources representing humanity's past and present. Its popularity has grown in the 21st century in parallel with the increased public desire to conserve and preserve heritage sites and growing global concern about the value of linking past, present, and future. From a tourism management perspective, the key to stimulating heritage tourism is to appeal to an interest in education and learning (Lee, 2015), as heritage incorporates educational aspects, emotional involvement, the search for individual and collective identities, and a desire for presumedly authentic experiences (Biran *et al.*, 2011). Heritage can thus connect tourism, economy, and culture with nation branding (Silverman, 2015), which is the use of symbolic resources of nationalist discourses to position the nation-state 'as a necessary frame of identity, allegiance, and affiliation' (Aronczyk, 2008). The reconstruction of the Ecuadorian Railroad system by the Correa regime fits this pattern perfectly. To rebrand the nation-state, Correa converted a disused and abandoned artifact of Ecuador's heritage into a symbol of a transformative political era.

Bhati *et al.* (2014, p. 122) state that, 'In many countries, the key decision maker on the interpretation of heritage and its impact on identity is the state; as both heritage and identity form a strong relationship with tourism, then heritage tourism inevitably becomes an issue of politics.' In the case of President Correa's railroad restoration project, industrial railway heritage tourism was used to boost a particular political ideology and contribute to the construction of the nation-state. The Ecuadorian railway became a highly contentious political project during Correa's regime. It remained so under successive administrations (Lenin Moreno, 2017–2021, and Guillermo Lasso, 2021–present), which let the project slowly die. While economic reasons, such as the fall of oil prices between 2014 and 2021 and, later, the global COVID-19 pandemic in 2020 and 2021, played a role in the demise of the renovated railway system, the intentional neglect to which it was subjected after Correa left power in 2017 speaks to how this project was seen as representative of his administration. To show why that came about, I will now focus on the railway project as developed by the Citizen's Revolution and its role in nation-building.

On 1 April 2008, through Agreement No. 028, President Rafael Correa Delgado declared the train system in its entirety a national heritage – more specifically, a Civil Monument and Historical, Testimonial, and Symbolic Heritage (*Monumento Civil y Patrimonio Histórico, Testimonial, Simbólico*) (Carrera, 2012). Agreement No. 028 also transformed EFE[3] into a public corporation, *Ferrocarriles del Ecuador Empresa Pública* (FEEP, Ecuador Railway Public Company). Thus, more than a hundred years after the rail system was first inaugurated, FEEP initiated a rehabilitation project with technical consultation from *Ferrocarriles de Vía Estrecha* (FEVE, Narrow Way Railways), currently a division of the *Empresa Española de Ferrocarriles* (RENFE, Spanish Railway Company). This massive undertaking required the reconstruction of 507 km of railway. The project was carried out using a three-pronged approach: rehabilitation of ways and stations, restructuring of the public corporation, and development of community-based tourism products, requiring an investment of US$340 million between 2008 and 2013 (Barreno Lalama and Brito-Henriques, 2015, p. 1924). The vast enterprise was concluded in 2013. The railway system, stations, wagons, and equipment were designated a Cultural and National Heritage that same year (Fig. 18.2).

The system showcased Ecuador's remarkable and varied landscape, which had earned the country the title of 'the Switzerland of the Americas' because of its 'lofty snow-clad peaks, deep ravines, and broad valleys enclosed by great mountain ranges' (Stabler, 1917, p. 241). The system's schedule included five 1 day thematic routes of 4–6 hour each through the Andean and coastal provinces and one

Fig. 18.2. The most beautiful train in the world.

multi-day excursion. At its peak in 2015, the railway system had more than 30 stations, 23 cafés, 14 handicraft markets, and 13 local museums, providing more than 5000 direct and 15,000 indirect jobs and involving 180 micro-enterprises. Depending on the route, the price of an expedition ranged between US$22 and US$69 (one way). Transportation, meals, sightseeing, cultural programs, and guided visits were all included.

The single-day routes were the *Tren Machachi*, the *Tren de la Libertad* (Liberty Train), the *Tren de los Lagos* (Train of the Lakes), the *Tren de la Dulzura* (Train of Sweetness), and the *Tren del Hielo* (Train of the Ice). The 4-day *Tren Crucero*, the system's signature voyage, took tourists through 456 km of mountainous landscapes and diverse communities and cultures in the Ecuadorian highlands and coastal areas (Fig. 18.3). It featured a reconditioned steam engine and four carefully restored carriages capable of carrying up to 54 passengers. This route, running between Quito and Durán, passed through six provinces, 22 municipalities, and 43 rural parishes. Prices to take the *Tren Crucero* ranged from US$1650 (luxury experience) to US$2300 (gold class), including transportation, accommodations, meals, sightseeing, and guide services.

The law that declared the railway system a cultural heritage also designated it a socially and economically sustainable tourism product whose main objectives included restoring the train system to the glory days of the prior century and making it accessible to as many Ecuadorian citizens as possible. To achieve this objective, the system was subsidized. There were special offers amounting to a 30% discount for children under 12, senior citizens, and people with disabilities. In addition, high school seniors from public institutions who graduated with the highest GPAs received a free trip on a student-oriented version of the *Tren Crucero*.

Heritage Tourism and the Train Voyages

Of course, tourism was not new to Ecuador. The Colón Archipelago (widely known as the

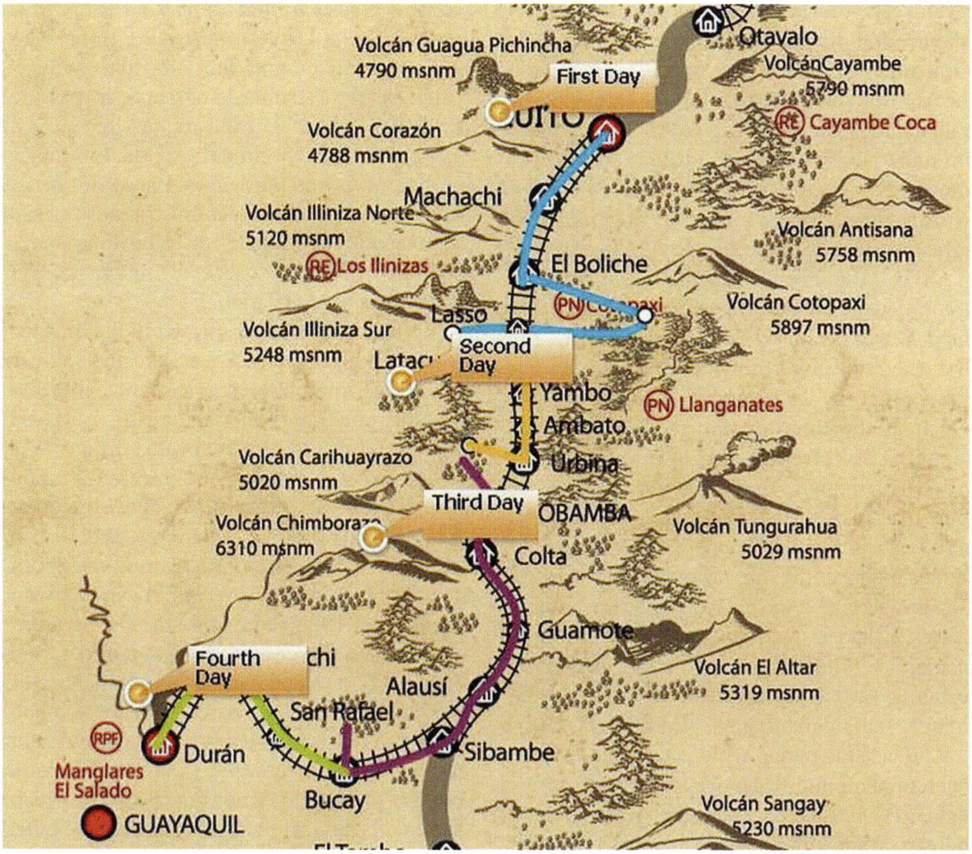

Fig. 18.3. Route of the *Tren Crucero*.

Galápagos Islands) has been a popular destination for naturalists and tourists since at least the 1830s and was protected from the excesses of tourism by Ecuadorian law as early as 1934. Ecuador's tourism industry grew around the 1950s and gradually became solidified over the next 40 years. The *Dirección Nacional de Turismo* (DITURIS, National Tourism Directorate) was formed in 1974, and the Ministry of Tourism in 1992 (Solís Carrión, 2007; Zurita, 2021). Ecuador has much to offer in the way of tourist attractions, particularly in terms of its ecological diversity. It was declared one of the 17 megadiverse countries in the world by the Convention on Biological Diversity, 'with one of the largest diversities of ecosystems and live cultures in a relatively small territory, with good internal connectivity, and strategically located' (Solís Carrión, 2007, p. 29). Some experts argue that Ecuador is the most biodiverse country in the world per km^2 because it holds 11% of the world's biodiversity (Solís Carrión, 2007, p. 29). Yet recognition of the tourism potential of other products, such as the railway system that was nicknamed 'the most difficult train in the world' only emerged in the 21st century, under President Correa.

In part, this happened for economic reasons. Tourism became an essential aspect of President Correa's policies under the coordination of the *Secretaría Nacional de Planificación y Desarrollo* (SENPLADES, National Secretary for Planification and Development). Income from tourism grew substantially during his administration, from US$492 million in 2006 to US$1,487.2 million 10 years later, a tripling of revenues (Cajo-Riofrío, 2020). One of the objectives of Correa's administration was to decentralize tourism away from the Galápagos Islands and the country's capital, Quito, to rural areas. With this in mind, he increased the Ministry of Tourism's budget from US$40 million annually in 2012 to US$150 million annually between 2013 and 2017 (Coloredo-Mansfeld *et al.*, 2018). President Correa wanted to transform Ecuador's tourism potential into a reality. In an interview for Travel Weekly in 2016, the President stated that tourism accounted for 6% of the country's GDP but that his goal was 'to have the tourism sector be the first, the most important source of revenue for the country, by 2025, more than oil exports', which at the time amounted to 12% of the country's GDP (Weissmann, 2016).

Even though the economic aspects of tourism were important to the government, social and political considerations provided a more significant incentive for promoting its growth. Tourism became a means to cement the government's political platform, thus establishing a memorable branding for the nation. According to Dinnie (2012, p. 15), nation branding is 'the unique, multidimensional blend of elements that provide the nation with culturally grounded differentiation and relevance for all of its target audiences' for different purposes. In the tourism industry, competition reigns supreme. Therefore, branding plays a fundamental role in establishing a destination as desirable. For example, one of Correa's boldest efforts to brand the nation was to purchase a 30s commercial promoting Ecuador during the 2014 Super Bowl broadcast, paying US$3.8 million for the spot. The commercial spotlighted Ecuador's natural resources and diverse cultures to the tune of The Beatles' classic 'All You Need is Love', tying into the country's 'All You Need is Ecuador' promotional campaign. This campaign featured the railroad system as an exemplary representative of Ecuador's cultural and natural diversity. The government's statistics estimated that this campaign generated a 14% increase in tourism visits.[4]

Railway routes can be desirable to tourists for various reasons. In some cases, the actual route is infused with significance, such as a trade route. In other cases, the traversed buildings or sites are the focal point. Some routes are short and regional while others are transnational (Timothy and Boyd, 2003). While there are only two railway systems that are part of the UNESCO World Heritage List (the Darjeeling Himalayan Railway in India and the Rhaetian Railway in the Albula /Bernina Landscapes across the Swiss Alps), historic railways are recognized tourist attractions around the world. In numerous discourses and interviews, President Correa argued that previous administrations lacked vision about the Ecuadorian railway system's potential to become a successful tourism product nationally and internationally.

In fact, tourists were drawn to the finished project. Though the revenue from the railway system was small, it is estimated that 200,000 domestic and international tourists traversed the Andes aboard the renovated trains. The *Ferrocarriles del Ecuador* also received several international accolades, including the Best Wider World Tourism Award by the British Guild of Travel Writers in 2013 and the South American Leading Luxury Train Award between 2014 and 2019 by the World Traveller Awards.

I started this case study with a quote from President Correa comparing the significance of the Eiffel Tower to the French with the symbolism of the railroad system for Ecuadorians. But Jorge Eduardo Carrera (J. E. C.), former General Manager of FEEP, finds one major area in which the Ecuadorian train system exceeded the Eiffel Tower as a representative icon of its country. The difference is that, for tourists, travelling on the train allowed them to meet Ecuadorians who are in the 'real country', not some 'beautiful tourist' spaces (interview J. E. C., 30 March 2023). They would encounter the Ecuadorian who lives in a rural area or the Ecuadorian who lives in a tourism destination but does not interact with visitors very often. Such a person might occasionally work in tourism-related activities but is not a 'tourate,' a local person in a tourist destination who interacts directly and frequently with tourists (Causey, 2003).

One of the objectives for reconstructing the railway system was establishing community-based tourism products (Ruiz Ballesteros and Solís Carrión, 2007; Echeverría Cedeño, 2018). Local communities not previously engaged in tourism were incorporated into the project as small-scale entrepreneurs offering food, handicraft products, guiding, and cultural entertainment. Carrera estimates that more than 6 million Ecuadorians in the 10 provinces and 33 municipalities where the railway system passed through benefited directly or indirectly from these newly created economic and social networks (Barreno Lalama and Brito-Henriques, 2015).

What did the estimated 200,000 tourists experience on these routes? The signature voyage, the *Tren Crucero*, focused on 'luxury experiences.' During its 4-day trip of 456 km between Ecuador's capital, Quito, and Guayaquil, Ecuador's largest city, tourists traveled through the Avenue of the Volcanoes and the infamous Devil's Nose – the most difficult stretch of the Trans-Andean Railroad. They visited traditional markets, a rose plantation, churches, and national parks, and stayed and dined at well-manicured former haciendas. The intention of the *Tren Crucero* was to attract international tourists who could afford this relatively expensive excursion and, in the process, subsidize the shorter 1-day routes intended for domestic tourists. By 2015, 60% of tourists in the *Tren Crucero* were international (interview J. E. C., 30 March 2023).

While the *Tren Crucero* and the five short routes had their particularities, a few overarching themes pertain to all. One example is the consistency of the experience. Visits to the different stations were structured to allow passengers sufficient time, 20–30 min at each station, to dine on regional dishes in the station's cafés and shop for local specialties in its handicraft stores. The idea was not only to let visitors experience local goods, cultures, and customs but also to fulfil one of the objectives of the reconstruction project by creating job opportunities for Ecuadorians in rural areas. In 2015, the Ecuadorian Central Bank calculated that rural areas endured a poverty rate of 39.9%, far above the 15.7% rate in urban areas (Cajo-Riofrío, 2020). It is estimated that the reconstructed railroad generated 21,000 jobs, significantly boosting local incomes. The emphasis on providing job opportunities in the countryside won the *Tren Crucero* the British Responsible Tourism Award in 2016 and the TO DO! Award in 2018 for being an 'innovative, socially responsible and sustainable tourism project which [...] created new tourism products along the travel route and a high-quality travel experience' (Giraldo, 2017, p. 3).

Another characteristic common to all the routes was that tour guides played a crucial role in curating the experiences offered to tourists, either by reading from a careful script (as in the case of the tour guides from the *Tren Crucero*) or by improvising and connecting with the tourists through a call-and-response approach (on all other routes) to reconstruct the past, reassemble national identities, and portray Ecuador's industrial/modern period. Except for the stations, some of which contained small museums or exhibit areas, most interpretations of the routes relied on trained tour guides. That is, much of the information provided about a destination occurred while the train was on its way there, leaving ample time for the tourists to explore on their own when the train stopped. In some stations, part of the entertainment included presentations by local folklore dance troupes (Fig. 18.4).

Navigating between following the script and improvising depended on the level of knowledge and presumed interest of the tourists. Tour guides managed the narrative, the photo opportunities, and the entertainment activities aboard the trains. On the short routes, guides asked numerous questions of the tourists and almost quizzed them about their general knowledge of Ecuador's history and geography. The absence of this constant interaction and its replacement with scripted descriptions of scenery and historical moments aboard the *Tren Crucero*, suggests that silence and restraint were viewed as appropriate ways to interact with more affluent tourists, who were more likely unfamiliar with details about Ecuadorian history and cultures. The time spent on route to a destination in the *Tren Crucero* was filled with soft instrumental music or silence.[5]

For domestic tourists in particular, one overarching theme was nostalgia for Ecuador's cultural heritage, including the train system itself. Karina (K. L.), a tour guide on the Train of the Lakes route, noted: 'Just like the train united the coast and the highlands in the 1900s, we are all integrated into the tourism railway project today' (interview K. L., 10 October 2014). Overall, the country on display was 'traditional', and the experience evoked was one of reminiscence. In fact, one of the mottos of Ferrocarriles del Ecuador was '*mira, siente, vive, recorre*' (look, feel, live, travel). On the short routes, information about the train catered to Ecuadorian tourists, and the guides offered a wealth of anecdotal

Fig. 18.4. Tourists and local tour guide in the *Tren de la Libertad*.

information about the places traversed, and how they came to receive their names and use. These descriptions invariably linked Ecuador's pre-and post-colonial history with the Inca empire. The theme of nostalgia was also evident in the food offered on the train and at the station cafés. Every item was considered a 'traditional' snack or dish, and many of them were no longer easily found in the modern supermarkets that abound in a country amid a national nutritional transition toward ultra-processed food (Gross *et al.*, 2017). Since most of the routes were in the highlands, typical snacks included maize on the cob with fresh cheese, fried fava beans, *tortillas de tiesto* (maize tortillas prepared in a clay pot), *café de chuspa* (coffee prepared in a fabric filter), cheese empanadas, toasted maize with lupines, or *dulce de leche*.

Conclusions

The concept of 'tourism regimes' (like the term 'heritage regimes') encompasses two ways of looking at tourism (Geismar, 2015). The first views tourism as a constantly constructed, contested, and constituted social phenomenon. The second relates to a commonsense view of tourism, generally applied through policies and practices by nation-states, where tourism is used to promote economic development and boost a country's international stature. But within both regimes, the idea of 'heritage' remains a central component, both in terms of attracting tourists, how representations of tourism are presented to tourists, and how those representations are used and understood by those living in that nation.

In the early 20th century General Alfaro's railroad system generated a discourse *about* a collective national project, a significant exercise in constructing the Ecuadorian nation-state (Clark, 2012). At the time liberals also used the railroad metaphor to call for a transformation of the nation-state through movement and connection (Clark, 2012). The construction of the railroad was framed as a redemptive work because it was believed to be a way to 'resurrect' and emancipate Ecuadorians morally. For liberals, the railroad system represented the 'light of progress' that the 'obscure' ways of conservatism had repressed. Ecuadorians were invited to depart from the past, represented by the Conservative party, and look forward to progress, represented by the Liberal party. The platform of the Liberal Party was 'movement, connection, growth, and expansion,' and the railroad was the capstone of its program (Clark, 2012, pp. 51–52).

Over a century later, President Correa's administration used the same metaphor. Correa molded his image after Eloy Alfaro in more ways than one. He called his political, economic, and social projects 'revolutionary' or the 'revolution of the 21st century'. He installed a new National Constitutional Assembly, which drafted a new constitution in Montecristi, Alfaro's birthplace.[6] I posit that his regime's focus on reconstructing and rehabilitating the railway system was not an accident. In an interview in 2016, Correa admitted the nostalgia he personally felt about the Guayaquil-Quito train route that he recalled was in operation when he was a child, 'and it was abandoned. When I came to office, one of my first tasks was to rescue this train, not for transportation but for tourism' (Weissmann, 2016). He intended to make the system accessible to most of the population, hence highlighting the value of the Citizen's Revolution by offering thousands of Ecuadorians the opportunity to see what the country had become under his regime.

The train inaugurated by Eloy Alfaro was viewed as a machine with the magical and symbolic power to transform Ecuador, to mobilize, connect, and unify its people. The train system inaugurated by Rafael Correa during his Citizen's Revolution was proposed as an instrument with the magical and symbolic power to transform Ecuador, to mobilize, connect, and unify its people, but also to share this view of the country with the world – a view of a hopeful small country that was demonstrating to the world what the application of the *Buen Vivir* policies could mean for tourism development, in line with the international branding campaign 'All You Need is Ecuador'.

As Coloredo-Mansfeld *et al.* (2018) aptly note, tourism growth, even if it is heritage-, community-, or conservation-based, runs the risk of being co-opted by the state when the state becomes its major champion. 'At the point when the state makes good on its investment promises, local people face enormous pressure to cede control of the distinctive economies of place amid regulations imposed by national ministries' (Coloredo-Mansfeld, 2018, p. 442). During Correa's administration, tourism policies aimed to link Ecuador's nation-branding with the more extensive public policies being put in place. One of President Correa's most visible actions was to grant rights to nature under its constitution (the concept of the *Buen Vivir*) while fostering an extractivist economic approach. Paradoxically, the revenues of the oil production boom of the early 2000s allowed the railway system to be reconstructed *and* subsidized. Other aspects of this nation-branding included the fact that 33% of Ecuador's territory was under environmental protection, including added protections for the Galapagos Islands, that the country boasted the best road system in South America, and that poverty was no longer an eyesore for tourists to endure. 'It is unpleasant for a visitor to come and see huge social inequalities. We have reduced poverty a lot and raised the standard of living. You won't see beggars on the streets, and this is very good for tourism' (Rafael Correa in Weissmann, 2016).

The railway system was a crucial actor in and witness to dramatic changes for over a century. It became a central actor once again after being rehabilitated. The Ecuadorian railway system is, in fact, 'a permanent actor in its territory and because of its role in community life, produce[d] numerous assemblages and generate[d] new networks and transformations' (Barreno Lalama and Brito-Henriques, 2015, p. 1923).

But the involvement of the state cuts both ways. In May 2020, the government of Correa's successor, Lenín Moreno (2017–2021), announced that FEEP would be eliminated due to the economic crisis caused by the global COVID-19 pandemic and because revenues from tourism did not cover the expenses incurred by the costly infrastructure. At US$25 per ticket for most 1 day routes, the system could not survive without subsidies (Ramos, 2019; Diario El Universo, 2020). By 2023, the government of Moreno's successor, Guillermo Lasso (2021–present), had allowed stations throughout the country to be left to decay, and FEEP was dissolved. There were rumors about having a private company run the railway system, but by July 2023, there was no indication that this process was in motion.

The reconstruction of the railway system for tourism and heritage purposes was an iconic component of Correa's *Buen Vivir* nation branding. Its vulnerability was profoundly tied to its emblematic nature for this regime, and the sounds and rattle of the most beautiful train in the world can no longer be heard traversing the Andes.

Notes

[1] Except for public officials, I have used pseudonyms to refer to my interviewees.
[2] The company received a financial influx from J.P. McDonald Company of New York (Loewenfeld, 1946, pp. 76–77).
[3] The previous step carried out by the National Congress in 2005 was to transform ENFE into EFE, *Empresa de Ferrocarriles del Ecuador* (Ecuador Railway Company).
[4] The US is the second largest tourism market for Ecuador (Colombia is the first).
[5] The exception to this rule occurred on the last day of the 4-day experience before the trip ended, when a live band joined the train and tourists were invited to dance and sing along to the tune of popular Ecuadorian and Latin American folk music.
[6] In fact, Alfaro's remains, which were deposited in a mausoleum at some point in the 1940s were exhumed and re-interred with honors in Montecristi on the initiative of President Correa.

References

Aronczyk, M. (2008) 'Living the Brand': Nationality, globality and the identity strategies of nation branding consultants. *International Journal of Communication* 2, 41–65.

Ayala Mora, E. (2002) *Historia de la Revolución Liberal Ecuatoriana*. Corporación Editora Nacional, Quito (Colección Temas, 5).

Barreno Lalama, A. and Brito-Henriques, E. (2015) La rehabilitación del ferrocarril en ecuador, como actor transformador del territorio. In: de la Riva, J. (ed.) *Análisis Espacial y Representación Geográfica: Innovación y Aplicación. Universidad de Zaragoza*. Zaragoza, Spain, pp. 1921–1928.

Bhati, A., Prcye, J. and Chaiechi, T. (2014) Industrial railway heritage trains: The evolution of a heritage tourism genre and its attributes. Journal of Heritage Tourism 9(2), 114–133.

Biran, A., Poria, Y. and Oren, G. (2011) Sought experiences at (dark) heritage sites. Annals of Tourism Research 38(3), 820–841.

Cajo-Riofrío, M.C. (2020) La Pobreza en Espacios Rurales del Ecuador y el Turismo Como Oportunidad de Desarrollo Sostenible. In: Pinilla Rodríguez, D.E., Yaulema, N.A. and Pulgar-Sánchez, K.S. (eds) Bienestar, Medio Ambiente y Turismo: Reflexiones desde la Experiencia Ecuatoriana y Chimboracense. Universidad Nacional de Chimborazo (UNACH) Riobamba.

Carrera, J.E. (2012) Incidencia de la rehabilitación del ferrocarril del Ecuador en el desarrollo local de las comunidades atravesadas por la vía férrea. Master of Arts, Universidad Politécnica Salesiana, p. 194.

Causey, A. (2003) Hard Bargaining in Sumatra: Western Travelers and Toba Bataks in the Marketplace of Souvenirs. University of Hawai'i Press, Honolulu.

Clark, K. (2012) *La Obra Redentora: El Ferrocarril y la Nación en Ecuador*. Universidad Andina Simón Bolívar, Quito (Biblioteca de Historia, 19).

Coloredo-Mansfeld, R. *et al*. (2018) Conflicts, territories, and the institutionalization of post-agrarian economies on an expanding tourist frontier in Quilotoa, Ecuador. World Development 101, 441–452.

Diario El Universo (2020) Ferrocarriles del Ecuador EP, Tras 10 Años de Creación, Será Eliminada. Available at: https://www.eluniverso.com/noticias/2020/05/19/nota/7845615/ferrocarriles-ecuador-eliminacion-personal-cierre (accessed 7 January 2021).

Dinnie, K. (2012) Nation Branding, Concepts, Issues, Practice. Butterworth-Heinemann, Oxford.

Echeverría Cedeño, M.F. (2018) Actitudes de los Visitantes en Áreas Rurales en Ecuador: Caso de Estudio Mindo. In: Muñoz Barriga, A., Osorio, M. and Guijarro, G. (eds) Tendencias de Investigación en Turismo en América Latina. Pontificia Universidad Católica del Ecuador, Quito, pp. 77–91.

Geismar, H. (2015) Anthropology and Heritage Regimes. Review of Anthropology 44, 71–85.

Giraldo, Á. (2017) Award Rationale Responsible Tourism on Board Tren Ecuador in Ecuador. Studienkreis für Tourismus und Entwicklung e.V, Seefeld, Germany.

Gross, J., Guerrón Montero, C., Hammer, M. and Berti, P. (2017) Creating healthy bodies in rural ecuador at a time of dietary shift. In: Sherwood, S., Arce, A. and Paredes, M. (eds) *Food, Agriculture and Social Change: The Vitality of Latin America* Routledge, New York.

Lee, Y.-J. (2015) Creating memorable experiences in a reuse heritage site. Annals of Tourism Research 55, 155–170.

Loewenfeld, E. (1946) The Guayaquil and Quito railway, Ecuador. The Southwestern Social Science Quarterly 27(1), 68–93.

Ramos, X. (2019) Con boletos desde $25 por pasajero, Ferrocarriles del Ecuador no cubre sus gastos y ha cerrado cinco rutas. Available at: https://www.eluniverso.com/noticias/2019/11/29/nota/7626533/tren-ecuador-precios-2019 (accessed 7 January 2021).

Ruiz Ballesteros, E. and Solís Carrión, D. (2007) Inroducción. In: Ruis Balleresteros, E. and Solís Carrión, D. (eds) Turismo Comunitario en Ecuador: Desarrollo y Sostenibilidad Social. Ediciones Abya-Yala, Quito, pp. 1–26.

Silverman, H. (2015) Branding Peru: Cultural Heritage and popular culture in the marketing strategy of Prom Perú. In: Encounters with Popular Pasts. Springer, New York, pp. 131–148.

Solís Carrión, D. (2007) De la resistencia a la sostenibilidad: el proceso histórico del turismo comunitario en ecuador y sus retos actuales. In: *Turismo Comunitario En Ecuador: Desarrollo y Sostenibilidad Social*. Ediciones Abya-Yala, Quito.

Stabler, J. (1917) Travels in Ecuador. The Geographical Journal 50(4), 241–254.

Timothy, D.J. and Boyd, S.W. (2003) Heritage Tourism. Pearson Education Limited, Edinburgh Gate, UK.

Weissmann, A. (2016) Ecuador's President Outlines His Vision for Tourism. Travel Weekly (online).

Zurita, E., *et al*. (2021) Sector turístico y su impacto en la economía ecuatoriana. In: *Universidad Nacional de Chimborazo (UNACH) Riobamba*.

19 Developing a Rural Tourism Route Through the Integration of Local Stakeholders: Challenges and Opportunities for Family Farmers in Porto Seguro – Bahia (Brazil)

Eduardo S. Sant'Anna, Verônica Feder Mayer and Osiris Marques
Tourism and Hospitality School, Universidade Federal Fluminense (UFF), Niterói, Rio de Janeiro, Brazil

Summary
Family farming plays a key role in global food production and economic development. Rural areas in Porto Seguro, Bahia, Brazil, possess a rich cultural heritage, sustainable agricultural practices, and traditional communities that have been overlooked by mainstream tourism. To unlock economic opportunities for small farmers and attract diverse tourists, the local government and university (Universidade Federal Fluminense) collaborated to enhance a rural tourism route. This case study explores the synergy among local government, the university, and residents, showcasing how it can lead to creative placemaking practices in rural tourism. The study focuses on the 'Motherland of Brazil' route, a sustainable tourism option, and emphasizes the importance of residents' involvement in destination planning through family farmers' participation.

The Value and Interest of the Case Study

This case study holds significant value in understanding the potential of rural tourism for small farmers, economic development, and cultural preservation. It highlights the overlooked territories of Porto Seguro/Brazil, known for its mass tourism impact, and addresses the need for diversification.

Background

This case study portrays an action research project implemented within a rural tourism route in Porto Seguro, Bahia, Brazil. The primary focus of this case is to explore how the design of a rural tourism route can foster the integration of local stakeholders and promote opportunities for creative placemaking practices. Placemaking refers to a broad range of activities that involve shaping and enhancing the character and identity of a given space (Lew, 2017). This process encompasses various actions, spanning from organic local community actions to more intentional and organized endeavours. This study examines three distinct rural tourism experiences as illustrative examples to showcase these concepts.

The 'Brazil Motherland' rural tourism route takes place in Porto Seguro (Fig. 19.1),

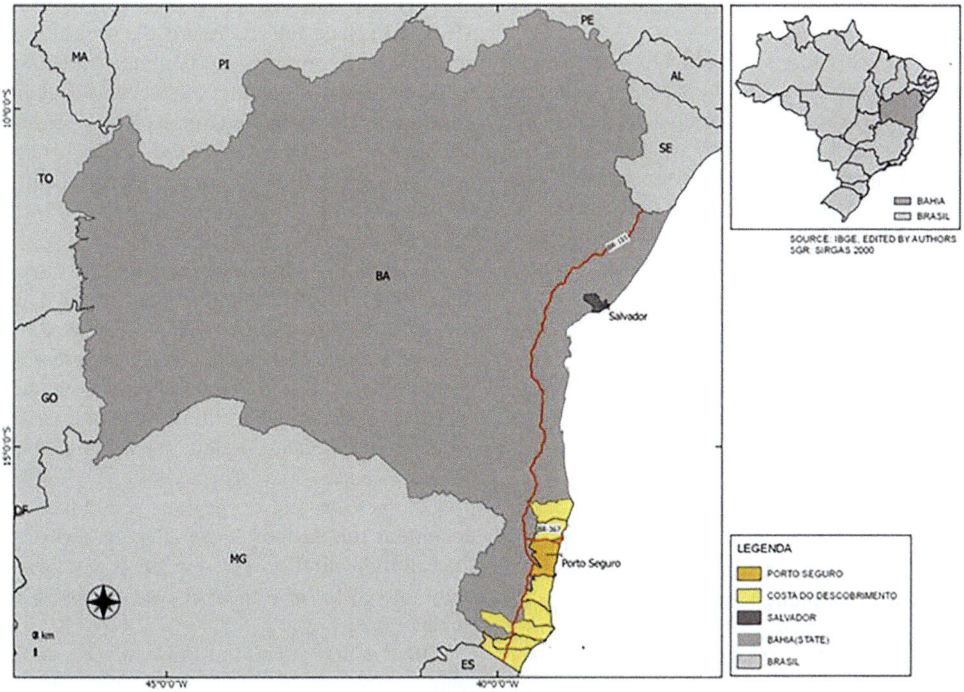

Fig. 19.1. Porto Seguro/Bahia within the tourist region of the 'Discovery Coast' (Source: Araújo and Abreu, 2022, p. 10).

a municipality situated in the southernmost part of Bahia State, Brazil, along the 'Costa do Descobrimento' or 'Discovery Coast'. It is accessible via the BR-101 and BR-367 highways, 656 km to the south of Salvador and approximately 1558 km from the Brazilian capital. Porto Seguro traces its origins back to the ancient captaincy of Porto Seguro, making it an important starting point in Brazil's history. The ancient captaincy of Porto Seguro refers to the historical administrative division during the colonial period in Brazil. Captaincies were territories granted by the Portuguese Crown to donatários (grantees) for the purpose of colonization. Over the past decades, it has experienced remarkable economic growth, particularly in the thriving tourism sector. The establishment of the BR-101 highway in the 1980s, coupled with the implementation of an international airport in the 1990s, solidified Porto Seguro as one of the largest mass tourism destinations in the north-eastern region of Brazil (Araújo and Abreu, 2022, p. 10).

Significantly, Porto Seguro gained considerable prominence in the first decade of the new millennium with the commemoration of the 500th anniversary of the 'Discovery of Brazil'. This milestone celebration further propelled the area's growth and development. Its unique geographical location, historical significance, and flourishing tourism industry create an appealing backdrop for conducting a case study.

The coastal area of Porto Seguro has been adversely affected by the negative consequences of mass tourism. These impacts include environmental degradation, overcrowding, particularly during major events like New Year's Eve and Carnival, and a decline in the authenticity of the local culture. To address these challenges and foster sustainable development, it is crucial to explore alternative forms of tourism, with a particular emphasis on rural tourism. Moreover, small farmers in rural areas often encounter limited economic opportunities and a decline in agricultural activities, which poses a threat to their livelihoods and the overall progress of the rural communities.

Thus, this case study aims to describe and analyse alternatives that harness the potential benefits of rural tourism for small farmers while bolstering the local economy and alleviating the negative impacts of mass tourism. In this sense, a national-scale action research project has provided an opportunity to address the integration of local stakeholders in planning a rural tourism route in Porto Seguro, Bahia.

Methods: Action research to design rural tourism experiences

The context for conducting action research in Porto Seguro was a national-scale project called 'Experiências do Brasil Rural' (Rural Brazil Experiences, RBE), which emerged as a collaborative effort between the Ministry of Tourism, the Ministry of Agriculture, and the Fluminense Federal University (UFF) in 2021 and 2022. RBE's core objective was to facilitate and bolster tourism development within rural areas by emphasizing gastronomic offerings intricately linked to tourism within tourism routes. The project centred its attention on travel routes encompassing family-owned agricultural enterprises actively involved in crop cultivation, livestock farming, agro-industrial pursuits, extractive industries, and tourism endeavours. Additionally, the project encompassed participation from small and medium-sized entrepreneurs, proprietors of lodging establishments, restaurants, and travel agencies.

The university's role encompassed the strategic planning, execution, and comprehensive evaluation of the outcomes from the action research process. This process unfolded across eight rural tourism routes spread across Brazil during 2021. During that year, social distancing guidelines associated with COVID-19 were in place, and, therefore, the project was predominantly implemented online. The study team consisted of eight professors and eight master's students in tourism. This case study discusses and reflects upon the research outcomes in one of the tourism routes, specifically the itinerary of Porto Seguro in Bahia, called Brazil Motherland.

A total of 13 family farmers and entrepreneurs from Brazil Motherland participated in the RBE activities during 2021. Participants went through four stages in this action research project (Fig. 19.2): (1) comprehensive evaluation; (2) professional training; (3) interactive tourism experiences design workshops; and (4) experiences trial. Throughout the entire project, pictures and screenshots were taken to document the various stages. It is noteworthy that all individuals involved in the project willingly agreed to have their images displayed in publications.

In the first stage, the university research team employed questionnaires and online group discussions to conduct a self-assessment, identifying strengths and weaknesses related to tourism in rural Porto Seguro, as well as hospitality, technical quality, management, and marketing aspects of the route. The goal was to get a clear view of the route's situation, paving the way for improvements and tailored interventions.

The second stage focused on route enhancement by offering online training classes and mentoring. The courses addressed challenges highlighted by members during the evaluation process, while mentorship supported family farmers in refining their services.

In the third stage, participants developed skills to craft tourist experiences that evoke emotions and meaning, fostering expertise in designing impactful experiences. Techniques

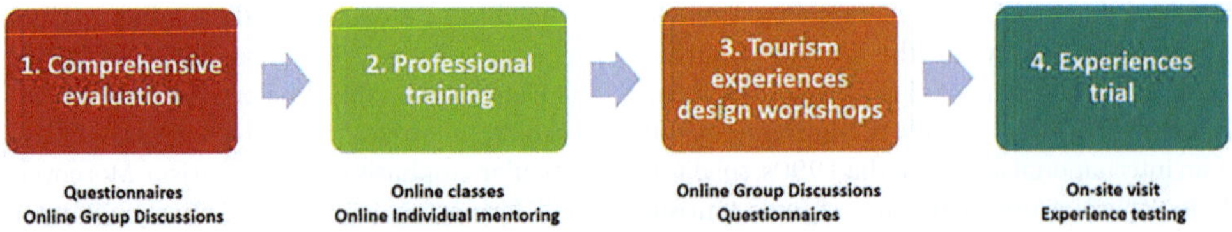

Fig. 19.2. Stages of the action research in Brazil Motherland.

used included online group discussions and questionnaires.

Testing the rural tourism experiences marked the concluding phase. Evaluation teams conducted on-site visits and carried out experience testing to evaluate technical and memorable aspects, which significantly contributed to enhancing the effectiveness for entrepreneurs and farmers along the rural tourism routes.

The following sections delve into the opportunities and challenges unearthed during the evaluation research. Furthermore, they present the results stemming from experience design and validation visits, spotlighting three illustrative instances of placemaking practices.

Challenges and Opportunities for Family Farmers in Rural Tourism

The participatory evaluation process with the family farmers and entrepreneurs resulted in an assessment of the challenges and opportunities for designing memorable and sustainable tourism experiences. This collaborative approach allowed the research team to gather insights directly from those involved in the project, capturing their perspectives on what made the route special and where improvements could be made. By engaging the key stakeholders in the evaluation, the research team ensured that their voices were heard.

The participatory evaluation encompassed various categories, with a particular focus on the opportunities available for tourists to learn, appreciate the natural beauty, be entertained, seek escapism, immerse themselves in local traditions, and the overall cohesion of the route. This thorough assessment provided valuable insights into the strengths and areas for improvement, ensuring that the subsequent development of the route would cater to the specific needs and desires of the community and visitors. Furthermore, since most of the entrepreneurs did not personally know each other, it was an important moment for building bonds with other members of the route (Fig. 19.3).

During the evaluation process, the route participants emphasized several opportunities for direct engagement with organic food production and local products as a major strength of the Motherland of Brazil tourist route. According to these locals, visitors may have the unique chance to immerse themselves in the production processes of cocoa, cupuaçu, açaí, buffalo mozzarella, and other locally sourced goods.

Fig. 19.3. Screenshot of one of the online meetings with local stakeholders from Porto Seguro, Bahia, Brazil (Source: Courtesy of Brazil Ministry of Tourism, 2021).

Furthermore, the hands-on connection with the origins of these foods provided an authentic experience that may foster a deep appreciation for local products and raise awareness about sustainable practices.

Another key aspect highlighted by the route participants is the opportunity to learn about agroecology and its best practices. This chance to learn about sustainability, environmental preservation, and responsible farming methods was potentially enlightening and may capture visitors' interest in adopting more conscientious practices.

The route participants also considered engaging with indigenous cultures to be an enriching experience. Tourists may learn about the traditions, language, history, and daily life of the indigenous communities present in the region. Additionally, tourists can actively engage in cultural activities such as traditional dances, body painting, and tasting indigenous cuisine. This immersive encounter with indigenous culture facilitates a deeper understanding and appreciation of Brazil's cultural diversity.

Besides the strengths, participants identified several weaknesses in the rural tourism route. First, they observed a need for better integration among the existing tourism experiences, which stemmed from the recent route implementation. These local actors emphasized the need for a more cohesive approach to establish a connection between tourism activities. Secondly, they highlighted the underutilization of sensory experiences as a missed opportunity. Despite the potential for engaging visitors' senses, these experiences needed to be adequately explored and structured. They also pointed out the poor conservation of buildings, inadequate infrastructure for accessing unique structures, low local awareness of the tourism potential, insufficient promotion of the region's rural beauty, and the absence of effective coordination between entertainment options. They urged collective efforts to address these weaknesses and proposed improvements, such as enhancing conservation efforts, improving infrastructure, engaging the local community, promoting the region's authentic rural beauty, and fostering better collaboration among route operators.

Tourism Placemaking: From Resources to Experiences

The planning of a tourism route in the rural area of Porto Seguro represents a strategic approach to tourism place making. Place-making is an inherent human behaviour that encompasses a spectrum of activities, ranging from spontaneous and unstructured actions of individuals (referred to as 'place-making' by Lew, 2017) to deliberate and planned actions undertaken by governments and tourism authorities (Lew, 2017). By designing rural tourism experiences, the focus is on creating a destination that not only appeals to tourists but also embraces and honours the local identity.

The rural tourism route serves as a platform to showcase the distinctive rural characteristics, cultural heritage, and natural beauty of the region, thereby shaping the destination's image and identity in a manner that resonates with the local community. Involving local stakeholders, particularly farmers, in the planning process ensures the integration of their valuable knowledge, expertise, and cultural practices. This fosters a sense of community ownership and pride and promotes sustainable tourism practices that respect and preserve the rural environment and local traditions. By considering the needs and aspirations of both visitors and the local community, the tourism route enhances the destination's identity and image while offering an engaging and authentic visitor experience.

To provide a focused analysis, we have chosen to highlight three noteworthy cases that exemplify the diverse range of experiences and the significant impact that family farmers have had on the development of the route. These three cases were chosen as representative examples of the diverse range of experiences offered by family farmers in Porto Seguro. Each one showcases a distinct aspect of the region's rural tourism offerings, providing visitors with memorable encounters that celebrate local traditions, sustainable practices, and the bountiful natural resources of the area. Together, they contribute to the overall appeal and authenticity of the Motherland of Brazil route.

Fig. 19.4. Four cyclists on their way to a farm (Source: Courtesy of Brazil's Ministry of Tourism, 2021).

Bahia Active: Bike tour through the rural world of Brazil Motherland

A major source of integration of the experiences was the work of one of the participants who designed a rural tourism biking trip (Fig. 19.4). Bahia Active is a cycle tourism company that specializes in promoting a healthy lifestyle. They excel in providing connections with nature and immersive experiences, whether it is through cycling, hiking, or boating. The company's carefully crafted routes are designed to seamlessly blend sporting activities with moments of joy, relaxation, social interaction, and immersion in the local surroundings.

Bahia Active provides an immersive cycle experience that offers participants the opportunity to appreciate the beauty of their surroundings, indulge in the intricate details of the rural landscape, and engage with the distinctive characteristics of the local residents in the villages they explore. This journey promises unique and transformative encounters, where individuals are encouraged to embrace emotions that unfold during 5 days of cycling. It offers a sustainable mobility proposal in the Costa do Descobrimento, integrating cycling, hiking, and canoeing into alternative tourist routes. This provides tourists with a unique local perspective, combining a relaxed history lesson on-site with the pleasure of exploring beautiful landscapes, rivers, and villages.

Embedded within the principles of place-making practices, Bahia Active's cycle expedition serves as a prominent illustration of collaborative and participatory place-making. It intertwines the aspirations of tourists, the expertise cultivated by the entrepreneur, and the distinct heritage intrinsic to the region. Collaborative place-making, as in this case study, often involves participatory change processes resulting from co-creation among residents, government authorities, non-governmental organizations (NGOs), and educational institutions (Lew, 2017). This kind of action can be in the middle of the spectrum between organic place-making, which transforms residents' attachment to the place into a substrate for a tourist experience and planned and intentional placemaking.

Organic barn (Celeiro Orgânico): Embracing a healthy lifestyle

Adriana Severino is a family farmer and leader of an association responsible for the certification of organic food production in Porto Seguro. She is originally from Rio Grande do Sul, in the southern part of Brazil. After enjoying a holiday in Bahia, she made the life-changing decision to settle there permanently, captivated by its exceptional climate and welcoming community. Adriana's expertise in organic certification was instrumental in facilitating her relocation, as it grabbed the interest of Porto Seguro's Department of Agriculture. Departing from her managerial role in a cooperative and agroindustry in Rio Grande do Sul, she embarked on a new journey in 2017, focusing on the certification of organic agriculture in Porto Seguro.

When Adriana was invited to take part in the Motherland of Brazil rural tourism route, she did not have any connection with tourism whatsoever. Yet she embraced the opportunity as a challenge and a way to align her values and causes with pedagogical tourism in her farm. She designed an activity called 'Embracing a healthy lifestyle' aimed at sensitizing locals and visitors to healthier eating habits.

In this experience, the visitors can engage in an immersive experience, delving into the daily routines of a collective of farmers who have made a conscious decision to cultivate organic crops, reaping the benefits of improved health and well-being. This agrarian endeavour is the birthplace of organic agriculture in Porto Seguro. The activity commences with a brief trail, in which Adriana guides the visitors through a scenic banana plantation (Fig. 19.5) and leads them to the verdant vegetable gardens. Along this path, visitors have the pleasure of savouring the flavours of the season through sampling Edible Wild Plants (PANCs). The visitors gain profound insight into the production methods of biofertilizers and natural repellents employed on the premises. They can also explore the meticulously tended vegetable gardens and learn about organic farming techniques.

Fig. 19.5. Adriana Severino (on the left) explaining the banana cultivation processes (Source: Courtesy of Brazil's Ministry of Tourism, 2021).

Besides advocating for organic foods and promoting healthy choices, Adriana is an advocate for a change in the narrative surrounding the 'Discovery of Brazil'. The teaching of Brazil's discovery in schools has long been a subject of debate, with prevailing narratives often presenting a misleading portrayal of harmonious interactions between indigenous peoples and the Portuguese.

> This issue of how they wrongly teach in schools about the discovery of Brazil, depicting a harmonious embrace between indigenous people and the Portuguese, is far from the truth. The real history was marked by struggle, conflict, and death. It was an enduring inequality that persists to this day. Now, there is a movement to change the narrative, transforming it into a story of the land and promoting tourism aimed at children from the age of 11, inviting them to explore the birthplace of Brazilian history. No longer the Coast of Discovery, but the Coast of Birth. This endeavour aligns perfectly with our efforts, offering a unique and immersive tourism experience unlike any other. (Adriana Severino)

Adriana's example vividly illustrates 'place-making', as expounded by Lew (2017). Not just a farmer and leader in organic food certification, Adriana is an active catalyst in crafting a unique tourist place. According to Lew (2017), 'place-making' involves imbuing spaces with meaning and identities, culminating in a distinct sense of place. Adriana embodies this concept, her emotional connection to Porto Seguro exemplifying how personal identification shapes authentic places. Adriana's profound bond with Porto Seguro underscores the emotional aspect of 'place-making', which translates into her design of a tourism experience. Her 'embracing a healthy lifestyle' (Lew, 2017) activity empowers visitors to interact with the place-making process, fostering an authentic sense of place.

Furthermore, Adriana challenges historical narratives. Her re-evaluation of the 'Discovery of Brazil' narrative, which often romanticizes indigenous-Portuguese interactions, is pivotal in shaping a place's authentic understanding. By revisiting the narrative and promoting realism, she contributes to constructing a place with a genuine and meaningful history, in line with place-making principles.

Trevo dos Búfalos: among cheese and buffaloes

Roberto Cangussu's family has been raising buffaloes for over 40 years. His father began managing these animals in the 1970s with two females and one male in Teófilo Otoni, a city in Minas Gerais. About 25 years ago, his family moved to Porto Seguro, where they have been exclusively involved in buffalo milk production and its derivatives. Around 10 years ago, they became aware of the tourism potential as the number of international flights to Porto Seguro increased. However, it was in 2021 that their involvement in the tourism industry became more tangible when they joined the rural tourism route of Porto Seguro and took part in the RBE project. Roberto states:

> I didn't see the potential that could be utilised and is still underrated, which is the countryside surrounding the tourist city. And with the advent of the work done by the Ministry of Tourism with you all, it opened up opportunities, and we started to get to know and visit properties that, as mentioned in one of the early meetings, could become your main business. Since that first meeting, I have been believing in and investing in this.' (Roberto Cangussu)

The experience offered by the family farmer provides an educational and engaging opportunity for visitors to connect with nature, water buffaloes, and the traditional craftsmanship of buffalo cheese production. Upon arrival, guests are welcomed into a serene environment where they can interact with the calm and gentle water buffaloes, feeling their skin and observing the source of the milk used for buffalo cheese. The experience includes a gastronomic journey, with a tasting session of freshly crafted products, allowing visitors to appreciate the unique flavours and textures of buffalo cheese. A guided tour of the farm's rustic shed offers insights into the meticulous techniques and craftsmanship involved in cheese production.

The journey continues through captivating landscapes, including partially flooded terrains and abundant foliage, providing an ideal habitat for the water buffaloes (Fig. 19.6). Guests witness the animals in their natural environment, deepening their understanding of their ecological

Fig. 19.6. Buffaloes swimming in a river (Source: Courtesy of Brazil's Ministry of Tourism, 2021).

significance. The experience culminates with a breathtaking sunset, where visitors can observe the water buffaloes cooling off in nearby bodies of water. Through this immersive experience, the family farmer fosters a profound connection with nature, appreciation for traditional craftsmanship, and an understanding of the cultural and ecological importance of water buffaloes and buffalo cheese production.

Roberto Cangussu and his family's journey in the tourism industry offers a compelling example of place-making. This case closely echoes Lew's (2017) notions, wherein the potential of a rural environment is recognized and transformed into engaging experiences. The Cangussu family tapped into the undervalued countryside surrounding Porto Seguro. With the rise in international flights, they seized the opportunity to intertwine their buffalo milk production with tourism, thereby highlighting the intrinsic value of their locale.

Their farm experience provides an immersive encounter. Visitors engage with water buffaloes, witness traditional buffalo cheese craftsmanship, and embark on a gastronomic journey. This authentic engagement cultivates a deeper connection with the place, fostering an appreciation for its cultural and ecological significance. Moreover, the guided tour through the landscape mirrors the observations of Zhou *et al.* (2023), offering visitors insights into the natural environment. The direct interaction with the terrain and the observation of buffaloes in their habitat align with the ethos of 'place-making', encouraging a profound connection with space.

Conclusions

The creation of the rural tourism route in Porto Seguro represents the engagement of family farmers, both those already involved in tourism and those seeking new opportunities. Also, the local government fostered collaboration and knowledge sharing among diverse stakeholders. This collaborative place-making approach promotes sustainable development while showcasing the region's rural resources and warm hospitality.

Online meetings and workshops organized as part of the RBE project provided a platform for farmers to connect and establish partnerships. However, one challenge encountered was the limited participation of farmers and entrepreneurs who successfully completed the project. The lack of interest, time, or staff capacity of some individuals hindered the full realization of the collaborative potential envisioned for the route. Indeed, many local entrepreneurs are quite involved with the mass tourism dynamics that take place in Porto Seguro, making it difficult for them to adopt a different mindset. Despite this limitation, the participating farmers formed a network of collaboration that enhanced the diversity and quality of the rural tourism experience.

The creation of the rural tourism route has opened economic opportunities for participating family farmers and local communities. Through the establishment of guesthouses, the sale of farm produce, and the provision of tourism-related services, families have the potential to generate income and improve their livelihoods. This initiative contributes to the economic development of the region, fostering inclusive growth and reducing dependence on a single sector.

The online nature of the project presented some limitations and challenges. The virtual format may have hindered the development of stronger personal connections and face-to-face interactions among participants. Additionally, the low number of farmers and entrepreneurs who completed the project highlighted the difficulties in ensuring full participation and commitment. These limitations underscore the importance of exploring alternative strategies to engage and motivate potential participants in future rural tourism initiatives.

The collaborative rural tourism route in Porto Seguro, developed within the RBE project, exemplifies the local government's dedication to sustainable development and cultural preservation. By engaging family farmers and promoting collaboration, the project has diversified tourism offerings, enhanced local economies, and showcased the region's natural and cultural assets. While limitations and challenges were encountered, the outcomes demonstrate the potential for inclusive growth and community empowerment through collaborative tourism initiatives. Future endeavours should consider addressing participation barriers and exploring hybrid approaches that combine online engagement with in-person interactions to maximize the project's impact.

References

Araújo, C.P.D. and Abreu, L.M. (2022) From fishing villages to international tourist destinations: Where are the citizens? The cases of Cabo Frio, Rio de Janeiro and Porto Seguro, Bahia-Brazil. *Sociedade & Natureza* 34, e62140.

Brazil Ministry of Tourism (2021) Projeto experiências do brasil rural: manual de implementação para desenvolvimento de experiências memoráveis em roteiros turísticos [Rural Brazil experiences project: Implementation manual for the development of memorable experiences in touristic routes.

Lew, A.A. (2017) Tourism planning and place making: place-making or placemaking? *Tourism Geographies* 19(3), 448–466.

Zhou, L., Cheng, X., Wall, G. and Zhang, D. (2023) Entrepreneurial networks in creative tourism placemaking: Dali village, Wuhan, China. *Tourism Geographies* 25(1), 220–242.

20 A City Developed for (and by) Tourism: Campos do Jordão (São Paulo, Brazil)

Bruna C. Mendes and Tatiana Marchetti Panza
Federal Institute of Education, Science and Technology of São Paulo, Campos do Jordão (SP), Brazil

Summary
Brazil provides several tourist destination options besides soccer, sun, and beaches. The cold weather and nature are the pillars of tourism in Campos do Jordão, a city that began to develop due to the tuberculosis pandemic that hit the country in the 18th century when it started hosting patients who sought a cure for this disease through life near nature, cold weather and resting. Over the years, it began to attract São Paulo City's elite, which sought isolated treatment for tuberculosis or private rest. This elite was the factor behind the city's development, which turned a city of the sick into a tourist city. The term 'Brazilian Switzerland' was adopted as a tourist differential in order to build the imagery of a European city in Brazilian lands. Known as one of the main winter destinations in the country, the city has about 48,000 inhabitants and attracts more than 4.9 million visitors on a yearly basis, on average. However, actions taken in the city often focused on changes to serve tourists, rather than its dwellers. The main challenges faced by this city comprise social contrasts between the local population and visitors, lack of social participation in the urban and tourist planning process, as well as poor public power performance in territory organization processes.

Methods

Information used in this case study derives from Mendes (2021) and Panza (2022), two doctoral theses focused on Campos do Jordão City (São Paulo State, Brazil). One of them addressed the local community's hospitality toward tourism and tourists. The other thesis focused on investigating the urban and tourist development process from the perspective of both dwellers and managers involved in the local territorial planning process. The aforementioned theses gathered quantitative and qualitative data via interviews, questionnaires and focus groups during 2020 and 2021. The authors' experience as local residents and workers should be seen as the factor defining their perception about tourism and its impacts on the city.

The Value and Interest of the Case Study

Campos do Jordão is a small city (48,000 inhabitants) located in São Paulo State's countryside, South-eastern Brazil. It is part of the Metropolitan Region of Paraíba Valley and North Coast (RMVale). It shares borders with certain municipalities in Minas Gerais State and its total area is protected by Conservation Units at different administrative scales (federal, state, and municipal). Fig. 20.1 presents a series of maps of Brazil, emphasizing the State of São

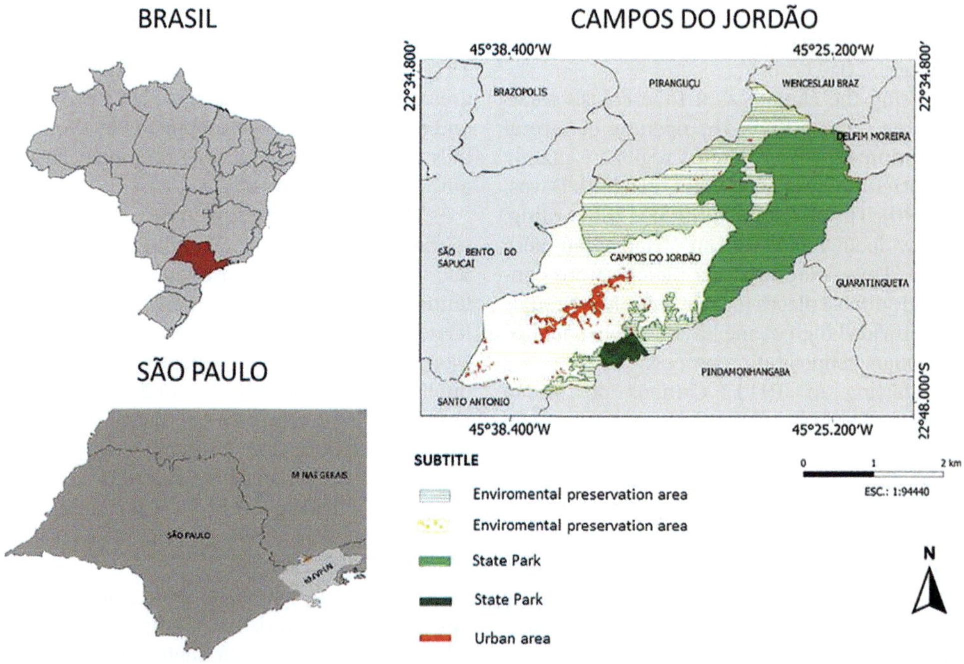

Fig. 20.1. The area of Campos do Jordão.

Paulo, the RMVale and its connection to the state of Minas Gerais and of São Paulo, and the city of Campos do Jordão, and its preservation areas, parks, and urban area.

The city of Campos do Jordão stands out for being located 1628m above sea level, in the area known as *Serra da Mantiqueira* (Mantiqueira Mountains). This is a mountainous area of extreme beauty and uniqueness, which required actions to preserve those natural spaces. Unlike much of the Brazilian territory, the investigated city attracts visitors searching for cold weather, nature and the feeling of traveling to an unusual place. Its image was built, over decades, from imagery based on glamour, sophistication and charm, which gave it the nickname 'Brazilian Switzerland'. Given its current slogan 'travel through the charms of the highest city in Brazil', it is considered to be the city inducing tourism in *Serra da Mantiqueira* Region. It welcomes more than 4.9 million visitors on a yearly basis, on average.

Campos do Jordão has complex features. Because tourism is the area's main productive activity, it represents an important scenario to help better understanding the development of tourist activity and its association with the built territory. Thus, the present study describes the city formation process and its transformation into a tourist site. It also explores aspects linked to its current tourism status, relationships with the community and challenges for territorial planning processes.

Historical Formation Process and Local Urban Development

Although Campos do Jordão's foundation dates back to 29 April 1874, its development is linked to the period of the tuberculosis pandemic (whose peak was between 1914 and 1940). Since there were no drugs to fight tuberculosis, attempted cures focused on the concept of climatic therapy. In this context, the city became a reference for climatotherapy treatments due to its cold and dry climate and its high solar radiation. This earned it the famous nickname of 'Brazilian Switzerland', by which it is still known today, since its climate is similar to that of Davos, Switzerland (Duarte *et al.*, 2007; Hammerl, 2011).

Using the climate as a tuberculosis treatment mechanism drove the opening of several sanatoriums in mountainous regions – mainly in Switzerland (e.g. Chamonix, in 1840; Davos, in 1890). This mechanism unified the 'resting, healthy food and clean air' triad (Hammerl, 2011). Thus, mountainous regions were considered suitable places to treat this disease, since they provided high ozone levels to help patients' pulmonary rehabilitation process.

Starting in 1911, Campos do Jordão received the first tuberculosis patients, who were stigmatized because this disease was a social challenge that required several health interventions. By 1913, the population was no longer made up only of people born in the town. Instead, for the most part, it consisted of people who had moved to the town in search of a cure for their disease. Houses, hotels and inns were mainly used to accommodate the sick. Because of this, the city became known as a health resort, both for its fresh air and for the radioactive water from the mountains (Academia, 1988; Hammerl, 2011).

Until 1970, when the treatment against tuberculosis became more efficient in Brazil, the process of urban formation in Campos do Jordão was characterized by healing activities. However, from 1940 onward leisure tourism was considered a likely factor that could guide local development. The construction of several hotels was promoted for this purpose by the state government and private groups interested in local tourism development. The growing number of jobs in construction, services and commerce attracted a large contingent of unskilled labor from other cities in the region. This helped to gradually transform the local area, but it was a disorderly occupation, as workers built their houses wherever they could find space, which favored the emergence of slums from 1940 onwards – the main sites being the slopes of the mountains in the central area of the city (Oliveira, 1991; Meneghini, 2018).

Despite rapid demographic growth and the expansion of commercial activities due to tourism, the first specific laws for local urban development did not emerge until the 1960s, focusing on the first master plan of 1962. According to Rosa Filho and Cortez (2010), despite the attempt to put things in order, from 1970 onwards there were large invasions of green and recreational spaces, land divisions, and private property developments in the local hills, whose occupation was promoted by the authorities themselves.

The city received a new incentive for tourism, which did not take into consideration its socio-spatial organization, namely: the launching of a highway, in 1978. This popularized the locality and generated strong real estate speculation. Consequently, it forced the public sector to make the urgent decision to stop approving land subdivisions, in 1979. Despite this attempt, a real estate boom took place in the mid-1980s, when thousands of houses and apartment buildings were built in Campos do Jordão (Academia, 1988; Oliveira, 1991; Silva, 2004; Abitante, 2016).

Due to its climatic, hydromineral and landscape characteristics, Campos do Jordão attracted people who left their usual place of residence for different reasons. Either they were looking for healing, a job or leisure. Whatever the reason, many people decided to make the city their new place of residence. Thus, from the very beginning, the studied city experienced a busy traffic of people, and the tourist activity was the factor that drove urbanization and local development processes. This shows the importance of this activity both for the city and the region.

Transforming the City Into a Tourist Site

Once the drug treatment for tuberculosis was discovered, it was expected that patients would no longer be sent to the city but would be treated in their place of residence. However, the city was already known for its ability to help treat disease, not just tuberculosis but any disease. A group of businessmen concerned about the unregulated growth of the city – which was leading to a deterioration in local health conditions due to the increasing number of patients and threatening the investments made to turn the city into a leisure destination – focused on improving the city's tourist image. The main objective was to transform the city´s image to focus on its natural beauty. This new development was encouraged when the state government became involved in the tourism sector. To achieve this, the process of guiding the use of space was begun, and specific areas were designated for the operation of hospitals and inns for the sick, as well as allotments to be used exclusively by tourists (Bertoli Filho, 2001).

A state decree defined the urban zoning process and divided the city into two specific zones: one for sanatoriums and treatment purposes and the other for tourism. The beginning of the city division process can be seen since the sanatoriums, which were built in donated places, were strategically located in the most popular areas. In comparison, the tourist area was reserved for the elite's summer houses and became a resting place, a fact linked to businessmen and politicians' interests and organizational processes (Hammerl, 2011; Prince, 2017).

Groups associated with the state and federal governments granted the region, mainly the areas where rich 'tourists' clustered, public and private benefits, such as installation of water and sewage networks, electrical grids and garbage collection. The public sector provided the mail service and undertook building a highway (Bertoli Filho, 2001). The urban organization of the tourist (and elite´s) area was based on European aesthetics. However, its development in the rest of the city, especially in the non-tourist areas, was generally done without planning. The improvisation on the basis of which the most popular houses were (and are) built has created a scenery typical of peripheral neighborhoods of large urban centers in the country. It has increased the illegal settlement of hillside areas that, according to environmental protection laws, should be covered with forests. This scenery does not meet tourist demand, but it characterizes the city (Fig. 20.2). The destruction of the place can be seen in the process of building Campos do Jordão, as the rather strong asymmetry between the working and tourist classes has contributed to the occupation of green areas by the working population for the

Fig. 20.2. Occupation of hillside areas.

construction of houses (Duarte *et al.*, 2007; Meneghini, 2018).

As the number of tourists increased, the city's landscape changed. As it has grown from a small city to a large one, it has faced problems such as the transformation of areas of no interest to the property market into slums. As flat land is highly valued and controlled, the poor population is confined to the hillsides, and the phenomenon of slums is very visible in Campos do Jordão, where buildings are prone to landslides, siltation, and sanitary problems. This situation is aggravated by the illusion of easy employability, as well as issues related to the seasonality of tourism activity and laws aimed at preserving the environment that end up restricting local improvements (Duarte *et al.*, 2007).

Despite these problems, the city has consolidated itself as a tourist option in the mountains, based on a European landscape, including typical architecture, and a cultural program different from that of other cities. Currently, music festivals with a diverse program are held in the summer (since 2021) and classical music festivals in the winter (since 1970). The latter is considered the largest classical music festival in Latin America. Both are organized by the São Paulo state government through the Secretariat of Culture and Creative Industries and by the Osesp Foundation in collaboration with the municipality of Campos do Jordão[1]. Tourists seek in this landscape not only the cool weather but also urban comfort and a variety of services (Fig. 20.3).

However, the logic driving the appropriation of spaces the social actors – in the current case, by tourists and dwellers – have quite different and, consequently, almost always conflicting natures. The difference between tourists and dwellers' goals drives the development strategies designed by the public and private sectors. Thus, Campos do Jordão City is featured by the contrast between wealth and poverty, since it strengthens the tourism economy without facing the social issues within a context of population growth

Fig. 20.3. European scenery.

in the midst of poverty (Cruz, 2003; Pinto and Ricci, 2006).

Overall, intense changes carried out in Campos do Jordão to turn it into a tourist destination were based on the political action of local landowners who implemented a socio-spatial segregation that reflected municipal occupation and spatial organization processes. The area known as Capivari, with its flat land, became a place occupied by tourists, making property speculation a challenge for residents who wanted to buy a house or land in the region, as financial values became prohibitive for the local population. The aforementioned area and other neighboring regions became privileged places, often occupied by tourists. Legal means were used to create a legislation to encourage tourists to buy land in these privileged areas, through tax exemptions or zoning. Local residents were left with areas on the slopes, without adequate study to build in places at risk of collapse – something common in areas where native trees were cut down to build popular houses, such as what occurred in the city of Campos do Jordão. It is also worth emphasizing the construction of an imagery to change the local image from a 'city of the sick' into a 'tourist city', based on using terms, such as 'Brazilian Switzerland', 'exuberant', 'untouched', and 'majestic' (Fig. 20.4).

Based on these investments, the city was acknowledged for its winter tourism, since it attracted tourists in June, July, and August. However, this tourism development process was not built based on clear guidelines focused on continuous and long-term planning. The city experienced a misdirected growth, which was concentrated in a single village (i.e. the tourist one). Although the local hotel structure and restaurant spaces have expanded, the city's urban and transport structure was neglected. This triggered conflicts during the tourist season, between tourists and the local community, especially with regard to the number of cars, congestion on the main roads, disregard for traffic rules, higher prices in supermarkets, and rubbish dumped all over the place.

Fig. 20.4. Planned tourist area.

Based on historical analysis applied to the city formation process, participation of the local population in decision making processes was neglected. Decisions about local tourism were made by the private (the city has a large number of external investors) and public (based on their own interests, they changed the guidelines every 4 years) sectors, that focused on maximum profitability. The economic impact was not analyzed beyond the number of tourists. The environmental impact was neglected, mainly due to the expansion of second homes. The social impact on the community was not considered as a point of reflection and the growing dissatisfaction of the local community with tourism and tourists was ignored. These analyses were postponed in order to develop local tourism in pursuit of profit. Tourism impacts on the city, be they positive or negative, were neglected.

Current Tourist Context

The city welcomed more than 4.9 million visitors in 2019 and more than 655,000 visitors in July this year alone. The number of visitors was 12 times higher than the number of permanent residents. Although the number of tourists has increased over the years, the annual growth rates of demand show fluctuations rather than a gradual increase. For example, a significant decrease in the pace of growth has been observed since 2017.

It is worth highlighting that while the number of tourists in the city is increasing, the profile of visitors has changed significantly. In the past, the city received elite tourists, based on the aspects presented previously. However, the current profile of tourists is that of day visitors with specific characteristics of mass tourism.

Nowadays, tourists stay in the city for approximately 2–3 days. Moreover, there is intense growth in the 1 day tourism market, which means that people do not have time to get to know the whole city, only its main points. Dissatisfaction with tourism and tourists, anxiety about the approaching high season, and irritation with traffic, are all mentioned by Andereck and Vogt (2000), and are already evident in the city, mainly in terms of traffic and overcrowding in certain places. These factors, in combination with low wages paid in the tourism sector, end up pushing the population away from its role as host. Another disruptive element lies in individuals with a second home, since they consider themselves hosts in the city. This factor shows a role-confusion process that directly affects the tourist-receiving and welcoming practice. In addition, the city's disorderly and unplanned growth led to the formation of several irregular neighborhoods that are susceptible to landslides at heavy rain times, as well as to heavy traffic and difficult access to places and services during the tourist season. Table 20.1 highlights the main differences between social actors.

Most of the hotel, restaurant, and tourist attraction businesses are concentrated in the tourist area. However, it is important to emphasize that this place is not visited by locals, as they consider it only a place for tourists. It is worth mentioning that tourists have to pass through other developed villages, where patients undergoing tuberculosis treatment were concentrated in the past, to reach the above-mentioned area. In the absence of spatial planning, these three villages are connected by a single road, which leads to heavy traffic during the tourist season, when it takes more than 90 min to travel distances that are at other times covered in 15 min. In addition to concentrated tourist flows, buses and delivery trucks contribute to the deterioration of local mobility. Given the stress caused by the time lost in traffic, retail workers are often treated poorly by tourists, which increases their dissatisfaction with the reception of visitors in their own city. Thus, it can be said that mass tourism causes inconveniences both for the local population and for the spatial planning process.

Despite the above issues, Campos do Jordão is the 10th most popular city in the country. It welcomes approximately 4.9 million tourists on a yearly basis (1.5 million of them visit the city in winter, alone) and wants to become one of the best tourist destinations in Brazil by increasing its tourist offering throughout the year. In order to achieve this goal, the Brazilian Tourism Engineering Agency (ABET – Agência Brasileira de Engenharia Turística) took accountability for the city's place branding process. This focused on defining its unique identity, based on elements seen as local strengths, such as its exuberant nature (symbolized by araucaria trees and the mountains), as well as on minimalist traits

Table 20.1. Urban dynamics in Campos do Jordão.

	Owner tourists	Dwellers
Financial status	High purchasing power (monthly income: clearly higher than that of dwellers)	Low purchasing power (monthly income is half of what is considered ideal[a]).
Risk areas	Some areas susceptible to flood events	3985 homes are located in risk areas (i.e. areas susceptible to landslides, floods, erosion, and subsidence)
Urban setting	Concentration of tourists in general regions showing internal homogeneity.	Concentration of dwellers in general regions showing less internal homogeneity. However, they inhabit the entire territory.
Sub-centers	They use all three centralities of the main road axis (Abernéssia, Jaguaribe, and Capivari Villages)	They use sub-centers close to their homes.
Accessibility	Enabled by their own vehicle	They have better accessibility due to the central location of their home.
Road system	Used by them	Used by them

Source: Abitante (2016).

[a]According to the Inter-Union Department of Statistics and Socio-economic Studies (Dieese), the ideal monthly income would be R$3.940,24 and the average income in Campos do Jordão is R$2.640.

that convey the style of the highest Brazilian city (1628 m above sea level) ABET, 2021.

Impacts on the Local Community

The so-called mass tourism developed in Campos do Jordão has distanced the local community from its role as host since it only reproduces the actions imposed by its employers. Despite showing satisfaction in receiving and coexisting with tourism and tourists, the fact that they always position themselves as workers, rather than as residents, emphasizes their distancing from the resident-host role. Losing the community's uniqueness of welcoming visitors in a competitive sector, such as tourism, leads to loss of competitiveness to other tourist destinations, and it requires taking intense actions to avoid inhospitality and hostility cases. Thus, one cannot refer to locals as resident hosts, as would be expected.

The sense of benefiting from tourism is of concern, as there is evidence that residents feel they do not derive personal benefit from this activity. This can be explained by the inequitable distribution of profits from tourism. The image of the city itself is criticized when it is portrayed as an ideal place for recreation and escape, which is not the city of the residents' daily lives. This factor leads residents not to recognize the space as theirs and use it for their pleasure, as the stigma of the tourists toward the residents' space is already established. Above all, the image of the city as a Brazilian Switzerland is considered outdated and not representative of the community. Therefore, it is necessary to reformulate it by taking advantage of what tourists are looking for, which is nature. However, the local community, which should play a key role in reformulating the city's image, does not know or use it. Actually, it acts as a service provider, rather than as a host.

Overall, residents emphasize the need of reorganizing the tourism activity in the city. The current tourism process leads to both chaos and disrespect. On the one hand, visitors do not respect its spaces and people. On the other hand, the city lacks proper structure to show visitors that they are welcome. This mass tourism maintenance process ignores that dwellers need to recognize themselves in that space since it belongs to them in the first place. Lack of such recognition breaks the link between dwellers and the city.

Residents do not show satisfaction with local urban policies associated with the main social functions of the city, such as transport, mobility, public security, social security, labor, health, and leisure. Most of the local population does not benefit from zoning and from access to public infrastructure in their places of residence. Moreover, low-income families are segregated in a process that creates imaginary and not so much imaginary borders between spaces designed for local dwellers and tourists.

The process to build a sustainable tourism activity must not only take into consideration attractions and infrastructure but also hosts (residents), since they will be the first ones to welcome the arriving tourists. However, given dwellers' distance from their role as hosts, which resulted from years of social alienation, previous relationships – influenced by mass tourism – left behind deep and negative 'scars', such as irritability and apprehension in the encounter between dwellers and tourists. This alienation process has isolated local dwellers in specific areas, as well as forced them to forbid themselves to visit certain spaces taken over by tourists and to distance themselves from the city. Therefore, before talking about welcoming, it is necessary to talk about the sense of belonging and the process of recovering the feeling that the city belongs to all. First of all, one must be able to answer the following question: What kind of tourism do we want for the city?

Challenges for Territorial Planning

The construction of a conceptual framework for the planning of cities and regions as territorial policies is still ongoing, and it is necessary to include the understanding of the concept of territory in the debates (Zanon, 2019). Therefore, in this case study, the term 'territorial planning' is used, taking into account the concept of territory in its integrative character, as referred to by Haesbaert (2004).[2]

In this sense, territorial planning is a process of formulating and implementing territorial planning policies that recognize the territory as a social space in its multiple dimensions and connections with the general context in which the place is embedded. Therefore, it is necessary to understand that the transformation of the socio-spatial reality of the territory is part of the historical and contemporary processes and the political and social struggles that have taken place in the territory, and from this to define cross-sectoral and articulated strategic actions that actually promote positive changes in favor of local and/or regional development.

Although Campos do Jordão has several attributes associated with its physical features, many of them represent real challenges for its territorial management process. Its rugged and sloped topography makes planning processes, as well as interventions to expand the road network and to implement affordable housing, hard to be accomplished, due to increased costs associated with these procedures. In addition, the lack of effective public policies focused on territorial organization and tourism development for more than 40 years has, and still has, several consequences for the local territory.

The socio-economic profile of the local population (approximately 50,000 inhabitants) and the socio-spatial distribution of the city demand special attention from the local public administration. More than 20% of the population living in Campos do Jordão in 2021 comprised low-income individuals. Many of them live in hillside areas with poor urban structure, which are at imminent risk of accidents. The high social and economic vulnerability experienced by the local population hinders the city's social development. Educational attainment is also a problem, with most of the population having only completed secondary education, which is compulsory in Brazil, by the age of 17. The lack of participation in professional training courses, at technical and higher levels, limits social participation, and access to new opportunities.

Lack of integration between public and private authorities, and the population makes it hard to promote tourism and to develop effective actions due to individualistic thinking based on personal interests. The municipal public administration has made efforts to draw up plans and specific public policies that did not exist until recently (dealing with culture, tourism, and urban mobility, for example), a fact that shows progress in this regard. However, few of them address the tourist issue. This hinders the application of public policies based on a territorial approach. Furthermore, lack of financial resources is the main factor stopping the city from complying with guidelines proposed in the aforementioned plans.

With respect to territorial planning and management, different administration levels (municipal, state, and federal) require constant dialogue between stakeholders to achieve integrated, shared, and decentralized territorial management. The interrelation among the aforementioned spatial spheres of public action, and their respective institutional capacities, can directly affect the local and regional development process, either in a positive or negative manner, depending on the articulation in place.

Thus, tourism can be used to promote local development, based on the elaboration of public policies, as well as of integrated and participatory planning policies that, above all, take into consideration the needs of the local population, rather than just focusing on increasing the city's financial resources. This integration must involve the urban and tourist public management systems, the process to strengthen relationships with the private sector, and effective social participation in planning and management processes.

Discussion and Conclusions

Tourism can provide several opportunities in places where it is implemented. However, it is necessary to fairly combine collective interests, as well as ensure benefits to all individuals involved in this process for it to happen. Campos do Jordão grew without the implementation of effective planning actions aimed at prioritizing permanent dwellers' interests and needs. Urban public policies have favored the interests of the real estate and tourism markets, to the detriment of the local community's welfare.

The loss of local control by the population is usually accompanied by unbridled growth of institutionalized tourism and resentment, as can be observed in cities such as Barcelona, Venice, Rome, Berlin, Amsterdam, and Paris. This study highlights a series of problems common to places characterized by massive tourism development without planning: spatial segregation, low-paying jobs, occupation of natural spaces by tourism/tourists without planning, disrespect for the place visited, and lack of territorial organization, all leading to dissatisfaction of the local population with tourism and tourists.

Based on the understanding that local development occurs when social change helps improve the quality of life for local people, change means planning for tourism, addressing the impact of this sector on local society, recognizing and understanding both the positive and negative impacts of this activity, developing partnerships with the private sector to encourage people to recognize the city as a place that can be shared and experienced by all, and taking advantage of and improving upon the actions currently undertaken by the city. However, it is a difficult task that cannot be carried out in isolation. It is necessary to involve different sectors so that planning can be carried out and actually practiced, as well as to monitor situations that may force changes along the path laid out. Overall, a process of rediscovery of the city is advocated, which allows the local community to look at everyday life from a new perspective, characterized by curiosity and interest.

Some interesting actions can already be seen in Rio de Janeiro and Porto Alegre. Costa, 2015 provides several references on how abandoned industrial spaces can be transformed into creative parks or habitats with residential and commercial spaces, following the example of New York with its Meatpacking District project. In Rio de Janeiro (state of Rio de Janeiro), the Porto Maravilha project in the Santo Cristo neighborhood, which aims to revitalize the port area, stands out. Fábrica Bhering, which closed in 1990, is rising to the challenge, transforming its six-story building into a rental space for film and soap opera shoots, as well as renting space to visual artists and for various events. In Porto Alegre (state of Rio Grande do Sul), the associations of residents and the private sector have enabled a transformation of the so-called fourth district, characterized by its industrial areas. Three private initiatives related to the creative industries and tourism emerged, giving rise to galleries, hostels and shopping centers, as well as the Refloresta project, which aims to integrate activities and people into a sense of community. Also worth mentioning is Project Recentro, a program to revitalize the city center of Recife (state of Pernambuco), which provides tax incentives for those who invest in the center, as well as cultural, janitorial and maintenance activities. Another highlight is the revitalization of the historic center of Aracajú (Sergipe state), which initially aimed to strengthen the traditional commercial activities of the place and later to incorporate tourism. In addition, through private partnerships, new functions were created for the different buildings of the town, in order to expand the leisure and cultural offer for both residents and tourists.

Considering these experiences already made in Brazil, there is a need to re-evaluate the use of certain spaces, to involve the community in the use of commercial spaces and to expand the offer of recreational and cultural spaces, especially for residents. The above examples show that without partnerships between public and private initiatives, nothing will be possible. There are several places in Campos do Jordão that could undergo a review of the use of space to reconnect residents with the place while creating new opportunities for tourists. For example, the old Santa Casa hospital is a place where many residents were born and is now a ruin. It would be a great place for a cafe, cultural area, and restaurant, among other options.

Despite the cited examples, one cannot ignore the point that there are few Brazilian cities that have an integrated vision for tourism planning. This vision needs to be expanded, updated, and integrated into public management practices. For example, tourist cities such as Curitiba (state of Paraná), João Pessoa (state of Paraíba), and Santos (state of São Paulo) are developing territorial planning policies based on the Sustainable Development Goals (SDGs), to ensure the social and economic development of the community and increase the quality of life of its inhabitants. In addition, the planning of these cities is driven by the intention to promote tourism through integrated cross-sectoral public policies.

Furthermore, the analysis of the tourism development process and the consequences of its activities, through the evaluation of its impact on the urban space, natural environment, and population, based on the perception of local residents, allow the identification of potentials, bottlenecks and changes promoted both in the analyzed place and its surroundings. It is necessary to develop more effective actions focused on community involvement and participation in the local tourism organization process. The local government and other entities involved in this process must also understand the importance of planning actions to be implemented in the short, medium, and long term, based on clear guidelines that do not depend only on policies. However, it should be emphasized that any decision making process must include an educational perspective in its actions, in order to raise awareness among the local population so that they understand that they are social actors responsible for the construction of the territory and the importance of participation in the local tourism planning process.

This process of raising awareness can be promoted through projects – whether formal or informal – that focus above all on tourism and its impacts, creating contexts that allow for greater participation by local businesses and the local population. In this way, a structure can be developed that provides recognition and appreciation of local work with actions that develop a sense of pride in working in a particular place, that instill a sense of belonging, recognizing the city as their own, that spread co-responsibility for local development and that encourage the welcoming of diverse visitors to the territory.

Tourism in Campos do Jordão is regional in nature, benefiting from its proximity to major urban centers. Therefore, further research should be conducted to gather information that can inform the tourism planning process – mainly basic information about tourist arrivals, number of nights, type of accommodation, expectations, and average spend. Tourism analysis should be linked to local public policies and the real estate sector to provide a comprehensive mapping of the city and its policies. Tourism observatories are an excellent option for this information organization. These groups are research institutions whose goal is to provide a solid information base for monitoring local tourism activities. In Brazil, there is the Brazilian Network of Tourism Observatories (Rbot), which, through a network of contacts, tries to support the different observatories in their organizational process. It is worth learning about the work of the Observatories of Goiânia, São Luis do Maranhão, and Minas Gerais, as they conduct annual surveys monitoring trends and following up on changes in employment, services provided, tourist arrivals, and visitor profiles. This helps facilitate (re) planning of tourism actions, developing partnerships with the public and private sectors and highlighting the necessary coordination needed for tourism to maintain its continuous development. Through the exchange of experiences with these observatories, the practice of continuous monitoring can be adopted by other places and institutions, as in the case of Campos do Jordão.

Campos do Jordão's territory management model should be revised, by taking into consideration more sustainable tourism development practices, based on the articulation of urban and tourism planning actions. Furthermore, the medium- and long-term impacts of tourism should be systematically and constantly predicted and monitored to enable reorganizing of the tourist activity development process whenever necessary.

Notes

[1] For more information, you may access the following websites: about the festivals, Available at: http://www.festivalcamposdojordao.org.br/o-festival/, and about Osesp Foundation, Available at: http://www.fundacao-osesp.art.br/home.aspx (accessed 17 November 2023).

[2] The territory is considered to be a 'hybrid space between society and nature, between politics, economy and culture, and between materiality and 'ideality', in a complex time-space interaction' (Haesbaert, 2004, p. 79).

Further Reading

Mendes, B.C. (2021) Hospitality in touristic cities: belonging and welcoming by the host-resident of campos do jordão (SP). Phd (Doctorate in Hospitality) – Pos Graduate Program in Hospitality, Anhembi Morumbi University, São Paulo (SP).

Panza, T.M. (2022) Tourism and public policies: analysis of their relations and influences in the territory of campos do jordão (SP). Phd (Doctorate in Urban and Regional Planning) – Pos Graduate Program in Urban and Regional Planning, Vale do Paraíba University, São José dos Campos (SP).

References

ABET (2021) COMTUR aprova estratégia de branding e nova marca turística de Campos do Jordão. Available at: http://www.abet.tur.br/comtur-aprova-estrategia-de-branding-e-nova-marca-turistica-de-campos-do-jordao/?fbclid=IwAR1N6KcsESE3Bpj5OcCbKWlaH4NHEblmCyNbkmyYPWoGMToB7_PzTnpHfxI (accessed 30 September 2021).

Abitante, J.C. (2016) *Segregação espacial e acessibilidade: uma construção para o desenvolvimento sustentável em Campos do Jordão – SP*. Dissertation (Master in Development, Technologies and Society)– Pos Graduation Program Development. Federal University of Itajubá.

Academia (1988) Academia de Letras de Campos do Jordão. *Imagens Jordanenses*: iconografia histórica de Campos do Jordão. Campos do Jordão (SP).

Andereck, K.L. and Vogt, C.A. (2000) The relationship between residents' attitudes toward tourism and tourism development options. *Journal of Travel Research* 39, 27–36.

Bertoli Filho, C. (2001) *História social da tuberculose e do tuberculoso: 1900-1950*. Fiocruz, Rio de Janeiro.

Costa, G. da (2015) Reconexão de espaços degradados à cidade por meio da reconversão de uso de vazios industriais: o caso do IV Distrito de Porto Alegre. Dissertation (Master in Arquitecture) – Pos Graduation Program in Arquitecture. Federal University of Rio Grande do Sul (UFRGS).

Cruz, R. de C.A. (2003) Introdução à geografia do turismo, 2nd edn. São Paulo, Rocca.

Duarte, R.F., Barbosa, A.S. and Bruna, G.C. (2007) O turismo e a transformação de cidades: O caso de campos do jordão – brasil e sintra-portugal. *Revista AE Ensaios* II (1), 01–26. Available at: http://

www.aedificandi.com.br/aedificandi/N%FAmero%203/3_SINTRA%20CAMPOS%20DO%20JORD%C3O.pdf (accessed 5 March 2021).

Haesbaert, R. (2004) *O Mito da desterritorialização: do 'fim dos territórios' à multiterritorialidade*. Bertrand Brasil, Rio de Janeiro.

Hammerl, P.C. (2011) *Campos do Jordão (SP): de estância de saúde à estância turística*. Dissertation (Master in Hospitality). Anhembi Morumbi University.

Mendes, B.C. (2021) Hospitalidade em cidades turísticas: pertencimento e acolhimento do anfitrião-residente de campos do jordão (SP). Thesis (Doctorate Degree in Hospitality) – Anhembi Morumbi University, São Paulo, Brazil.

Meneghini, R. (2018) *Turismo e a busca por um pedaço de chão em Campos do Jordão – SP*. Novas edições acadêmicas, Saarbrucken (Germany).

Oliveira, J.O.S. de (1991) *A quem interessa a urbanização clandestina? Estudos sobre Campos do Jordão*. Dissertation (Master in Arquitectute and Urbanism) – São Paulo University.

Panza, T.M. (2022) Turismo e políticas Públicas: análise de suas relações e influências no território de Campos do Jordão (SP). Thesis (Doctorate Degree in Urban and Regional PLanning) – Vale do Paraíba University, São José dos Campos, Brazil.

Pinto, L.A.S. and Ricci, F. (2006) Qualificação profissional e turismo: uma discussão sobre as questões sociais e o espaço turístico construído na cidade de Campos do Jordão. In: *IV Seminário de Pesquisa Em Turismo Do Mercosul*. Caixas do Sul.

Prince, A.E. (2017) *O Estado de SãO Paulo e a luta contra a tuberculose no Século XIX e meados do Século XX*. Editora e Livraria Cabral Universitária, Taubaté, SP.

Rosa Filho, A. and Cortez, A.T.C. (2010) A problemática socioambiental da ocupação urbana em áreas de risco de deslizamento da 'Suíça Brasileira'. *Revista Brasileira de Geografia Física* 3, 33–40.

Silva, M.G.L. (2004) *Cidades turísticas*: identidades e cenários de lazer. Aleph, São Paulo.

Zanon, E.R. (2019) Planejamento Territorial: Reflexões sobre uma cultura em construção no Brasil. *ENANPUR*. Natal.

21 Placemaking Perspective and Determinants of Tourism: A Case Study in a Creative Design City in Brazil

Elias José Mediotte[1], Magnus Luiz Emmendoerfer[1] and Thiago Duarte Pimentel[2]

[1]Universidade Federal de Viçosa, Viçosa, Minas Gerais, Brazil; [2]Federal University of Juiz de Fora, Juiz de Fora, Minas Gerais, Brazil

Summary

This case study delves into the realms of creativity, exploring the crossroads of digital-era territorial redefinition, urban marketing, and tourism planning. Its objective is to comprehend how placemaking takes shape in a Brazilian Creative City, focusing on tourism dynamics, potentials, and limits for redeveloping urban spaces to promote new social interactions and economic opportunities. Employing a qualitative methodology, we address the study site's nuanced social, cultural, and historical underpinnings. Carried out in Brasília (DF) in 2022, a UNESCO Creative City of Design since 2017, the study reveals that placemaking requires tourism elements to comprehend resignified places, despite the intricate revitalization of spaces through heritage listings endorsed by local urban design. Additionally, we highlight the vital link between resources, meanings, and creativity in tourism recovery, notably in the wake of the COVID-19 pandemic.

The Value and Interest of the Case Study

This study contributes by framing and applying placemaking guidelines in analysing tourism for local tourism management and development in an empirical case, as well as developing analytical tools for that purpose. It also highlights the significance of placemaking knowledge in shaping public policies and initiating projects that distinguish areas, like creative cities, in terms of revitalization and resignification of places. This highlights the potential of placemaking as a practical tool for managing tourism.

Background

Tourism is a deliberately planned and market-driven endeavour primarily focused on economic goals. Consequently, even in challenging circumstances, it can offer a feasible remedy to promote territorial development (Fayos-Solà *et al.*, 2014).

Destinations, especially those with a tourist focus, go beyond being just "products," "goods," or "business ventures." They hold a deeper significance – they embody the essence of places where "people live, work, and celebrate. If we truly aim to foster sustainable communities, it is essential to recognize them as intricate tapestries woven from complex relationships and interwoven networks" (Hall, 2004, p. 216). This perspective enables them to create opportunities and reap benefits for the societies in which these activities flourish, primarily driven by creativity.

Cultural dynamics have changed significantly since deindustrialization, stemming from the capitalist crisis in the 1970s. The former dominance of spectacle and imitation in daily life (Ferreira, 2017) has yielded to a fresh socio-economic landscape propelled by creativity. According to Landry (2013), creativity permeates life, acting as a resource with implications and applications in thought, action, and pursuit of economic and socio-cultural outcomes.

In this context, the focus is on making life better and adding more fun to leisure activities. The effort to reshape cities' images by merging business with popular celebrations, coupled with the global collaboration among leaders, has captured researchers' attention, prompting them to consider transformative shifts in the realm of tourism (Richards and Duif, 2018).

Concerning this matter, the United Nations Educational, Scientific and Cultural Organization (UNESCO) launched an international policy named "Creative Cities" in 2004. This policy categorizes cities as creative zones, connecting them within the structured UNESCO Creative Cities Network (UCCN). The network encompasses seven creative sectors: crafts and folk art, design, film, gastronomy, literature, media arts, and music (UNESCO, 2017), each providing fertile ground for the growth of the creative economy.

Creative cities go beyond formulating spaces for efficient governance, with the goal of promoting development, including inclusive, safe, resilient, and sustainable tourism. They also tap into innovation, culture, connectivity, tourism, and urban transformation to encourage the spread of skills and abilities. Their uniqueness, rooted in their originality, drives the creation of entrepreneurial capital (Emmendoerfer, 2019). Additionally, these cities harness the potential of the creative economy to shape goods and services from both tangible and intangible aspects, breathing vitality into specific locations within their territories.

Moreover, the evolving significance of socially constructed spaces has promoted and nurtured a redefinition of territories, surpassing physical boundaries and embracing inclusiveness and creativity. Amid this transformative shift, placemaking emerged as a key tool or approach. In essence, it involves reinvigorating, co-creating, and co-producing spaces and locales, catalysing the rebirth of cultural customs through imaginative interventions (Richards and Duif, 2018). Through placemaking, local development can be propelled, particularly by revitalizing public spaces, boosting their strength, inclusiveness, security, and sustainability (Mediotte *et al.*, 2022).

William H. Whyte coined the term placemaking in the 1970s, with the Project for Public Spaces (PPS). As they delve into the concept, Heemann and Santiago (2016, p. 11) explain that placemaking "involves observing, listening, and asking questions to the people who live, work, and visit a particular space to discover their needs and aspirations." They propose that successful placemaking in city areas, whether public or private, encompasses:

- spaces for cultural festivities, connecting people;
- hubs of active economic and social interactions;
- gathering places for diverse individuals of various ages, genders, and cultures; and
- stages for public life, facilitating shared experiences and participation in daily life.

Given this context, we view placemaking as more than just building spaces. It is a vital tool for imaginative urban and territorial planning. It acts as a tourist framework that lets us explore the city, considering many dimensions.

Placemaking: Guidelines for Contextual Analysis of Tourism

The concept of placemaking goes beyond city size. We firmly believe that any city can harness its potential to foster an identity rooted in local culture. The goal is not only superficial marketing

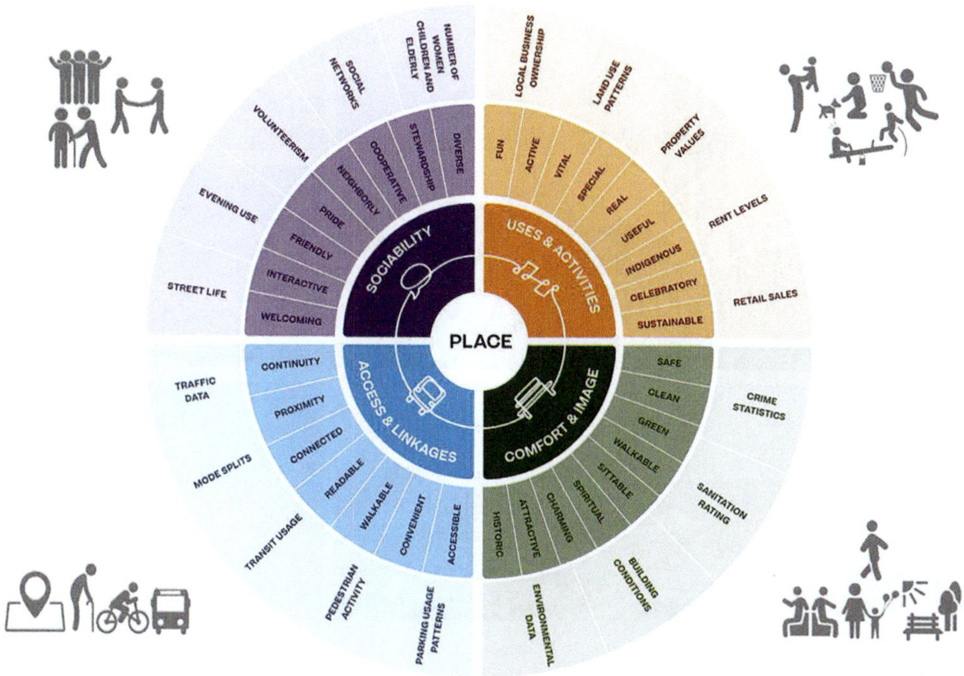

Fig. 21.1. Placemaking guidelines. Source: Adapted from Ferreira (2017) and Vasconcelos (2017).

or branding but also fostering inclusivity and creativity. This underscores the importance of events that showcase a city's cultural identity through value building, attracting and keeping talents, and consolidating reinforcements to shape the city's desirable tangible and intangible resources (Zitcer, 2018).

In this regard, creating cultural hubs within a creative cities network could open doors for a new tourism model that promotes economic and socio-cultural growth. It is important to emphasize that local residents are a key element in upholding economic cycles, especially in countering gentrification's effects, as highlighted by Harvey (2006).

Thus, the fundamental elements that define cities as successful placemaking spaces can be outlined by the indicators shown in Fig. 21.1. These indicators function as tools to navigate the changing dynamics, swift market transformations, and the impact of new tech, as well as the relationships between residents, tourists, and governance structures (Heemann and Santiago, 2016) that unfold within cities.

Based on Fig. 21.1, PPS highlights in its guidelines that most spots, such as a town square in the city centre, a small neighbourhood park, or even a street, have the potential to apply placemaking by utilizing its core indicators:

- *Accessibility and connectivity*: They are simple to reach and conveniently linked to other important nearby places.
- *Comfort and image*: They provide comfort and project a positive image.
- *Activities and uses*: They attract people to engage in the activities held in the area.
- *Sociability*: They are sociable environments where people naturally gather and visit frequently (PPS, 2016, p. 4).

In our approach, we take inspiration from the research conducted by Mediotte *et al.* (2022). They offer a clear structural framework for implementing placemaking in both public and private spaces.

Fig. 21.2. Brasília – Creative city of design.

This involves collaboration between specialists such as managers, public agents, academics, and non-specialists, including members from the private sector and civil society. Thus, following their placemaking implementation framework (Mediotte *et al*., 2022), we selected Brasília as our case study focus. This city, also known as the Federal District and the capital of Brazil, earned the UNESCO Creative City title for Design in 2017.

Case Study: Brasília as UNESCO Creative City of Design

The construction of Brasília, intended as the administrative centre or capital of Brazil, began in 1955 and lasted until 1960, covering planning, design, and construction (IBGE, 2022). The city is renowned for its monumentality, where various scales merge to forge an innovative landscape heritage, anchored by Lucio Costa's urban plan and Oscar Niemeyer's architecture (IPHAN, 2022).

Brasília is turning to tourism to boost its creative economy, responding to 21st-century global demands. In 1987, UNESCO declared it a world heritage site for its historical, cultural, natural, and urban importance (IPHAN, 2022). Additionally, the city achieved the highest Ministry of Tourism (MTur) recognition (concept A) via the Brazilian Tourism Map: Categorisation of Municipalities in Touristic Regions (BRASIL, 2022). In this regard, thanks to its architectural emphasis on innovative urban design, Brasília was named a UNESCO Creative City of Design in 2017 (Fig. 21.2).

The data collection for this case study took place in 2022, specifically during July and August. The research employed a combination of observation, fieldwork, and in-depth interviews with key figures representing the city's governance. These individuals play a pivotal role in securing and preserving the UNESCO Creative City title within the UCCN. The interviewees were labelled as research participants (RPs) and the interviews followed ethical guidelines according to CNS No. 01/2013. Notably, all the entities and institutions represented by the RPs are based in Brasília, Distrito Federal.

To select and reach out to interviewees (research participants), we employed a mapping strategy based on the city's UNESCO Creative City of Design application dossier.

Our initial goal was to identify the main actors leading governance and the "Brasília, UNESCO Creative City of Design" label. We reached out via e-mail or available commercial phone lines, leading us to the primary figures within Brasília (DF). Thanks to the network governance aspect, we assumed robust communication ties among the label representatives. When meeting an interlocutor, we asked for contacts of other pre-selected individuals, using a snowball technique. Beyond local governance, regional and national-level actors were also identified, detailed in the research participant description:

- RP1 – UNESCO Representative in Brazil (National level)
- RP2 – Representative of the Ministry of Tourism (MTur) Brazil (National level)
- RP3 – Representative of the Business Chambers at the Federation of Commerce of Goods, Services, and Tourism of the Federal District (Fecomércio) DF (Regional level)
- RP4 – Representative of the State Secretariat of Tourism (SETUR) DF (Regional/Local Level)
- RP5 – Civil Society Representative of Brasilia Creative City of Design (Local level)
- RP6 – Artisan and Independent Market Vendor (Local level)

Results

We applied Mediotte *et al.* (2022) placemaking framework due to its relevance to tourism within a creative city, especially concerning design. Additionally, we used a placemaking checklist, developed by the case study authors, who are research members of the UNESCO Chair in Creative Economy & Public Policies (catedraunesco.org). This checklist, shown in Fig. 21.3, served as our conceptual framework for this case study.

We assessed the placemaking checklist in Brasília, focusing on the prime tourist attractions highlighted by the RPs and observed on-site by one of the authors of this case study. Our attention was drawn to places frequently visited by tourists, Instagram-worthy spaces, public squares, and events. The goal was to witness instances that captured the essence of the UNESCO Design endorsement.

Tourist Complex 01

- Metropolitan Cathedral Nossa Senhora Aparecida
- National Museum of Brasilia
- National Library
- Claudio Santoro National Theatre
- Brasília Conjunto Nacional Shopping Mall Plaza

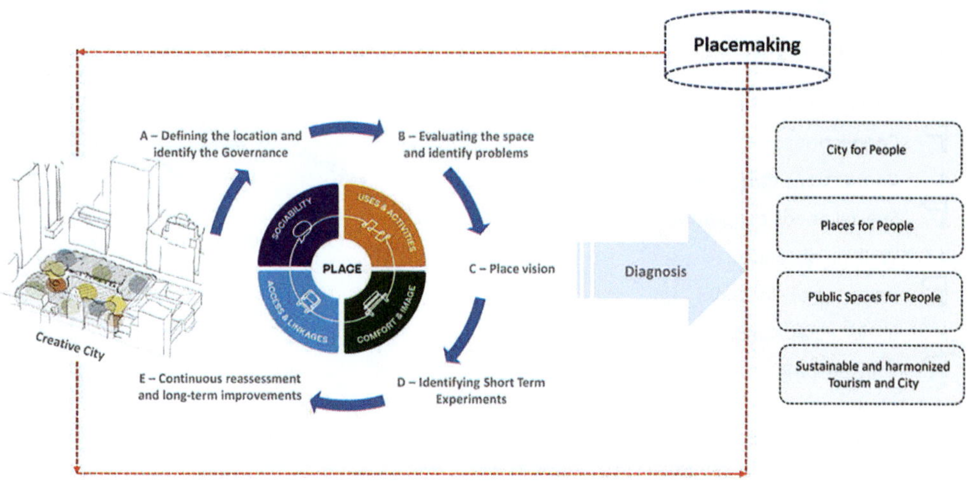

Fig. 21.3. Placemaking roadmap for local tourism management and development. Source: Own elaboration based on Fayos-Solà *et al.* (2014).

Tourist Complex 02

- Esplanade of Ministries
- National Congress
- Itamaraty Palace
- Three Powers Plaza

Tourist Complex 03

- Brasilia TV Tower
- TV Tower Craft Fair
- City Park Plaza

Tourist Complex 04

- Church of Nossa Senhora de Fátima
- Plaza in front of Renato Russo Space

Tourist Complex 05

- Instagrammable – Graffiti on the old Cine Cultura
- Instagrammable – Infinu Space
- Instagrammable – Renato Russo Space

We sorted the four indicators into specific blocks on the checklist, followed by key questions. Each section presents options: "Yes" or "No. The placemaking checklist consists of 58 questions that assisted us in conducting this case study.

- Block 1 – Accessibility and connections of the public space.
- Block 2 – Activities and uses of the public space.
- Block 3 – Comfort and image of the public space.
- Block 4 – Sociability of the public space.

We consider tourism complexes to be visited by individuals of varying ages, physical abilities, and financial circumstances. People interact and participate in these areas for tourism or community events. It is important for these spaces to be welcoming to all, catering to different requirements and encouraging social interactions. Here, we provide the finalized checklist from our on-site research, commencing with the Accessibility and Connections block. See Figs 21.4 and 21.5.

We focused on synthesizing the responses that obtained the most significant "YES" and "NO" applications for this case study. However, the tables relating to each block with questions used are available in the ZENODO repository (Available at: https://doi.org/10.5281/zenodo.8383576, accessed 22 May 2024) for complete consideration, which can be used for future applications, including elsewhere.

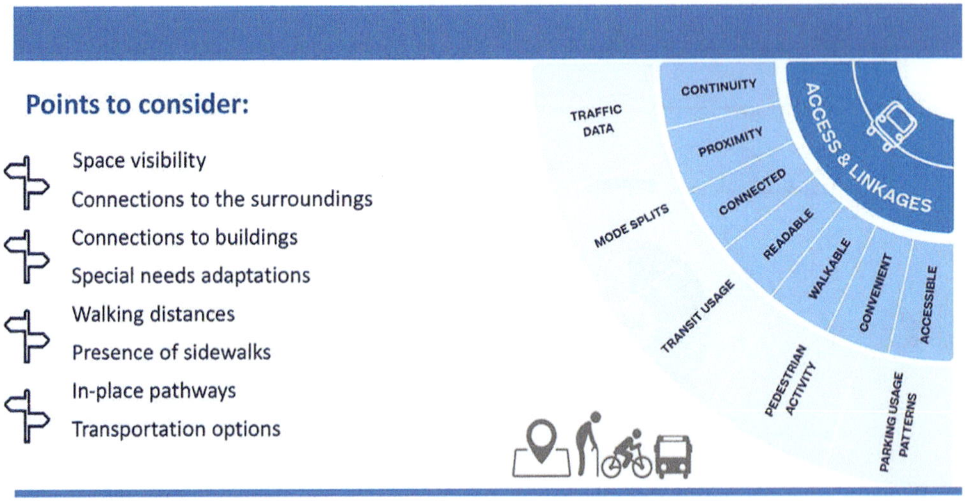

Fig. 21.4. Block 1. Accessibility and connections. Source: Own elaboration, based on PPS (2018) and Vasconcelos (2017).

Fig. 21.5. Accessibility and connections of the spaces designated for the study.

In Block 1, Accessibility and Connections, we recorded that the most critical and evident items relating to complexes 01, 02, 03, 04, and 05, with a concentration of "No" responses, were issues like use of space by occupants or residents of adjacent buildings; the presence of shops and commercial services within a short distance from the space, and transportation options (bus, train, subway, car, and bicycle) to reach the space.

Urban mobility stands out as one of the key aspects that demands special attention when it comes to placemaking implementation.

> I just miss a lot a more aesthetic look at the urban development of the city [...] it is much more than a City of Design; it is a Smart City, where the **transport service works, and it is the Design that provides all that functionality** [emphasis added] that exists within the transport service of Curitiba [Creative City of Design in Brazil, since 2014], which is an example to the world, right? Model from Brazil to the world. This is Design, and I think Brasilia does not seize it very well (RP5).

Another noteworthy concern is the limited involvement of local residents in the observed areas. According to some RPs, this lack of interaction between residents and the city is attributed to cultural factors. None the less, Brasília has taken steps, particularly amid the COVID-19 pandemic, to motivate residents to partake in the city's offerings. Despite these efforts, the prevailing sentiment is that residents appropriate from the urban space occurs in a voluntary and spontaneous way, not realizing all the work undertaken by the authorities. This suggests that even more efforts need to be made in order to improve and create facilities for the residents and tourists, fostering a sense of collective involvement rather than individual experiences of urban space appropriation.

> Today, in Brasília, if you visit the TV Tower on the weekend or our natural spaces, you will see that they are being increasingly used as tourist attractions. This is a result of the government's efforts to diversify the offerings, to move away from the administrative and political aspects...

and to show that Brasília is also about nature, culture, architecture, and even rural areas. **We have worked hard to change the residents' perception of their own city. However, there is still much room for improvement** [emphasis added] (RP4).

Next, we focus on the activities and uses block of the spaces designated for the study. See Figs 21.6 and 21.7.

In Block 2, activities and uses, we recorded that the most critical and evident items relating to complexes 01, 02, 03, 04, and 05, with a

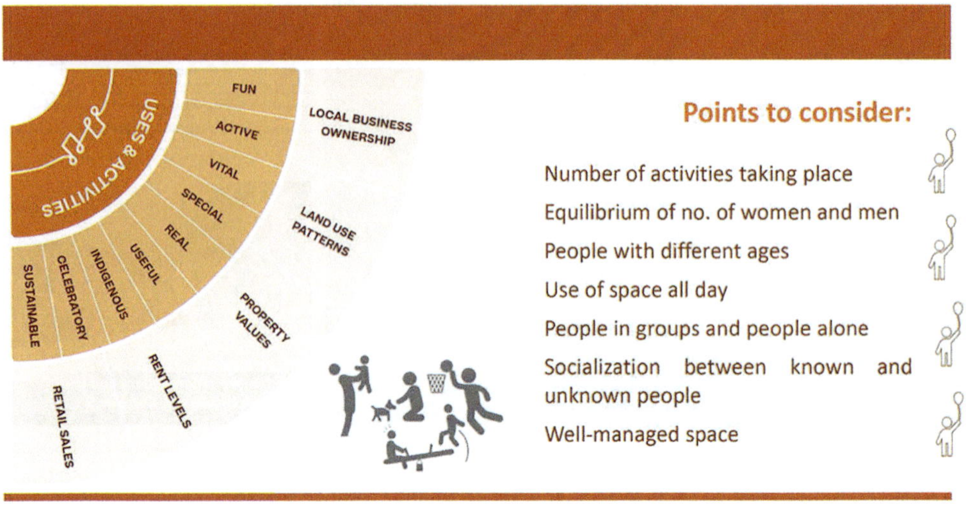

Fig. 21.6. Block 2. Activities and uses. Source: Own elaboration, based on PPS (2018) and Vasconcelos (2017).

Fig. 21.7. Activities and uses of the spaces designated for the study.

concentration of "No" responses, were issues about the use of the space by people (including your diversity) in all parts of this space, with different activities taking place such as walking, eating, engaging in sports, relaxing, and reading.

We view this indicator as potentially the most debated. The *use of spaces* and their activities seem irregular and do not fully encapsulate the essence of sites for immersive or innovative tourism experiences.

> This pandemic came as a punch in the stomach for everyone. Moreover, it was precisely through creativity that the best projects, the best ways of new coexistence came about... of belonging to a territory again. [...] I do not even like to say 'legacy of the pandemic,' which is complicated, but it really was the use of public areas. So, in Brasília, we did not have that habit, even though we live with the Cerrado[1]. We have a highly rich Cerrado area, but people did not use it (RP5).

Undoubtedly, one of the factors is tied to the city's unique character in preserving its *architectural heritage*. As a result, many of the city's complexes are primarily meant for appreciation and contemplation.

> People from Brasília, who have a high consumption pattern and historically did not even look around and would travel abroad, well... they had never paid attention to their surroundings before ... [now] I believe that, maybe not exactly, but with the pandemic, this might also be due to a decline in the [socio-economic] standard ... they started to look around and today there are a series of experiences offered. (RP1)

Several sites hold significant potential for collective, inclusive, and resilient use. However, the prices of goods and services, which do not always reflect the design label, particularly in sectors like gastronomy and crafts, may not be affordable for all audiences, revealing a *segregating* trend within the city.

> I cannot really answer that [your question on if *the prices of the products and services on offer are affordable for locals and tourists alike*] because I think it is a bit more personal. Unfortunately, or fortunately, compared to Brazil, Brasília has a higher per capita income, and a higher monthly income, compared to other states, right? So, it is difficult to compare because of this issue of disparities. Salary disparities. So, **some [products and services] will not be affordable, some will be affordable for tourists** [emphasis added] (RP1).

While the RP highlighted a few instances, the practical outlook of perceiving these places mainly as consumption spots signifies a form of exclusive tourism.

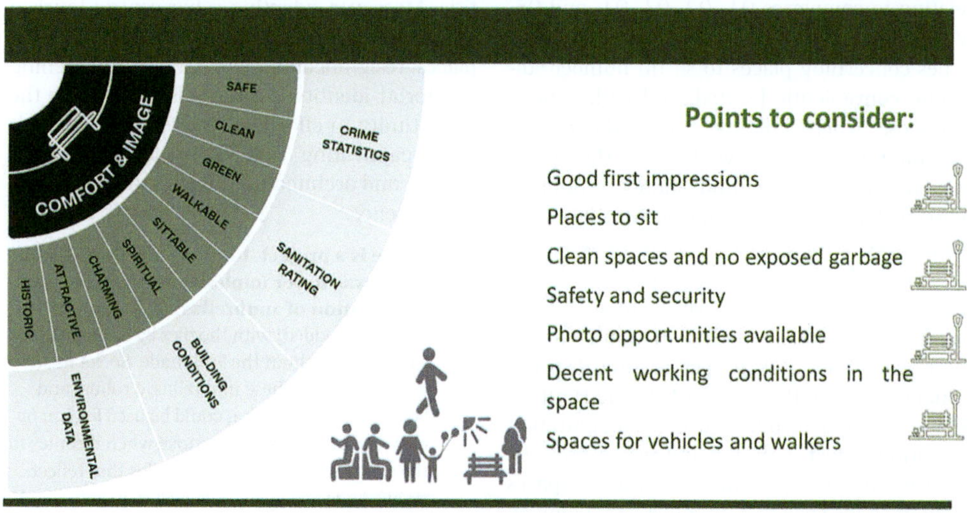

Fig. 21.8. Block 3. Comfortability and image. Source: Own elaboration, based on PPS (2018) and Vasconcelos (2017).

Fig. 21.9. Comfort and image of the spaces designated for the study.

Next, we address the comfort and image block of the spaces designated for the study. See Figs 21.8 and 21.9.

In Block 3, Comfortability and Image, we recorded that the most critical and evident items relating to complexes 01, 02, 03, 04, and 05, with a concentration of "No" responses, were issues concerning places to sit (in number sufficient, conveniently located, and with options to choose, whether in the sun or the shade), dominant presence of vehicles in the space, the offering products (and shops or outdoor markets) have consistent images that align with the identity of the space.

This indicator shows a creative city's tourist effectiveness hinges on two key factors. The first factor is the limited seating in tourist areas. Many complexes lack proper seating distribution, often due to their UNESCO World Heritage Site status and resulting urban configuration limitations. Even with ongoing urban projects, tourists frequently navigate these spaces without shaded benches or convenient points to buy water or find protection during rain or other weather conditions.

Furthermore, RP3 points out that tourists are more attracted to Brasília's bars than outdoor beauty. This viewpoint contradicts the core values of placemaking, a concern amplified by the city's UNESCO Creative City of Design title. Moreover, adopting a bar-centric tourism approach contradicts the inclusive *ethos* of places resignification based on strengthening territorial identity. Visitors also yearn for the opportunity to effortlessly experience the city's most captivating features – its landscapes, nature, and architecture – each symbolizing the local essence.

> **There is a project, for example, that I loved, but it was never implemented [...] It is the distribution of umbrellas, all colourful** [emphasis added] with the tiles by Athos Bulcão, spread throughout the Esplanade. So, let us say you rent one of these umbrellas, a robust and beautiful umbrella that could be used for sun or rain, because Brasília is no joke when it comes to the sun... So those silver umbrellas that reflect, right? And then you would pick it up at the Cathedral, and return it wherever you wanted, because there would be collection points all over the city. This is a really cool solution that I think

> would be beautiful. Just imagine the tourists, all with Athos Bulcão umbrellas, colourful and spread out across the Esplanade... [...] You do not necessarily have to try to change the city; maybe it is about changing the tourists' perspective of the city (RP3).

Apparently, the urban areas frequently visited by tourists seem best explored by car. These regions are typically administrative zones and lack thoughtful tourist-friendly features. This is particularly evident when navigating the primary route showcasing the city's architectural design. While striking government structures grace this path, there is a noticeable shortage of amenities catering to tourists: limited shaded seating (important due to the region's semi-arid climate), minimal options for food and drink, an absence of activities that reflect local traditions, and a shortfall of spaces that encourage integration, especially through a design-oriented creative economy.

The prevailing impression portrays a city focused on business, where tourism is gradually taking shape, but with limited opportunities for contemplation, engagement, and creative experiences that narrate the local history through the design lens. This process should be resilient and inclusive, catering to diverse ages and socioeconomic backgrounds, aside from nighttime festivals and events. Unfortunately, these events often lack inclusivity, primarily targeting specific social groups. This approach does not align well with sustainable development goals and leans more towards economic interests.

Another crucial factor in this indicator is the establishment of open-air markets, particularly in supporting artisans selling handmade goods. Notably, a lack of a united body or association to oversee their welfare, environmentally and socially, poses a significant shortcoming. Besides, from the interviewees point of view, their association does not take part in the creative city because it does not generate income for them.

> We have permission from the Priest... just to work here. However, as for an Association, we do not have one. We have thought about pursuing some kind of benefit for ourselves... we even thought about starting an Association, but not everyone is fully on board... It would be much better if we had a representative. During the rainy season, it is really tough here. We have to lift up the stall, find shelter, and the rains here are quite heavy, you know? (...) We **do not participate in the Creative City, but if there was something that brought benefits to us, for sure, yes** [emphasis added] (RP6).
>
> [regarding the Cathedral vendors exposed to the weather] When I was the deputy secretary, I also tried to address the issue [to put them there on the sidewalk of the Cathedral], because I do not even think that is the ideal place for them

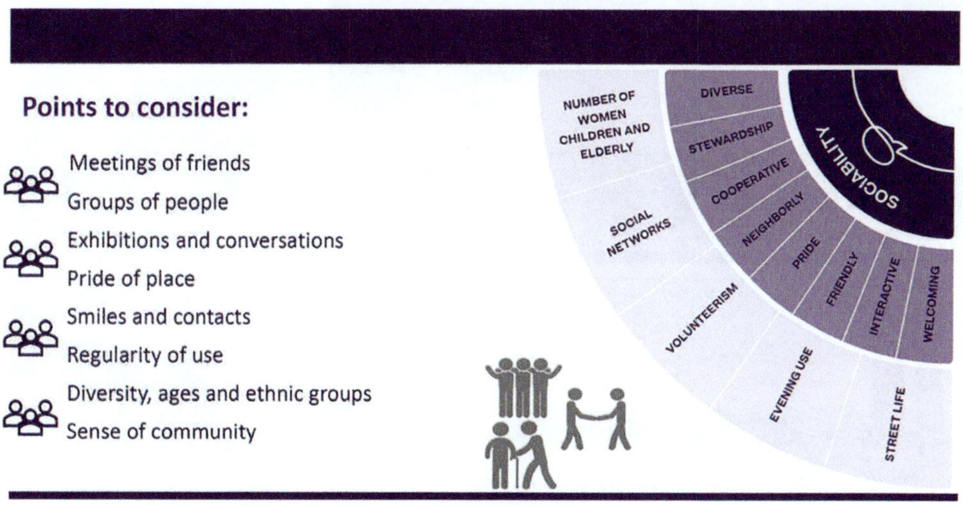

Fig. 21.10. Block 4. Sociability. Source: Own elaboration, based on PPS (2018) and Vasconcelos (2017).

> to be [...] Because there is the TV Tower right above. But then, tourists also find it far away. [...] the issue of heritage protection, it is a limiting factor within these aspects, but at the same time, it is what gives us legitimacy to continue being who we are: a preserved city, a heritage city, and a creative city (RP3).

Even with local government efforts, the challenge persists. Tourists and vendors still utilize these spaces without adequate support or amenities to guarantee a positive experience for both sides.

Next, we address the sociability block of the designated study spaces. See Figs 21.10 and 21.11.

In Block 4, Sociability, we recorded the most critical and evident items relating to complexes 01, 02, 03, 04, and 05, with a concentration of "Yes' responses, indicating mainly that people use the space in group interaction. In determinate spaces, such as complexes 01 and 04, the people use the space for sociability in more numbers only during events or festivals. We also recorded that people were not communicating or interacting with people from other groups.

Looking at how people interact in the studied areas leads us to think about how tourist spots are set up. Except for a few cases, people do not really talk to each other much in these places. That is because the spaces are not really designed for people to connect. Whether it is families or friends, most people just go to these places, take pictures, and leave. Not many things are happening there regularly to make them comfortable or interesting. This recalls the idea that "If a place isn't welcoming for the local residents, it won't be appealing to tourists either" (RP2).

The findings of this study emphasize that the creative city designation does not align closely with the fundamental factors that foster tourism through this creative label, nor does it adhere well to the guiding principles that should be associated with this designation. The data suggests that, in Brasília, the design element needs to be more in line with the desired outcomes linked to the creative economy.

Our analysis underscores a stronger connection between the creative economy and sectors like gastronomy and crafts. However, we understand that this aspect is fundamental

Fig. 21.11. Sociability of the spaces designated for the study.

for a creative city, as the design status alone is not enough to fuel the tourism established through this global creative label; the mix of different cultural expressions matters the most. Brasília has important elements across all these aspects. Nevertheless, the design label falls short in forging the needed connections to nurture a vibrant and enduring creative economy in tourism that goes beyond surface-level appreciation.

The design label points to a major challenge: the city's hands are tied when it comes to altering its urban structure due to its World Cultural Heritage status. However, we have also scrutinized other areas that could be revitalized to promote tourism that is inviting, strong, and creative, and offers distinctive experiences to tourists. This involves acknowledging how essential resources, meanings, and creativity connect during the revival of tourism, especially after the COVID-19 pandemic. Thus, the city's governance must go beyond the role of directing tourists through areas with different visitor flows. Instead, it should actively encourage them to spend time in and fully embrace these spaces.

Conclusions

The active engagement of stakeholders in discussions about the urban potential that validates a UNESCO Creative City is crucial at the public and collective level. Therefore, understanding the role of a UNESCO Creative City within this framework presents a dual nature: it poses a challenge that needs addressing and stands as a fundamental necessity when considering the growth of local tourism. This pertains to both the conceptual and practical dimensions of reflection and implementation.

This means that the way stakeholders interact and the potential of a label that shapes and connects the city of Brasília are important factors, especially through the designated spaces where we carried out the placemaking diagnosis. These spaces are not solely about their physical boundaries. Instead, they are the products of people's perceptions, experiences, emotions, and representations tied to a label that ignites creativity.

As a result, the city needs to adopt actions beyond just connecting and aligning with the public (or private) agenda, to incorporate collaborating across sectors, and embracing holistic strategies based on placemaking principles. These measures aim to meet the demands of the thriving and vibrant tourism in this city and the residents. This also considers the beliefs and genuine interests of all these people in the native traditions that shape the images and symbols of the cultural and creative evolution of this community.

Finally, the COVID-19 pandemic has imposed numerous challenges on society, particularly on local development, whether in a creative city or not. Given this, we propose that placemaking can be a first step to restore social unity, fortify identity, boost inclusivity, resilience, and resignification of places. This hinges on merging labels with untapped prospects for thoughtful, experiential, and creative tourism.

Note

[1] *Cerrado* is a Brazilian savannah biome, characterized by seasonal forest, typical of a seasonal tropical climate defined by two seasons: one rainy and the other with a long biologically dry period.

Acknowledgements

Thanks to Brazilian National Counsel of Technological and Scientific Development (CNPq) and the Minas Gerais Research Foundation (FAPEMIG), Brazil.

References

Brasil, Ministério do Turismo (MTur) (2022) Mapa do Turismo 2022. Available at: https://shorturl.at/zKP48 (accessed 5 May 2022).

Emmendoerfer, M.L. (2019) *Inovação e Empreendedorismo no Setor Público*. ENAP, Brasilia.

Fayos-Solà, E., Alvarez, M.D. and Cooper, C. (2014) *Tourism as an Instrument for Development: A Theoretical and Practical Study*. Emerald Group Publishing, Bingley.

Ferreira, V.M.S. (2017) A rede de cidades criativas da unesco: Uma perspectiva das cidades brasileiras (dissertação de mestrado). In: *Universidade Federal de Goiás (UFG)*. Goiânia, GO.

Hall, C.M. (2004) *Planejamento Turístico: Políticas, Processos e Relacionamentos*, 2nd edn. Contexto, São Paulo.

Harvey, D. (2006), *Condição pós-Moderna: Uma Pesquisa Sobre as Origens da Mudança Cultural*, 6th edn, Edições Loyola, São Paulo.

Heemann, J. and Santiago, P.C. (2016) Guia do Espaço Público: Para Inspirar e Transformar. Available at: https://shorturl.at/iotO3 (accessed 2 May 2022).

Instituto Brasileiro de Geografia e Estatística (IBGE) (2022) IBGE Cidades: Brasília. Available at: https://shorturl.at/guOV6 (accessed 5 May 2022).

Instituto do Patrimônio Histórico e Artístico Nacional (IPHAN) (2022) Brasília (DF). Available at: https://shorturl.at/lvzl8 (accessed 5 May 2022).

Landry, C. (2013) *Origens e Futuros da Cidade Criativa*. SESI-SP, São Paulo.

Mediotte, E.J., Silva-Junior, A.C. and Emmendoerfer, M.L. (2022) Placemaking sob Orientação Empreendedora do Effectuation: Metodologia Para (re)Vitalização e Desenvolvimento Local. XLVI Encontro da ANPAD. Available at: https://shorturl.at/lxRV1 (accessed 15 August 2023).

Project for Public Spaces (PPS) (2016) Placemaking: What if we Built our Cities Around Places? Available at: https://shorturl.at/ntJY9 (accessed 5 August 2023).

Project for Public Spaces (PPS) (2018) What Makes a Successful Place? Available at: https://shorturl.at/ayBG5 (accessed 5 May 2022).

Richards, G. and Duif, L. (2018) *Small Cities with Big Dreams: Creative Placemaking and Branding Strategies*. Routledge, New York.

United Nations Organization for Education, Science and Culture (UNESCO) (2017). Available at: https://shorturl.at/GHQ25 (accessed 5 May 2022).

Vasconcelos, P. (2017) What Makes a Great Place? A Project for public Spaces Diagram. Available at: https://shorturl.at/elmQ6 (accessed 5 May 2022).

Zitcer, A. (2018) Making up creative placemaking. *Journal of Planning Education and Research* 405, 1–11.

Section 5: Economic Development and Policy in Tourism

22 Fiscal Incentives for Hotels in the City of Buenos Aires, Argentina

Carolina I. Garcia, Christof Pforr and Michael Volgger
School of Management and Marketing (Tourism Research Cluster), Faculty of Business and Law, Curtin University, Western Australia, Australia

Summary

This case study focuses on the Hotel Sector Promotion Scheme, a programme implemented in the City of Buenos Aires, Argentina, for the period 2019–2023. The scheme provides fiscal incentives in the form of tax credits for developing and improving hotel infrastructure in the city. To explore this case, we use secondary data from the City Legislature and media outlets specializing in tourism, as well as primary data from interviews with tourism stakeholders. By reviewing the programme and its particularities, this contribution sheds light on the interface between policy making, evidence use and partnerships.

The Value and Interest of the Case Study

Public policies focusing on hotel incentives are not uncommon in tourism destinations, but the Hotel Sector Promotion Scheme in Buenos Aires presents some features that stand out. The case illustrates how evidence can play an important role in tourism policy design. It also highlights that partnerships between government agencies can lead to relevant tourism policies. Overall, the case shows the interface between policy making, evidence use and partnerships in the tourism sector.

Context

The Autonomous City of Buenos Aires (referred to as Buenos Aires henceforth), Argentina's capital, is a key destination for global tourism. The cosmopolitan city offers a broad range of attractions of which culture, history and gastronomy stand out. It represents the gateway to tourism experiences in Argentina and ranks highly in tourism popularity, has many award-winning restaurants and is often found on global lists of best destinations (United Nations World Tourism Organization, UNWTO, 2019). In 2019, it welcomed a record 2.9 million international tourists (Tourism Agency, 2020).

The importance of tourism for Buenos Aires has long been formally recognized. In 2001, a law determined that tourism was of 'public and cultural interest' for the city (Law N° 600, 2001). The regulation also established sets of guiding principles, among them the generation of suitable conditions for investments in the sector. In 2007, an autonomous Tourism Agency was created within the city government (Law N° 2627, 2008). This agency aims to 'design and implement policies and programmes for promoting, developing and advancing tourism as a strategic economic activity' (Law N° 2627, 2008).

The Tourism Agency has traditionally focused on promoting tourism and attracting visitors, especially international tourists. Nonetheless, the organization has many other functions and powers (Law N° 2627, 2008). It has developed policies in collaboration with other government areas. This is also the case for the Hotel Sector Promotion Scheme, a policy developed in 2018 by the Tourism Agency in partnership with InvestBA, the investment agency in Buenos Aires.

Aim and Methodology

This case study focuses on the Hotel Sector Promotion Scheme, a programme implemented in Buenos Aires for the period 2019–2023. By exploring this initiative and its particularities, this contribution sheds light on the interface between policy making, evidence use and partnerships, which can help develop better policies in the tourism field. To explore this case, we use both primary and secondary data. Primary data was collected by means of online interviews with 13 tourism stakeholders (including tourism public and private sector representatives) using purposeful and snowball sampling between April and August 2023. Secondary data was retrieved from the City Legislature (law files, laws and decrees) and local media outlets specializing in tourism.

Policy Making in Action – Hotel Sector Diagnosis

The policy making process for the Hotel Sector Promotion Scheme started in 2018 with an analysis of hotel developments in Buenos Aires. It considered evidence concerning several aspects of the hotel sector: (i) performance and forecasts, (ii) quality and profits, and (iii) tax scenario and competitors (Fig. 22.1).

Hotel performance and forecasts

In 2018, tourism in Buenos Aires showed a strong economic performance. That year, the sector contributed US$6.7 million to the economy, and hotels created more than 21,000 direct jobs. The hotel occupancy rate in the first half of 2018 reached 57%, the highest in 7 years. International tourist arrivals and nights also grew during this period (Legislature Archives, 2018) (Fig. 22.2).

One contributing factor to this rapid expansion of tourism was a growth in air travel connections, both domestic and international. Considering the overall new seats available, there was a potential for an 18% increase in international tourism arrivals and expenditure. In particular, in the first 8 months of 2018, flight bookings for trips between September 2018 and August 2019 from main cities worldwide increased by 32%. Additional air routes planned for the end of the year were expected to contribute further to future tourism growth (Legislature Archives, 2018). Moreover, compared to many other cities, Buenos Aires was competitive in its prices because of a recent devaluation of its currency and the exemption of international tourists from paying accommodation taxes (Hosteltur, 2019a).

The increased arrivals suggested that more tourists would book hotels in Buenos Aires. The Tourism Agency worked on forecasts for different scenarios (conservative, moderate, optimistic). Even in the conservative and moderate scenarios, international tourists staying in hotels were expected to grow steadily from 2.1 in 2017 to 3.5 million in 2022. The overall occupation rate was also expected to increase towards 2022 for all types of hotels (5, 4 and 3 stars, apart hotels and boutique hotels), exceeding 85% (Legislature Archives, 2018).

The forecasted demand led to estimations on the number of hotels needed towards 2022. These estimates highlighted an insufficient accommodation supply to cater for expected tourists. Estimations indicated that between 34 and 80 new hotels, depending on the scenario, were required (Legislature Archives, 2018).

However, there were relatively few projects planned that would add new hotel rooms in Buenos Aires. In the medium term (3–4 years ahead) only 712 new rooms were expected to be added, a figure considered low in light of the forecasted increase in demand. It was also low compared to accommodation projects in other city destinations in the region: Lima in Peru (1800 planned rooms), São Paulo in Brazil

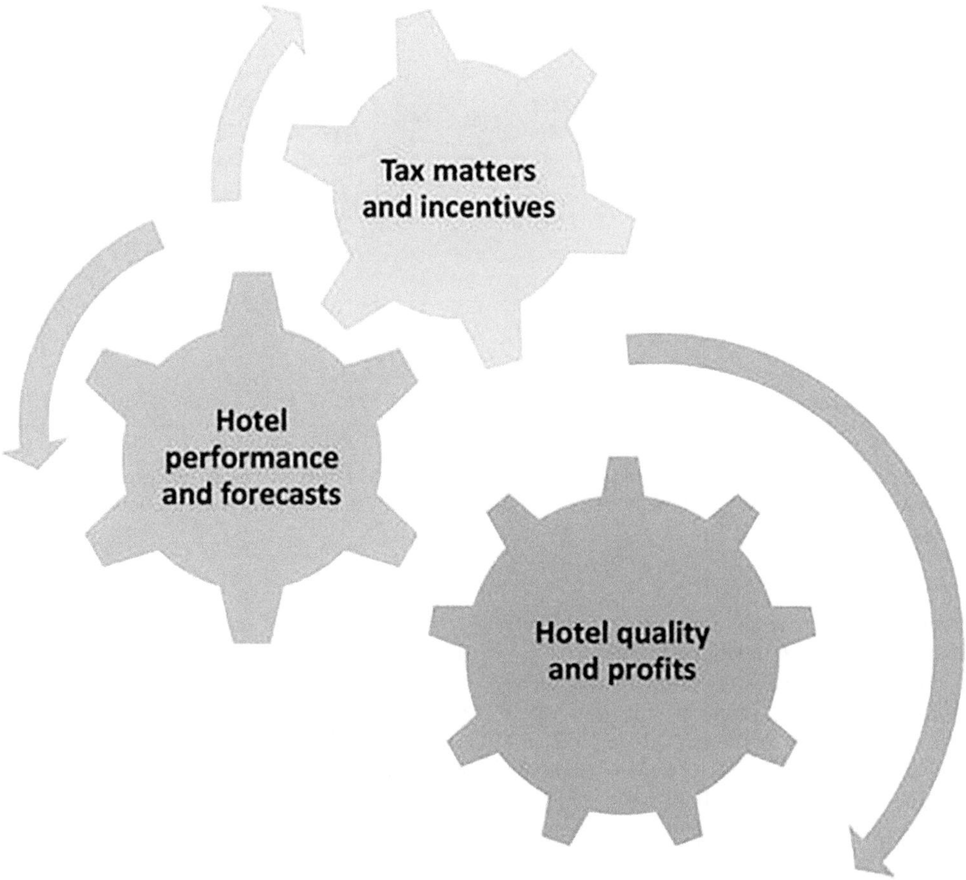

Fig. 22.1. Evidence about the hotel sector in Buenos Aires.

(1385), Santiago de Chile (1340) and Bogotá in Colombia (1228) (Legislature Archives, 2018).

Hotel quality and profits

A department within the Tourism Agency was devoted explicitly to market intelligence and monitoring. This department started a big data project in 2017, making Buenos Aires a pioneer in these efforts in South America (Surname *et al.*, 2018). The project involved analysing prices and hotel quality, tracking price evolution and travel forecasts, and benchmarking these findings with other city destinations in the region. As a result of this work, relevant insights concerning the quality and profits of hotels in Buenos Aires emerged.

Regarding quality, analysts collected over 150,000 comments by tourists rating accommodation in Buenos Aires. They compared these perceptions with their equivalent for accommodation in Lima (Peru) and Santiago de Chile (Chile), both of which were identified as competitor destinations for Buenos Aires. They used data mining sentiment analysis. In other words, they studied a large data set to identify relevant insights on tourists' feelings towards accommodation offerings. With this technique, analysts identified opportunities to improve accommodation facilities, infrastructure and maintenance, especially in 4-star hotels (Legislature Archives, 2018). The Tourism Agency's President highlighted at the time that 'our competitors, Santiago de Chile and Lima, have emerged into international tourism in recent years and have a

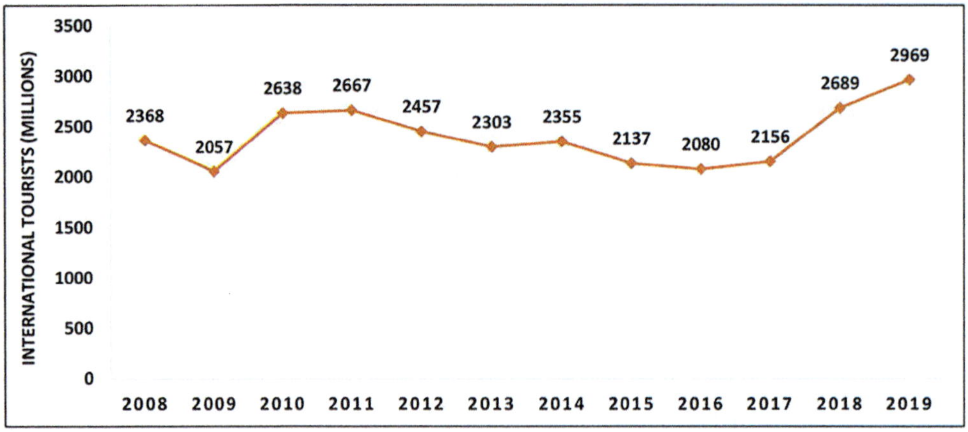

Fig. 22.2. International tourist arrivals in Buenos Aires, 2008–2019. Notes: Figures expressed in millions of tourists. A change of methodology in measurements was introduced in 2018. Source: Tourism Agency (2020).

much newer, more modern and more up-to-date hotel infrastructure' (Hosteltur, 2019a).[1]

Quality concerns were also voiced by hotel groups like Hilton, a company with eight hotels in Argentina, two in Buenos Aires. In 2017, Hilton's Director of South Cone[2] Development claimed that Argentina had some excellent tourism providers, but also some in need of change. He highlighted a quality deficit in the international tourism offer and that many hotels required improvements (Hosteltur, 2017a).

Concerning profits, the Tourism Agency tracked the evolution of hotel prices in Buenos Aires. It found that the rate in 4-star hotels was below that of competing cities, a situation that had a negative impact in terms of business opportunities and investment returns for hoteliers. On the other hand, the private sector complained about the relatively high costs faced by hotels in Buenos Aires. The Vice President of the Argentinian Federation of Hotel and Food Service Professionals (FEHGRA in Spanish), an institution representing tourism companies, commented:

> We have a tax, social security and labour cost that what it does is that a hotel in Argentina of the same category with respect to Brazil, Chile, Mexico or any country in South America, and even some in Europe, costs more. This is because we have a much higher cost floor than these countries. (Reportur, 2017)[1]

The President of the Argentinian Tourism Hotels Association (AHT) voiced similar concerns, highlighting consequences for hotel quality.

> We have been talking informally, but continuously, with the Tourism Agency about the problems we are going through; about how difficult it is, from the lack of profitability due to tax pressure, to modernise our hotels. (...) The truth is that a contradiction was generated, because on the one hand it [the Tourism Agency] is making a promotional effort and the airplane revolution to bring tourism from abroad, and on the other hand with the fiscal stranglehold we have, it was impossible for us to have the structure up to date and consistent to receive this tourism. (Hosteltur, 2018)[1]

Tax scenario and competitors

Further evidence was sought to understand the tax scenario for hotels in Buenos Aires. Argentina has different taxes at the national, provincial and local levels. For hotels and restaurants, taxes represent around 35% of the service's final price (Argañaraz *et al.*, 2016). One of the most relevant provincial taxes applies to businesses' gross income. In Buenos Aires (an autonomous city with the constitutional status of a province), hotels contributed 0.5% of all gross income tax in 2018. Hoteliers paid a gross income tax rate that varied according to annual turnover. In 2018, there was a fiscal agreement between the federal government and the provinces to reduce this tax rate and, as a result, companies paying 5% in 2018 would pay 4.5% in 2019 and 4% from 2020 onwards.

Hotels in Buenos Aires had not received any tax incentives for nearly 50 years at the time of these discussions. The last tax incentives for hotels were granted in the 1970s when Argentina was about to host the Football World Cup. On this occasion, there were income tax deductions for new hotels located in seven areas around the country, the City of Buenos Aires among these (Law N° 17752, 1968). Several hotels in the city were built with this incentive, and many had not undergone any renovations since then (Legislature Archives, 2018). This helps to explain why some hotels were identified as low quality or poorly maintained.

In comparison, competing destinations had implemented incentives via fiscal schemes in recent years. Colombia and Uruguay, for example, had successful national tax programmes for the sector (Legislature Archives, 2018). Furthermore, 18 out of 23 provinces in Argentina provided tax benefits for hotels, primarily focusing on gross income taxes. In this regard, Buenos Aires presented a less competitive hotel investment and operation setting.

The Policy Shape – Fiscal Incentives for Hotels

The comprehensive analysis of the hotel sector influenced the subsequent steps in the policy making process. Two government agencies developed partnerships and designed a fiscal incentives scheme for hotels. A bill was then tabled and discussed in the City Legislature, and the Hotel Sector Promotion Scheme was approved and launched.

Developing a policy instrument: Tax incentives

The formulation of the Hotel Sector Promotion Scheme was initially led by InvestBA, Buenos Aires' Investment & Trade Promotion Agency. InvestBA is a department within the Ministry of Economic Development and Production that aims to encourage investments and position Buenos Aires as an attractive business market. InvestBA's General Director stated at the time of these discussions that 'basically this project [the Hotel Sector Promotion Scheme] started within the area that I lead [InvestBA].... Concerning tourism, although the Tourism Agency does a very good job in what has to do with promoting and attracting tourists to the city, there was no investment incentive or tax incentive' (Legislature Archives, 2018).[1]

The project then became a collaborative initiative between InvestBA and the Tourism Agency. InvestBA was the promoting body, while the Tourism Agency became the implementing authority because it had direct competencies in tourism development. The Hotel Sector Promotion Scheme was discussed at several meetings in the City Legislature. The first step was for commissions (i.e. representatives from different political parties) to examine the project. Three commissions on economic development, tourism and sports, and budget contributed to discussions held during September and October 2018 and tabled an incentives bill in the City Legislature.

The partnership between the two government organizations, the Tourism Agency and InvestBA, was important in the policy making process. Not only did the Tourism Agency have direct competencies in the tourism field, but it also had tourism-related knowledge and hotel-specific evidence. Meanwhile, InvestBA had expertise in understanding the city's tax scenario and designing incentive schemes. The two worked in partnership and, when the bill was tabled, they jointly presented the Hotel Sector Promotion Scheme proposal in the City Legislature.

The private sector welcomed the initiative, with the President of the Tourism Hotels Association highlighting that in complex economic scenarios, having a fiscal incentives scheme 'offers a huge opportunity for development.... [The] uncertainty [we have] at an economic level is not something to which you can stop paying attention if you are thinking of putting a *peso* [a unit in Argentinian currency] either to remodel or build' (Hosteltur, 2018).[1]

The idea of introducing fiscal incentives in the sector was also discussed in tourism-specialized media. Hilton's Director of South Cone Development, for example, presented his views on boosting hotel investment in Argentina:

The Hotel Sector Promotion Scheme was approved during a Legislature session in November 2018 (Law N° 6038, 2018) without much controversy. No questions were asked after the project was presented during the session, and the project attracted 38 positive and only ten negative votes with 12 abstentions. The scheme started in February 2019 with the approval of the law guidelines (Decree N° 74/19, 2019).

The fine print of the tax incentives scheme

The Hotel Sector Promotion Scheme was a scheme that provided fiscal incentives for hotel investment projects in Buenos Aires. With this scheme, hoteliers could apply for credit equivalent to a percentage of their investment and use this credit to pay for the city's gross income tax. The scheme had two target beneficiaries: new hotels and existing hotels. The construction of new hotels had been identified as necessary with the forecasted increasing tourist demand. It was also aligned with the Tourism Agency's goal in its strategic plan to double the impact of tourism on economic development in Buenos Aires. On the other hand, the improvement of existing hotels responded to a need for increased infrastructure quality to enhance competitiveness, translating into higher profits.

Potential beneficiaries had to present their investment projects to the Tourism Agency, which assessed their eligibility. For new hotels, suitable projects included the construction and equipment of hotels without authorization to operate before the programme started. For existing hotels, eligible projects comprised the revamp, extension, improvement and purchase of additional equipment. In both cases, beneficiaries had to demonstrate technical and financial capabilities.

The scheme presented different conditions for new and existing hotels (Table 22.1). These were updated for new hotels shortly after the law took effect. By July 2019, there had been no demand for these investment projects and their base and maximum credit was therefore almost doubled (Hosteltur, 2019b). A new law was proposed and ratified (Law N° 6186, 2019), with the conditions listed below reflecting this update.

Table 22.1. Conditions of the Hotel Sector Promotion Scheme for target beneficiaries.

	Construction of new hotels	Improvement of existing hotels
Base credit	20%	40%
Maximum credit	50%	60%
Characteristics considered to determine maximum credit	• Historical building (10%) • SME (10%) • Location (4–10%)	• Historical building (10%) • SME (10%)
Timeline from project approval until end of construction/ improvements	Up to 5 years	Up to 3 years
Timeline from credit allocation until end of credit	Up to 15 years	Up to 10 years
Additional benefits	• Fiscal stability • Deferral of payment and credit for other (minor) taxes	• Fiscal stability
Additional requirements	Land purchases not eligible for credit	Minimum investment per hotel category

Note: Percentage figures indicate ratios of hoteliers' investments.
Source: Own elaboration based on Law N° 6186, 2019.

The base credit was 20% of the investment for new hotels and 40% for existing hotels. In each case, credit could reach a maximum of 50–60% depending on specific hotel characteristics. New and existing hotels could receive an extra 10% in credit if they were located in the Historic Protection Area, a Protected Building, or a National Historical Monument. This recognized the higher cost of investment projects in heritage areas where buildings, for instance, require higher quality construction materials. Both types of hotels were eligible for an additional 10% in credit if they were a micro, small or medium enterprise. Finally, new hotels could increase their credit by 2% or 5% depending on the suburb in which they were located. Timelines were also different for each target beneficiary reflecting that new hotel constructions take longer than hotel improvements.

There were additional benefits and requirements for investment projects in the scheme. All projects were guaranteed fiscal stability, meaning that the city government committed to no increase in tax burden for beneficiaries until they had entirely spent the granted credit. New hotels were also eligible for a deferral in paying other city taxes but could not claim any credit for land purchases. Existing hotels had the additional requirement of a minimum investment per hotel category in order to be considered eligible beneficiaries.

A key feature of the Hotel Sector Promotion Scheme was that the benefit (i.e. credit to pay gross income tax) would be received after the investment project was finished. This guaranteed that the projects were indeed carried out. As highlighted by InvestBA's General Director at the time:

> This means that the economic incentive is given today, but it can start to be used once the work is finished. For example, if a new hotel takes one and a half or 2 years to be built, you can only start using the benefit once the Government verifies that it has actually made the investment and the hotel begins to operate. What is the logic of this? We want to generate a positive externality today and we want investment and employment to be generated today. We want all the demand for supplies – faucets, cement, etc. – to be generated today, because that also pays for Gross Income [Tax]. We want the fiscal cost to be applied only when the work is finished, within a year and a half or two' (Legislature Archives, 2018)[1]

Policy Progress and Outcomes

Approved in late 2018, the Hotel Sector Promotion Scheme was expected to generate US$150 million in hotel improvements and US$500 million in new hotels (La Agencia de Viajes, 2019). We used primary and secondary data to assess the policy progress and outcomes. Secondary data was retrieved from legislation and various media outlets. Data collection faces some limitations because hotel project submissions were classified as confidential and are therefore not publically available (Ruling N° 2/ENTUR/19, 2019). An additional challenge to assessing the programme's success stems from the COVID-19 pandemic that began only a year after the scheme started. Given these limitations, we complement the analysis of secondary data with a qualitative assessment using primary data collected via semi-structured interviews with tourism stakeholders from Buenos Aires.

Considering the information available in specialized media, by mid-2019, 15 new projects had been submitted to enhance existing hotels, for a total value of US$8.66 million (Hosteltur, 2019c). These included Palacio Duhau – Park Hyatt, Four Seasons, Ibis Congreso and Amerian Buenos Aires Park Hotel proposals. Hotels applied for different improvement projects, including incorporating new shared spaces, renovating building façades and improving customer service with specialized software. In December 2019, the city government announced that more than 40 projects proposing new constructions or improvements had been submitted (Buenos Aires Government, 2019).

Interviewed stakeholders made an overall positive assessment of the Hotel Sector Promotion Scheme. Several highlighted the scheme as one of the most relevant tourism policies in Buenos Aires in the last 10 years. In fact, government officers ended up meeting with their counterparts in other Argentinian provinces interested in developing a similar initiative. The fact that the policy design involved considerable evidence was also widely acknowledged. Various stakeholders emphasized that private associations were involved in the policy process and had the opportunity to provide feedback. At the same time, some claim this involvement existed only at a relatively late stage of the process, providing input in an already rather 'polished' project.

The implementation stage of the scheme was not exempt from challenges, with interviewed stakeholders highlighting it was a complex policy. As mentioned above, the scheme's conditions had to be changed after the law was approved. In addition, interviewees pointed to weaknesses in policy monitoring. Intra-government communication was also difficult, particularly between the Tourism Agency and government agencies dealing with taxes, supervision of commercial premises and the environment. Lastly, bureaucracy was a constraint mentioned by several interviewees, as many projects could not be carried out because applying hotels were not complying with all rules or had tax debts.

Interviewees expressed mixed views about policy outcomes. During 2019, the first year the scheme was in place, there was much interest in the initiative, as mentioned above. Nevertheless, few projects were approved, with an interviewee claiming these were mostly from bigger hotels when in reality the smaller hotels would have needed the incentive most. Indeed, some stakeholders said the policy's main target group were not the more prominent hotels but smaller ones, but the latter did not end up benefitting from the scheme. Another interviewee pointed out that it was natural for big players to react first and present hotel improvement projects, and these were relatively easier and more straightforward to deliberate on for the decision makers.

Early into the scheme's 2nd year, the COVID-19 pandemic started. As a result, the Hotel Sector Promotion Scheme was modified to respond to the crisis. For hotel improvements, the minimum investment amounts per hotel category were dramatically reduced (La Agencia de Viajes, 2020). Additionally, a mixed use of properties was allowed so a building could host a hotel and another business, like a restaurant (La Agencia de Viajes, 2020). According to one of the interviewees, these changes had negative impacts on the ability to pursue the initial intention of the scheme as they resulted in a 'patched' law. Another interviewee, however, claimed that thanks to these modifications hotels could play an important role in hosting COVID-19 patients when hospitals were at capacity.

The Hotel Sector Promotion Scheme is scheduled to end in 2023, but stakeholders claim there is still a significant lack of hotels in Buenos Aires. Recent discussions provide hints concerning a potential future of the initiative. In December 2022, the President of the Tourism Agency highlighted the relevant investment encouraged by the scheme and claimed the government would try to expand it to establish incentives targeted explicitly at new hotels with MICE (Meetings, Incentives, Conferences & Exhibitions) venues (Mensajero, 2022). Interviewees also mentioned that work towards a new law is in progress, incorporating the experience from the implementation of the original scheme.

Final Remarks

This case study focused on the Hotel Sector Promotion Scheme in Buenos Aires. The scheme provides fiscal incentives through tax credits for developing and improving hotel infrastructure in the city.

First, we identified the evidence considered in the initial steps of the policy making process. This evidence focused on different aspects of hotel development in Buenos Aires: (i) current and forecasted performance, (ii) quality and profit, and (iii) taxes and competitors. These areas of investigation were identified by two government agencies, the Tourism Agency and InvestBA, who partnered to table a bill in the City Legislature. The private sector also voiced concerns about the state of the hotel supply in the city and expressed support for the proposed measure.

Then, we presented the policy shape of the Hotel Sector Promotion Scheme. This involved understanding the tax scenario and the role of both government agencies in developing the scheme. The fine print of the scheme showed the conditions that existing and new hotels could benefit from depending on several characteristics, including location and business size.

Finally, we focused on the policy's progress and outcomes. Overall, the scheme was well received, with many projects submitted. Furthermore, the Tourism Agency and InvestBA used a wide range of supporting evidence to design the policy, as was shown with both primary and secondary data. However, the policy implementation was not exempt from challenges, including monitoring of results, intra-government communication and bureaucracy. At the same time, the need for new hotels in Buenos Aires continues and might in the future be addressed with a new law that is currently in discussion.

Overall, unravelling the process behind the Hotel Sector Promotion Scheme in Buenos Aires illustrates how policy making, evidence use and stakeholder partnerships interlace in tourism policy.

Notes

[1] First author's translations of what was originally stated in Spanish.

[2] The South Cone, a subregion in South America, consists of varying countries based on the author or organization's definition, but typically includes the southernmost countries on the continent. For the Hilton, the South Cone includes Argentina, Paraguay, Uruguay and Bolivia.

References

Argañaraz, N., Mir, A., Devalle, S., Costamagna, P. and Marcó, M. (2016) *Estudio integral de la carga tributaria argentina sobre el sector de hoteles y restaurantes*. Fiscal Analysis Argentinian Institute IARAF. Available at: http://ahgbahiablanca.org.ar/wp-content/uploads/2016/01/La-carga-tributaria-argentina-sobre-restaurantes-y-hoteles-2015.pdf (accessed 22 March 2023).

Buenos Aires Government (2019) La Ley de Promoción para el Sector Hotelero supera los $1100 millones de inversión. buenosaires.gob.ar, 9 December. Available at: https://buenosaires.gob.ar/hacienda-y-finanzas/desarrollo-economico/noticias/la-ley-de-promocion-para-el-sector-hotelero-supera (accessed 21 March 2023).

Decree N° 74/19 (2019) City of Buenos Aires.

Hosteltur (2017a) Hacia la profesionalización de la inversión hotelera en Argentina. Hosteltur.com.ar, 7 December. Available at: https://www.hosteltur.com/lat/118970_profesionalizacion-inversion-hotelera-argentina.html (accessed March 2023).

Hosteltur (2018) AHT: 'Los incentivos permitirán que los hoteles se aggiornen a la demanda internacional'. Hosteltur.com.ar, 17 August. Available at: https://www.hosteltur.com/lat/120300_aht-incentivos-permitiran-hoteles-se-aggiornen-demanda-internacional.html (accessed 21 March 2023).

Hosteltur (2019a) Buenos Aires busca inversores para afrontar el crecimiento turístico. Hosteltur.com.ar, 25 February. Available at: https://www.hosteltur.com/127028_buenos-aires-busca-inversores-para-afrontar-el-crecimieno-turistico.html (accessed 21 March 2023).

Hosteltur (2019b) Buenos Aires duplicará beneficios fiscales para construir hoteles. Hosteltur.com.ar, 18 June. Available at: https://www.hosteltur.com/lat/129428_buenos-aires-duplicara-beneficios-fiscales-para-construir-hoteles.html (accessed March 2023).

Hosteltur (2019c) Buenos Aires: Hay 15 hoteles con proyectos de reforma y se sumarán otros 30. Hosteltur.com.ar, 30 June. Available at: https://www.hosteltur.com/lat/129690_buenos-aires-hay-15-hoteles-con-proyectos-de-reforma-y-se-sumaran-otros-30.html (accessed 21 March 2023).

La Agencia de Viajes (2019) Los incentivos fiscales atraen $ 368 millones en reformas hoteleras y buscan captar establecimientos 0Km. argentina.ladevi.info, 1 August. Available at: https://argentina.ladevi.info/actualidad/los-incentivos-fiscales-atraen-368-millones-reformas-hoteleras-y-buscan-captar-establecimientos-0km-n4023 (accessed 21 March 2023).

La Agencia de Viajes (2020) Hoteles: Buenos Aires modifica los incentivos para inversión. argentina.ladevi.info, 27 July. Available at: https://argentina.ladevi.info/hoteles/hoteles-buenos-aires-modifica-los-incentivos-inversion-n23930 (accessed 21 March 2023).

Law N° 17752 (1968) Argentina.

Law N° 600 (2001) City of Buenos Aires.

Law N° 2627 (2008) City of Buenos Aires.

Law N° 6038 (2018) City of Buenos Aires.

Law N° 6186 (2019) City of Buenos Aires.

Legislature Archives (2018) Expediente 2691-J-2018 (law N° 6038 files. City of Buenos Aires Legislature, Argentina.

Mensajero (2022) Casi 7 millones de viajeros llegaron a BUE durante 2022. mensajero.com.ar, 30 December. Available at: https://mensajero.com.ar/2022/12/30/casi-7-millones-de-viajeros-llegaron-a-bue-durante-2022/ (accessed 21 March 2023).

Reportur (2017) Feghra preocupada por la presión tributaria y la baja rentabilidad. Reportur.com.ar, 23 March. Available at: https://www.reportur.com/argentina/2017/03/23/feghra-preocupada-por-la-presion-tributaria-y-la-baja-rentabilidad/ (accessed March 2023).

Ruling N° 2/ENTUR/19 (2019) City of Buenos Aires.

Tourism Agency (2020) Turismo Internacional en la Ciudad de Buenos Aires Informe Anual 2019. Available at: https://turismo.buenosaires.gob.ar/sites/turismo/files/Turismo%20Internacional%20en%20la%20CiudaddeBuenosAires%20-%20Informe%20Anual%202019.pdf (accessed 22 March 2023).

United Nations World Tourism Organization (UNWTO) (2018) In: *UNWTO/WTCF City Tourism Performance Research*. UNWTO E-Library. DOI: 10.18111/9789284419616.

United Nations World Tourism Organization (UNWTO) (2019) UNWTO/WTCF City Tourism Performance Research. Report for Case Study: '*Buenos Aires, Argentina*' Working Paper. Available at: https://webunwto.s3.eu-west-1.amazonaws.com/s3fs-public/2019-11/buenosairescasestudy.pdf (accessed 25 March 2023).

23 Impacts of Capital Investment in Tourism Infrastructure: The Case of the Kuélap's Cable Car System

Cecilia Rivas Medina[1] and Sandra Cherro Osorio[2]
[1]School of Hospitality and Tourism Management, Universidad Peruana de Ciencias Aplicadas, Lima, Peru; [2]Business and Construction, Melbourne Polytechnic, Melbourne, Australia

Summary
Located in the north of Peru, the Archaeological Complex of Kuélap – an ancient fortress nestled in the cloud forest – is considered a significant cultural resource representative of one of the country's most important pre-Inca cultures: the Chachapoyas. While being a main destination on the north-eastern tourism circuit, this site required considerable development in order to attract tourism to the area. This case study investigates the relationship between capital investment in tourism infrastructure and improvements to tourism services. A combination of public and private partnerships allowed for the improvement of accessibility and tourism information facilities. The implementation of a cable car system to the fortress, the first of its kind in the country, has significantly reduced the trip distance from 2 h of hiking down to 20 min. The inclusion of signposting, infography and resting areas has also contributed to tourism satisfaction. Furthermore, through the development of community-based products and services, the local population is expected to benefit from the increase in visitation. As a result of these developments, tourist numbers showed a positive increase, contributing to the local economy. This case study examines current impacts and provides an analysis of other factors related to the management of heritage sites and their implications in the long term.

The Value and Interest of the Case Study

The fortress of Kuélap is an important archaeological site in the north of Peru. The site was legally protected in 1998 when it was declared a Cultural Heritage of the country. This important cultural resource consists of an area of 460 ha and contains various ancient buildings. Yet, tourism services and infrastructure were underdeveloped until recent years. In order to increase Peru's destination competitiveness, a series of actions were taken to improve tourism in various cultural sites, including Kuélap. This case study explores the results of the investments made through a public and private partnership to develop tourism in the fortress of Kuélap. It also identifies other tourism opportunities and future challenges in the context of a heritage site in a country from the Global South.

Context

This case study takes place in the Archaeological Complex of Kuélap, located in the province of

Fig. 23.1. Location map of Kuélap (courtesy of El Mundo – Spanish newspaper website).

Luya in the Amazonas region in north-east Peru (Fig. 23.1). This important archaeological site, declared a Cultural Heritage of Peru in 1998, is currently registered in UNESCO's tentative list to be selected as a Cultural Heritage of Humanity. This site hosts the remains of one of the most important pre-Inca cultures, the Chachapoyas. It covers an area of 460 ha that comprises a set of archaeological sites including the impressive Fortified Citadel of Kuélap, which is made up of 420 circular buildings divided into two sectors: Pueblo Bajo and Pueblo Alto (Low Town and High Town). The highlights of this Citadel are the stone friezes, the wall and the circular houses.

Considering its cultural and historical importance, the Ministry of Foreign Trade and Tourism – MINCETUR in Spanish – identified Kuélap as one of the most important sites for tourism development in the north-east of Peru. Previous literature has indicated that the relationship between heritage and tourism is contentious and even contradictory (Nuryanti, 1996; Aas *et al.*, 2005). Planning the management of heritage places requires involvement and cooperation among stakeholders including private and public sectors, non-profit organizations and locals (Nuryanti, 1996). Importantly, in order to minimize any issues, an understanding between all those involved needs to be reached so resources can be preserved and tourism activities can carry benefits to all, particularly local communities (Aas *et al.*, 2005).

Within this context, this case study examines the impacts of a public-private partnership to improve infrastructure and mobility to the archaeological site of Kuélap. The construction of a cable car system has facilitated access to the fortress, resulting in an increased tourism demand, improvement of visitor satisfaction and generation of income through the provision of tourism services at the local and regional level. While this operation has certainly brought benefits in the early stages of the development, other long-term challenges need to be considered such as site conservation issues, inadequate institutional capabilities and the level of involvement of local people in decision making.

Public-Private Partnerships for Tourism: Kuélap

Since 2003, public and private stakeholders have cooperated in identifying ways for the preservation of the archaeological site and development of tourist initiatives at Kuélap. These stakeholders comprise MINCETUR, the National COPESCO Plan (technical public body specialized in investment projects for tourism infrastructure at a national level), the Ministry of Culture, the Regional Government of Amazonas, cooperative agencies and private companies.

The first intervention took place in 2004 with a project that prioritized activities from the Master Plan for Management and Conservation of Kuélap and its surroundings. The focus was on the development of tourism activities so efforts prioritized the improvement of infrastructure to facilitate accessibility to the site. A project conducted in 2005, however, identified that access to the site, located on the top of a mountain, was difficult and lengthy. The government saw the need to engage in Public-Private Partnerships to address these issues and improve tourist services.

According to the World Tourism Organization (WTO), Public-Private Partnerships (PPPs) 'bring together stakeholders with different objectives and skills, and resources in a formal or informal voluntary partnership to improve the attractiveness of a regional destination, its productivity, associated market efficiency, and the overall management of tourism' (World Tourism Organization (WTO), 2015, p. 12). Rogerson (2016) further highlighted the importance of PPPs as vehicles for infrastructure development in the tourism sector. In Peru, PPPs are defined as a form of participation of private investment through long-term contracts where the state intervenes via a public entity to partner with one or more private investors. In the Peruvian context, PPPs ensure that risks and resources are distributed in the development of public infrastructure projects, with optimal levels of services for final users. Under this modality, the ownership of the investments can be maintained, reversed or transferred to the state, depending on the nature and scope of the project and the provisions in the contract. The PPPs may take the contractual form of concession, operation and maintenance, and management, as well as any other contractual modality permitted by law (El Peruano, 2018). PPPs cover investment in public infrastructure in general, public services, infrastructure-related services and public services, applied research projects and technological innovation projects.

A large-scale project at the local, regional and national level executed under the modality of PPPs was the 'Improvement of accessibility to the monumental archaeological zone Kuélap' (Lahura and Sabrera, 2022). Additional projects focused on providing support to the tourists visiting the archaeological site, such as the 'Implementation of Signaling and breaks in the Circuit of Tourist Walk of Tingo Viejo towards the Archaeological Complex of Kuélap' and the 'Construction of the Tourist Parador and Implementation of Tourist Signage of the Archaeological Complex of Kuélap'. All of them had the purpose of improving the conditions for the development of tourism (MINCETUR, 2016; Proinversion, 2017, 2021, 2023).

The 'Cable cars of Kuélap' project, also developed through a PPP, is within the scope of investment in public infrastructure and support for the tourism sector to contribute to the economic and social development of the country. This project emerged as a co-financed state initiative and was awarded in 2014 to the Kuélap Telecabin Consortium formed by a French and a Peruvian company. This partnership oversaw the design, construction, equipping and operation of the cable car system. The contract was granted for a concession period of 20 years. The construction started in August 2016 and concluded in November of the same year, resulting in an investment of US$21 million (Proinversion, 2017).

Before the execution of the project, Kuélap was only accessible by car or bus through a winding and uncomfortable road followed by a long walk. The construction of the cable cars significantly reduced the transit time from the district of Tingo to Kuélap from approximately 2 hour drive and walk to only 20 min. The 4-km route has overcome a difference of approximately 1000m between the town of Tingo Nuevo and the Fortress of Kuélap. The project includes a boarding station located in the centre of Tingo Nuevo, a road between the boarding station and the departure platform, a departure platform, the cable transport system with a capacity for eight people and the arrival platform (Fig. 23.2). The implementation of this project resulted in an improved access to the site allowing for a greater number of visitors.

Rodriguez *et al.* (2018) indicated that the main benefits from the installation and operation of cable car system include:

1. An increase in the number of domestic and international visitors to the fortress. It is expected that by the 10th year of operation (2027), the number of tourists will have increased at least by 50% in comparison to 2016, prior to the start of the operation of the cable car system.
2. An increase in the level of income among tourism service providers, including restaurants, accommodation and local businesses within the scope of the project.
3. Expectations that in the 10th year of operation, at least 50% of tourists will feel very satisfied with the services received during their visit to Kuélap.
4. By the end of the concession, the area available for the provision of tourist services will have increased by at least 250%.

Fig. 23.2. Results of the project showing the ease in accessibility to Kuélap (courtesy of Proinversion website).

Analysis of the Improvement of Infrastructure to Kuélap

Increase in tourism demand

Tourist arrival data starting in 2001 show that the total number of visitors to Kuélap has increased by approximately 10%, on average, each year. In 2016, 1 year before the commercial operation of the cable cars began, visitor numbers were around 56,000. By the end of 2017, 103,000 visitors were reported, which is equivalent to an increase of 84% in the number of visits during its 1st year. The highest number of visitors occurred in 2018, with 110,000 visitors. That is a 96% increase in visits compared to 2016. These encouraging numbers show that visits to the Kuélap Fortress had almost doubled as a result of the improved infrastructure that has contributed to time savings and comfort in the journey of visitors. The cable cars also became part of the attraction as they are the first of this type in the country. Visitor numbers maintained similar levels until 2019. However, the COVID-19 pandemic caused a substantial reduction in visitation to 28,377 (−49%), with the majority being domestic visitors (87%). In 2021, the lifting of travel restrictions and the reopening of destinations resulted in more encouraging numbers, with 59,613 visitors (6% more than in 2016). These visitors were mainly from the domestic market (Fig. 23.3).

These data highlight the achievement of the project's general objective: to provide a comfortable, fast and modern transportation to access Kuélap. The major outcome has been an increase in demand resulting in a high influx of visitors to the site. Statistics from 2017 to 2022 show that visitation is dominated by domestic tourists but the number of international tourists to the site was certainly significant from 2017 to 2019 (MINCETUR, 2020). It should be noted that the Peruvian tourism sector is still recovering from the effects of the COVID-19 pandemic. By the end of 2022, international tourism to the country was still 54% below 2019 levels (Thopson, 2023).

Studies of the relationship between infrastructure development and tourism have investigated the effects of policies on infrastructure construction (Deng *et al.*, 2019), transportation improvements as perceived by local communities (Kanwal *et al.*, 2020) and the impacts of airport infrastructure in regional development (Doerr *et al.*, 2020; Chow *et al.*, 2021). Lahura *et al.*'s (2019) study focused on the evolution of visits to Kuélap as a result of investment in infrastructure and identified that it had greatly exceeded

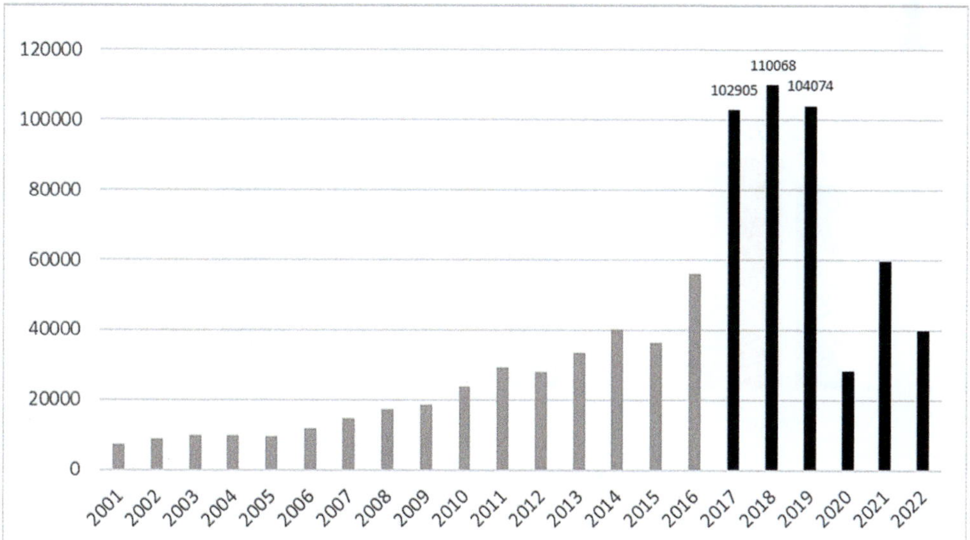

Fig. 23.3. Number of tourist arrivals to Kuélap (courtesy of MINCETUR website).

expectations. A follow-up study by Lahura and Sabrera (2022) further corroborated that the tourism potential of sites like Kuélap can be enhanced through investment in infrastructure. This provides significant indirect benefits to the economy of the region. As the country continues to recover from the negative economic impacts associated with the COVID-19 pandemic, PPPs for infrastructure development could become a useful tool to further develop tourism to other less-known archaeological, cultural and natural attractions. Alongside the development of infrastructure through PPPs, site conservation should also take priority particularly when higher revenue could assist in covering the costs involved in conservation. While the Peruvian government has classified Kuélap as a part of the cultural heritage of the country, conservation and maintenance of the site have been overlooked for many years. As an example, in 2022, the site was closed for a few months due to the collapse of a 15-m-high wall. The Minister of Culture at the time declared that the disaster was caused by the abandonment the site had experienced in previous years. Worryingly, it was also noted that other side walls were at risk due to deterioration (Los Angeles Times, 2022). The regional government spokesperson also reported that despite requests for effective intervention to preserve the site, Kuélap has been neglected by the national government (InfoBae, 2022). These issues highlight that while increased tourist numbers may seem to be positive, funding and enforcing conservation is also important if the site is expected to continue receiving visitors.

Improved tourist satisfaction

MINCETUR conducts surveys about visitor satisfaction with tourist sites across the country. After the implementation of the cable cars, visitor satisfaction from 2017 to 2019 showed an improvement from 77.7 to 90.7 points (out of 100 points in which over 80 points is high satisfaction, 70–80 points medium satisfaction and under 70 points low satisfaction). Domestic tourists' satisfaction, in particular, increased from 74.9 in 2017 to 97.7 points in 2019 (Fig. 23.4). The addition of the cable car system, signposting, infography and resting areas on the trekking path from Tingo Viejo to Kuélap resulted in the improvement of the visitor experience (Fig. 23.5). The Peruvian Export and Tourism Promotion Board (PROMPERU) also developed promotional material for the site specifically targeted at the overseas market.

In terms of specific criteria measured in tourist satisfaction, the latest data available

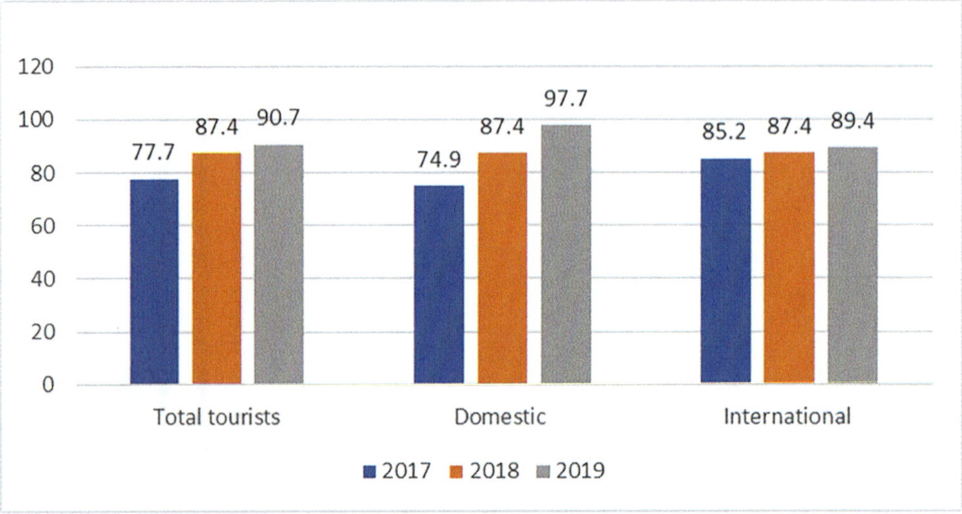

Fig. 23.4. Visitor satisfaction among domestic and international tourists to the fortress of Kuélap (courtesy of MINCETUR website).

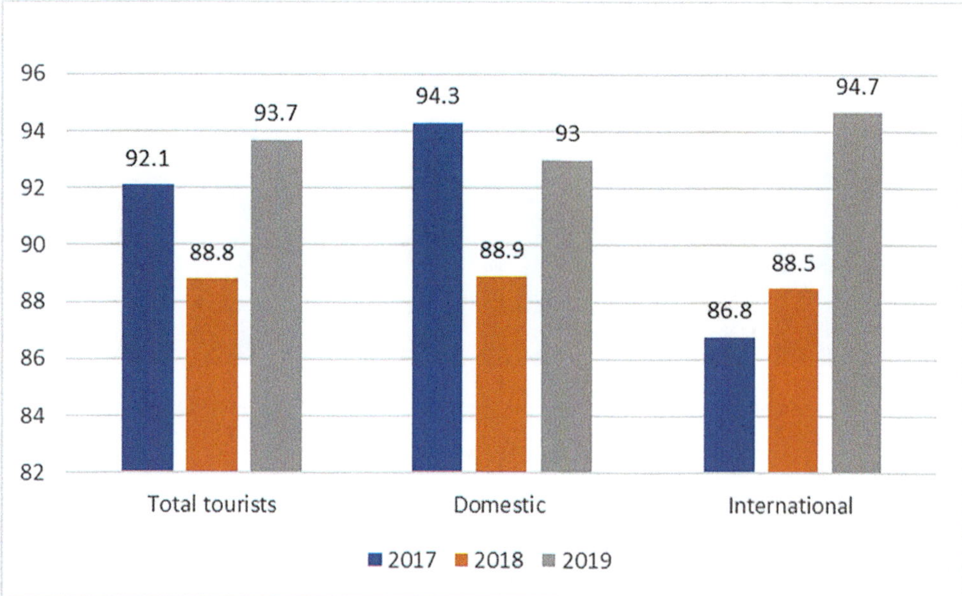

Fig. 23.5. Visitor satisfaction with the cable car service among domestic and international tourists to the fortress of Kuélap (courtesy of MINCETUR website).

correspond to 2019 when tourists indicated that security (90 points) and cleanness (90 points) of the site were the main aspects of satisfaction. Transportation and the time to arrive at the destination also scored high, with 88.4 points and 87.3 points, respectively. The lowest score, at 79.6 points, was for information available at the site (MINCETUR, 2023).

Regarding the degree of satisfaction with the cable car service among national and international visitors, it increased from 92.1 points to 93.7 points, and in the case of international tourists, Fig. 23.5 shows a notable increase from 86.8 points in 2017 to 94.7 points in 2019.

Data available from the pre-pandemic period show an increase in visitor satisfaction among domestic and international tourists. In addition, the indicators of Level of Satisfaction from NPS (Net Promote Score), available at MINCETUR (2018, 2019), indicate that in 2018, the goal set for the 10th year of operation of the cable car system, which was that at least 50% of tourists would feel very satisfied with the services received during a visit to Kuélap, had already been exceeded. Fig. 23.6 shows that in 2018, 57.4% of visitors to the site were very satisfied and this percentage increased to 66.4% the following year.

Direct and indirect benefits to the local population

The improvement in infrastructure and tourism services at Kuélap generated direct and indirect economic opportunities for the local populations, particularly for Tingo. Direct benefits were related to the number of visitors and their average expenditure on various services, whereas indirect benefits were calculated based on the number of visitors and the length of their stay in the Amazonas region and in Chachapoyas, Luya and Tingo, the closest towns to the site (Araujo, 2019).

In 2017 and 2018, out of 222 dwellings, 42 were businesses related to the provision of tourism-related services: restaurants (36%), supply stores (33%), lodgings (19%) and others such as equipment rental (12%). A survey indicated that 73.81% of those businesses considered that the operation of the cable car system

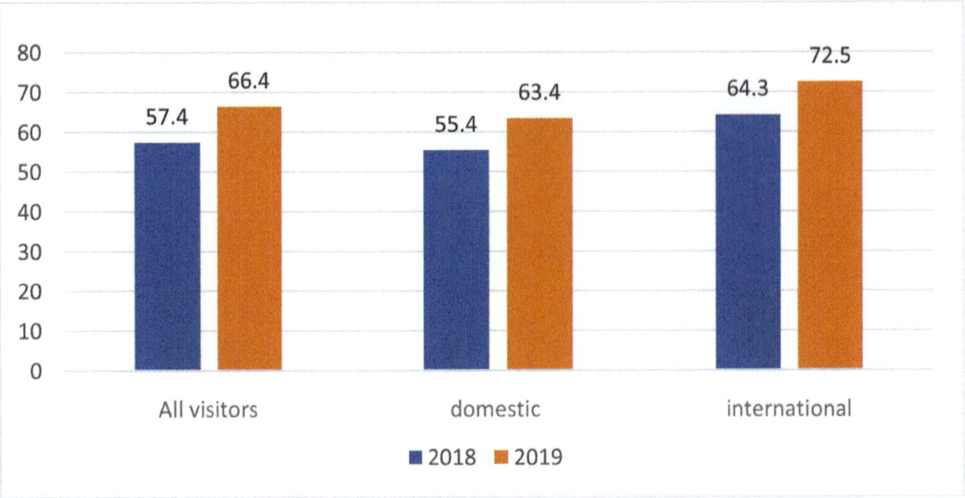

Fig. 23.6. Level of satisfaction among domestic and international visitors to Kuélap (courtesy of MINCETUR website).

has had a positive effect on their business. In 2019, on average they received 37 daily customers with an average expenditure of US$18/day. This is equivalent to an annual income of US$274,083. The average annual investment of the 44 businesses together was estimated at US$10,014 (Araujo, 2019).

Considering the increasing number of visitors and their expenditure in local businesses, this demand has resulted in the need to hire workers from the local community who either work in the town providing services at hotels or restaurants (Araujo, 2019) or maintaining the site (Rodriguez *et al.*, 2018).

Collaboration with local communities in decision making related to the management and operation of heritage sites has been identified as a critical part of the process of heritage management (Aas *et al.*, 2005). In developing countries, however, structural limits and a lack of a participatory development approach may impede such participation (Tosun, 2005). Visitor numbers data up to 2019 have shown mainly positive economic benefits in nearby towns to the fortress. However, no studies have focused on potential social benefits and challenges that may arise from the increase in visitor numbers.

Conclusion

Previous studies have argued that lack of infrastructure can hamper the tourism sector in less developed countries. Investment through partnerships between public and private sectors could be a key way to overcome this challenge (Liese and Beirsheim, 2011; Saner *et al.*, 2019; Kanwal *et al.*, 2020). This case study presented evidence of the encouraging initial results generated from the PPP undertaken for the improvement of infrastructure and tourism services in the Peruvian cultural attraction fortress of Kuélap. The 1st year of the implementation of the cable car system showed a growth in visitor arrivals of around 84%, a goal that was expected for the 10th year of operations. Studies by Lahura *et al.* (2019) and Lahura and Sabrera (2022) identified that the increase was not accidental and have proven a causal relationship between the investments made and the increase in demand of visitors to the archaeological site.

The greater number of visitors has contributed to local businesses, directly and indirectly, related to the provision of tourism services and products. Locals could invest those profits to enhance their products or services but perhaps more significantly, the additional source of income could assist their quality of life. This is relevant for regional communities that lack economic resources. In effect, investment in infrastructure for tourism purposes could also present benefits for local communities through the creation of services that were absent before, like roads and sanitation, among others. Aside from the current positive contributions of the cable car system, this case study has raised some issues that may become problematic in the long term, namely the need for conservation and collaboration with local communities.

Being the first cable car system developed in Peru, it has become an asset in itself, attracting a great number of domestic visitors. The cable cars are an innovative element that allows visitors to experience great scenic views but, more importantly, contributes to the decrease of distances, particularly in an area that has uneven terrain and requires trekking at high altitude. In a country like Peru with a rich history and culturally significant archaeological sites located in demanding geographical settings, the opportunity to implement services like the cable cars offers a great potential for tourism development. This is particularly important as Peru continues to rebuild its tourism industry after the devastating effects of the COVID-19 pandemic.

The authors acknowledge that while the PPP showed positive results from its inception until 2019, visitor numbers to the site have decreased from 2020 to 2022 as the country seeks to rebuild its tourism offering to encourage higher visitation, particularly from international tourists. In the years to come, further studies on the site will allow for exploration of additional dimensions of the effectiveness of the cable car system and its benefits for tourism services, especially those provided by the local population.

References

Aas, C., Ladkin, A. and Fletcher, J. (2005) Stakeholder collaboration and heritage management. *Annals of Tourism Research* 32(1), 28–48. DOI: 10.1016/j.annals.2004.04.005.

Araujo, W. (2019) Analysis of the *economic impact of the Kuélap gondola lift system on tourism in the Amazonas Region, Period 2017-1018.* Master's Thesis in Business Administration. Neumann Business School. Tacna. Available at: https://repositorio.epneumann.edu.pe/bitstream/handle/20.500.12892/120/TRABAJO_DE_INV_MAN_ARAUJO_ERICK.pdf?sequence=1&isAllowed=y (accessed 28 March 2023).

Chow, K.W.C., Tsui, H.K.W. and Wu, H. (2021) Airport subsidies and domestic inbound tourism in China. *Annals of Tourism Research* 90, 103275. DOI: 10.1016/j.annals.2021.103275.

Deng, T., Hu, Y. and Ma, M. (2019) Regional policy and tourism: A quasi-natural experiment. *Annals of Tourism Research* 74, 1–16. DOI: 10.1016/j.annals.2018.10.001.

Doerr, L., Dorn, F., Gabler, S. and Potrafke, N. (2020) How new airport infrastructure promotes tourism: evidence from a synthetic control approach in German regions. *Regional Studies* 54(10), 1402–1412. DOI: 10.1080/00343404.2020.1714022.

El Peruano. (2018) Legislative Decree Regulating the Promotion of Private Investment through Public-Private Partnerships and Active Projects. Available at: https://www.mef.gob.pe/es/normatividad-sp-29923/por-instrumentos/decretos-legislativos/17955-decreto-legislativo-n-1362-1/file (accessed 28 March 2023).

InfoBae (2022) Kuelap: Collapses, Actions, Losses in Tourism and Everything You Need to Know About the Emergency in the Pre-Inca Fortress. Available at: https://www.infobae.com/en/2022/04/12/kuelap-collapses-actions-losses-in-tourism-and-everything-you-need-to-know-about-the-emergency-in-the-pre-inca-fortress/ (accessed 16 October 2023).

Kanwal, S., Rasheed, M.I., Pitafi, A.H., Pitafi, A. and Ren, M. (2020) Road and transport infrastructure development and community support for tourism: the role of perceived benefits, and community satisfaction. *Tourism Management* 77, 104014. DOI: 10.1016/j.tourman.2019.104014.

Lahura, E. and Sabrera, R. (2022) The effect of infrastructure investment on tourism demand: A synthetic control approach for the case of kuelap, peru. *Empirical Economics* 65, 443–478. DOI: 10.1007/s00181-022-02334-1.

Lahura, E., Puscan, L. and Sabrera, R. (2019) *Effect of Infrastructure Investment on Tourism Demand: Evidence from the Kuélap Archaeological Complex.* Working Paper 2019-015. Central Reserve Bank of Peru. Available at: https://www.bcrp.gob.pe/docs/Proyeccion-Institucional/Encuentro-de-Economistas/2019/ee-2019-lahura-puscan-sabrera.pdf (accessed 20 April 2023).

Liese, A. and Beirsheim, M. (2011) Transnational public-private partnerships and the provision of collective goods in developing countries. In: Risse, T. (ed.) *Governance Without a State?: Policies and Politics in Areas of Limited Statehood*. Cambridge University Press, Cambridge, UK.

Los Angeles Times (2022) Part of the Wall of Pre-Inca Fortress Fall. Available at: https://www.latimes.com/espanol/internacional/articulo/2022-04-11/cae-parte-de-muralla-de-piedra-de-fortaleza-preinca-en-peru (accessed 16 October 2023).

MINCETUR (2016) Kuelap: experience beyond your imagination. In: *Project to Enhance the Understanding of the Factors That Explain Destination Competitiveness*. Progress Report to the UNWTO Committee on Tourism and Competitiveness, p. 33.

MINCETUR (2018) Level of Satisfaction of National and Foreign Tourists Visiting Amazonas 2017. Available at: https://www.gob.pe/institucion/mincetur/colecciones/577-peru-nivel-de-satisfaccion-del-turista-en-regiones (accessed 8 May 2023).

MINCETUR (2019) Level of Satisfaction of National and Foreign Tourists Visiting Amazonas 2018. Available at: https://www.gob.pe/institucion/mincetur/colecciones/577-peru-nivel-de-satisfaccion-del-turista-en-regiones (accessed 8 May 2023).

MINCETUR (2020) Level of Satisfaction of National and Foreign Tourists Visiting Amazonas 2019. Available at: https://www.gob.pe/institucion/mincetur/colecciones/577-peru-nivel-de-satisfaccion-del-turista-en-regiones (accessed 8 May 2023).

MINCETUR (2023) Statistical Information System. Available at: https://datosturismo.mincetur.gob.pe/appdatosturismo/content3.html (accessed 11 November 2023).

Nuryanti, W. (1996) Heritage and post modern tourism. *Annals of Tourism Research* 23(2), 249–260. DOI: 10.1016/0160-7383(95)00062-3.

Proinversion (2017) Kuelap Cable Car System. Available at: https://www.investinperu.pe/es/app/procesos-concluidos/proyecto/5464 (accessed 22 April 2023).

Proinversion (2021) Project APP Telecabinas Kuelap Doubled the Number of Tourists of the Archaeological Center. Available at: https://www.investinperu.pe/es/app/detalle-noticia/proyecto-app-telecabinas-kuelap-duplico-el-numero (accessed 22 April 2023).

Proinversion (2023) Telecabinas Kuelap has the Capacity to Move Up to 4000 Visitors a Day. Available at: https://www.investinperu.pe/es/pi/detail-news/telecabinas-kuelap-tiene-capacidad-para-trasladar (accessed 22 April 2023).

Rodriguez, M., Ricart, J.E. and Fageda, X. (2018) *PPP for Cities Case Study. Kuelap Gondolas (Peru)*. IESE Business School University of Navarra, Specialist Centre on PPPs in Smart and Sustainable Cities (PPP for Cities) and Development Bank of Latin America (CAF). Available at: https://www.iese.edu/wp-content/uploads/2019/03/ST-0469.pdf (accessed 12 May 2023).

Rogerson, C.M. (2016) Public-private partnerships for tourism infrastructure development: evidence from the cradle of humankind, south africa. *Tourism—An International Interdisciplinary Journal* 64(4), 419–430.

Saner, R., Yiu, L. and Filadoro, M. (2019) Tourism development in least developed countries: Challenges and opportunities. *Sustainable Tourism: Breakthroughs in Research and Practice*, 94–120.

Thopson, S. (2023) Foreign Tourism in Peru Has Not Recovered to Pre-Pandemic Levels. Available at: https://www.tourism-review.com/tourism-in-peru-not-successful-as-expected-news12974 (accessed 22 May 2023).

Tosun, C. (2005) Stages in the emergence of a participatory tourism development approach in the developing world. *Geoforum* 36, 333–352. DOI: 10.1016/j.geoforum.2004.06.003.

World Tourism Organization (WTO) (2015) Affiliate members global reports, volume eleven – public-private partnerships: tourism development. UNWTO, Madrid.

24 Conditional Subsidies for Domestic Tourists: The Case of PreViaje in Argentina

Carolina I. Garcia[1], Natalia Porto[2] and Matías O. Ciaschi[2]
[1]School of Management and Marketing (Tourism Research Cluster), Faculty of Business and Law, Curtin University, Western Australia, Australia; [2]Instituto de Investigaciones Económicas, Universidad Nacional de La Plata, Provincia de Buenos Aires, Argentina

Summary

This case study explores PreViaje (in English, PreTrip), an unprecedented tourism demand program launched in 2020 in Argentina as a response to the COVID-19 outbreak.

PreViaje is a public policy at the national level that subsidizes the selling of tourism services to residents travelling within the country. Its *sui generis* nature primarily lies in two characteristics: (i) the selling of tourism services takes place before travelling and (ii) the government subsidizes (up to) half of the tourist spending. The program features conditionalities that determine eligibility for the subsidy.

To study the emergence and evolution of the program, we use the three policy strategies suggested by the UNWTO to deal with the negative effects of COVID-19: (i) managing the crisis and mitigating the impact, (ii) providing stimulus and accelerating recovery and (iii) preparing for tomorrow.

The Value and Interest of the Case Study

The case of PreViaje shows how one single dynamic policy addresses the three strategies suggested by the UNWTO (2020) to deal with the consequences of COVID-19 on tourism. PreViaje is a one-of-a-kind program that has shown adaptability over time and seems to be here to stay. It has the potential to influence tourism policies in other Latin American countries.

Background

Argentina is well positioned as a world tourism destination. According to the United Nations World Tourism Organization (UNWTO , 2023), the country led international arrivals in South America in 2019 with almost 7.4 million international tourists, followed by Brazil with close to 6.4 million. Argentina is also one of the destinations with the highest international tourism receipts in the region, with more than US$5.2 million in 2019 and after only Brazil with close to US$6 million in the same year. Domestic tourism is even more relevant in Argentina. The country had 45 million domestic tourists in 2019, most of whom travelled for holiday and leisure (54%) or visiting friends and relatives (48%) (Argentinian Tourism and Sports Ministry, 2020).

The strong overseas and domestic tourism demand has its counterpart in a growing tourism supply. There were 41,650 registered tourism businesses in June 2021, most in the gastronomy sector (Argentinian Tourism and Sports Ministry,

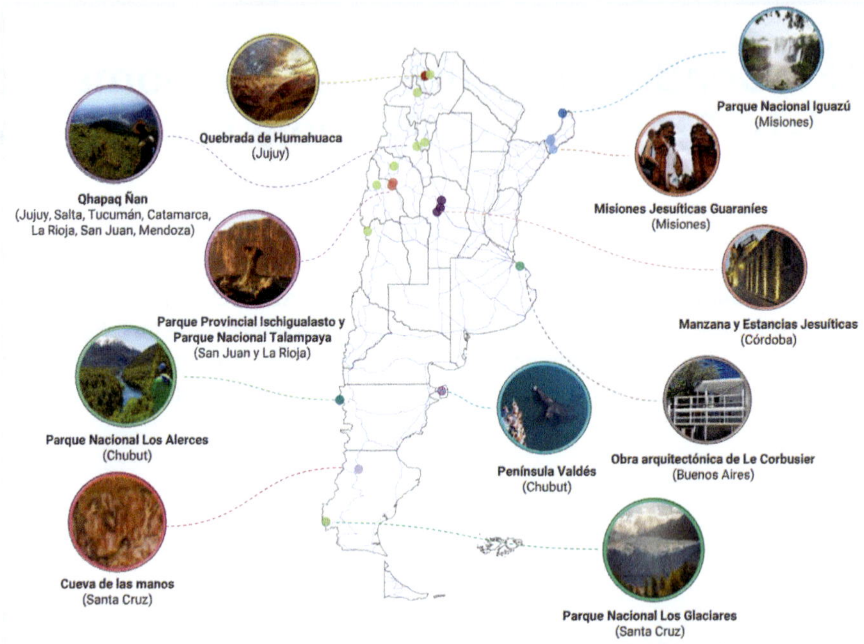

Fig. 24.1. UNESCO World Heritage Sites in Argentina. Source: Argentina Productive 2030 (2023), based on UNESCO (2023).

2022). Focusing on accommodation, around 20,000 lodgings featuring nearly 800,000 beds were available in 2021 (Argentinian Tourism and Sports Ministry, 2023b).

The endowment of tourism resources in Argentina is based on its unique natural and cultural assets. These are reflected in the 11 UNESCO (United Nations Educational, Scientific and Cultural Organization) World Heritage Sites the country has distributed throughout its territory (Fig. 24.1), showing part of its diverse cultural expressions and highly varied landscape. Nature-based tourism has been lately leading Argentina's tourism promotion strategies. *Ruta Natural* program (Natural Route) reflects these efforts by organizing the country's tourism offer in 17 routes that share a common natural identity (Fig. 24.2) (Argentinian Tourism and Sports Ministry, 2023a). The routes range from jungles and waterfalls in the north-east to mountains and glaciers in south-west Patagonia.

Tourism is a highly relevant sector in the Argentinian economy. It accounted for almost 7% of the country's exports and was a source of jobs for around 6% of all those employed in 2019 (Argentinian Statistics and Census National Institute INDEC, 2022).

Within this context, the impacts of COVID-19 in the year 2020 were devastating for the country and particularly for the tourism sector. Moreover, Argentina was among the countries with heavier mobility restrictions during the pandemic, with both international and national travel completely banned for several months. The country developed various programs to counter the negative effects of the pandemic. The most relevant tourism-specific public policy was called PreViaje and involved the promotion and subsidy of anticipated tourism sales for domestic visitors.

Aim and Methodology

This case study explores *PreViaje*, an unprecedented tourism demand program launched in 2020 in Argentina as a response to COVID-19.

Fig. 24.2. Natural Route program map. Source: Argentinian Tourism and Sports Ministry (2023a).

The case examines the emergence and evolution of the program following the three policy strategies suggested by the UNWTO (2020) as a response to COVID-19, which settle the bases for the policy to consolidate as a long-term scheme of domestic tourism promotion.

UNWTO policy strategies consist of actionable recommendations to help countries mitigate the short-term impacts of COVID-19 on tourism, provide a stimulus for recovery, and guide the long-term development and resilience of the sector (UNWTO, 2020). They are based on two principles: first, that supporting travel and tourism is supporting jobs and livelihoods, and second, that tourism has a proven capacity to bounce back and drive the recovery of other sectors (UNWTO, 2020).

Our contribution matches each UNWTO strategy with a different stage of implementation in the PreViaje program. For each stage, we first introduce the UNWTO strategy as a foundation. Then, we link this foundation with the specific policy approach adopted in *PreViaje*. Finally, we discuss the characteristics and conditionalities of each policy stage.

One Policy, Evolving Strategies

The start of PreViaje policy: Managing the COVID-19 crisis and mitigating the impact

Foundations

The first strategy to counter the adverse effects of COVID-19 was the management of the crisis and mitigation of its impacts. According to the UNWTO (2020, p. 15):

> Millions of livelihoods across the world are directly and indirectly dependent upon travel and tourism. As such, these measures are of the utmost urgency and should be implemented immediately, especially those related to providing timely, targeted, and temporary relief to the most affected people and firms for as long as the emergency lasts. UNWTO also emphasises the importance of supporting businesses economically and fiscally and of restoring confidence in the sector. Above all, support should be given to sustain jobs. Where this is not possible, every effort should be made to protect workers and create new opportunities, particularly for vulnerable groups.

Policy approach

The design and launch of PreViaje in October 2020 responded to strategies aligned with this first recommendation. While a broader package of measures was implemented to counter the effects of the pandemic and several initiatives looked after tourism businesses, PreViaje was a large-scale tourism-specific public policy.

PreViaje was a national-level policy that subsidized the selling of tourism services in advance to residents travelling within the country. The program tackled many of the UNWTO's (2020) first strategy targets, particularly sustaining jobs, supporting tourism businesses and restoring travel confidence with a demand-driven approach. This is because the program was created with two main goals in mind: (i) boosting domestic tourism demand and dynamizing the economy by achieving genuine income in the sector through the advanced selling of services, and (ii) promoting the federal distribution of resources to the different actors that make up the national tourism chain.

PreViaje was designed to subsidize up to half of the domestic tourist spending. The program was structured in a series of steps for the tourism demand and supply. It also featured conditionalities that determined eligibility for the subsidy (Fig. 24.3). These conditionalities were related to provider registrations, sales and travel windows, claiming credit, and using credit.

Provider registrations

Any tourism business willing to take part in the program had to be formally registered and apply to be part of *PreViaje*'s list of providers. Only specific economic activities were eligible: accommodation, travel agencies, transportation, car rentals, parks and museums.

Sales and travel windows

Tourists had to make travel-related purchases in specific periods of time and travel during eligible dates. In the first edition of the program, visitors could make purchases until the end of 2020 for trips taking place at any point during 2021.

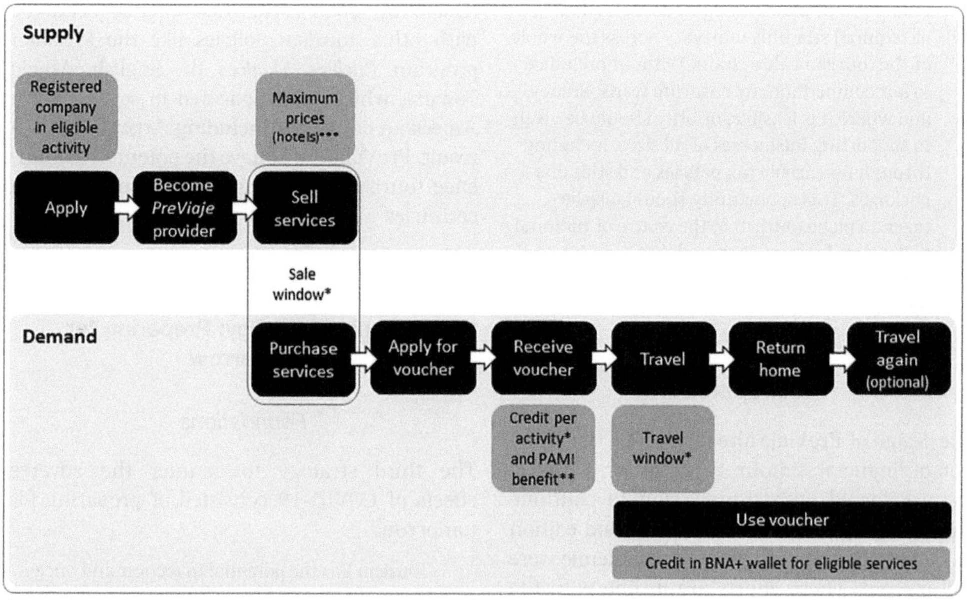

*Updated for each edition and incorporating seasonality concerns in the program's 3rd edition.
** Incorporated in the program's 2nd edition.
*** Incorporated in the program's 3rd edition.
PAMI is the Spanish acronym for the Comprehensive Medical Care Program for retired people over 70 years old without health insurance.
BNA+ is the national bank's digital wallet.

Fig. 24.3. *PreViaje's* steps and conditionalities. Source: Based on Garcia *et al*. (2024, Forthcoming). Figure originally developed for 'Wicked contexts pushing for change in Argentina: Tourism, labour informality and PreViaje policy' in *Tourism Policy-Making in the Context of Contested Wicked Problems* edited by Christof Pforr, Markus Pillmayer, Marion Joppe, Nico Scherle and Harald Pechlaner.

These dates would be assessed to determine if they were eligible PreViaje purchases.

Claiming credit
In order to receive a reimbursement in the form of a credit, visitors had to submit their purchase receipts on an online platform. They had a minimum expenditure per purchase and overall submission. As mentioned before, purchases had to take place within a specific timeframe, for certain travel services, and to be consumed in a pre-determined period of time. The credit went up to 50% of the purchase amount, with maximum money returns depending on the economic activity.

Using credit
Once the reimbursement was approved, beneficiaries received a credit via Argentina's state bank's, *Banco Nacion*, digital wallet (BNA+). This credit could be used during their planned trip or in subsequent trips. It was suitable for purchases related to the economic activities mentioned above (accommodation, travel agencies, transportation, car rentals, parks and museums) and some additional ones: excursion trips, tourism centres, attractions, gastronomy, regional products, entertainment and other tourism services.

Crisis = Opportunity? Providing stimulus and accelerating recovery

Foundations

The second strategy to counter the adverse effects of COVID-19 was the provision of stimulus and acceleration of recovery. Following the UNWTO (2020, p. 23):

> Due to the sudden, unprecedented fall in demand triggered by the COVID-19 pandemic, the travel and tourism sector will require

> financial stimulus to support recovery. Levels of required stimulus may vary across the whole of the tourism value chain. Without prejudice to fair competition, by ensuring transparency and where it is feasible, priority should be given to supporting businesses of all sizes, including through favourable tax policies and stimulus packages. This opportunity should also be taken to place tourism at the centre of national policies and ensure sustainability is an integral part of the sector as it recovers and grows again.

Policy approach

The design of PreViaje already tackled the provision of financial stimulus to the sector. This idea was reinforced with the decision to continue with the program and launch a second edition in 2021, when the effects of the pandemic were still present. These efforts are in line with the UNWTO (2020)'s second recommendation.

Not only did the government decide to continue with the program, but they also expanded it. The second edition featured a special benefit for retired people over 70 years old without health insurance. In Argentina, they are members of the government's Comprehensive Medical Care Program (PAMI is its acronym in Spanish). This group was eligible to claim up to 70% of their tourism expenditure as credit, 20% points above the 50% for all other PreViaje beneficiaries.

Additionally, the program *50 Destinos* (in English, 50 Destinations) was strengthened after the launch of *PreViaje*, with the idea of complementing its efforts. This program focused on the development of tourism infrastructure throughout the country and was a relevant driving force in terms of financial stimulus.

Another initiative launched as a complement to PreViaje was *Ruta Natural* (in English, Natural Route). This program targeted the development and promotion of nature-based tourism in Argentina. *Ruta Natural* introduced 17 natural routes focusing on enhancing tourism following sustainability concerns.

The consolidation of PreViaje generated international interest in the program. The Argentinian government provided specific training explaining PreViaje to other countries in the region, including Ecuador, Peru and Paraguay (Hosteltur, 2021). This had already occurred with other tourism policies like the Mexican program *Pueblos Mágicos* (in English, Magic Towns), which was replicated in several Latin American countries, including Argentina. As a result, PreViaje could have the potential to influence tourism policies in other Latin American countries.

Learning in policy: Preparing for tomorrow

Foundations

The third strategy to counter the adverse effects of COVID-19 consisted of preparing for tomorrow.

> Tourism has the potential to recover and once again establish itself as a key part of national economies and of the wider sustainable development agenda. This crisis may also offer a unique opportunity to shape the sector to ensure it not only grows but it grows better, with inclusivity, sustainability and responsibility prioritized. Furthermore, to build for the future, special attention should be placed on building resilience and on promoting sustainability at all levels. (UNWTO, 2020, p. 33)

Policy approach

PreViaje consolidated as a tourism policy in Argentina with subsequent third and fourth editions in 2022 and 2023. In these editions, some changes were introduced. This is aligned with the UNWTO's (2020) recommendations about preparing for tomorrow via tourism policies. Changes in PreViaje included off-peak season travel windows, compulsory price caps for hotels and increases in maximum money returns.

Off-peak season travel windows

The main policy change concerned eligible travel windows. These were restricted to very specific dates that coincided with the off-peak season in Argentina. As a result, PreViaje shifted from a broader demand program to a more tailored initiative focusing on reducing seasonality and distributing the benefits of tourism both in space and time, with implications for domestic tourism development. This approach is relevant considering that the asymmetric development of tourism in the country is partly due to the concentration of tourists in the high season, especially in January, February and July (Argentinian Tourism and Sports Ministry, 2023a).

Compulsory price caps for hotels
Price caps were determined for hotels in each destination. This involved discussions between the national government, local governments and hoteliers to determine maximum rates for hotel rooms. The aim of this change was to prevent unjustified hotel price increases as a result of PreViaje.

Increases in maximum money returns
As up to 50% of their purchases, with maximum money returns depending on the economic activity. These money returns were revised and updated considering the overall price increases in the country.

Challenges in policy implementation

PreViaje is a program featuring unique characteristics that has shown adaptability over time. Still, its implementation was not exempt from a diversity of challenges. In a recent presentation, the National Director of Marketing and Tourism Promotion identified some of these challenges and how they were overcome (Escuela de Economía y Negocios – UNSAM, 2023). Table 24.1 systematizes these learnings, complemented by statements by media outlets (La Agencia de Viajes Argentina, 2023), and groups them under common themes: communication to the private sector, communication to potential beneficiaries, development of web platform, validation of submissions, credit reception and use, and assistance and response to complaints.

Final Remarks

PreViaje is considered a highly successful policy. Since its implementation, the program has benefited approximately 7 million tourists (Argentinian Government, 2023), boosting the tourism sector and creating around 54,000 jobs (Argentinian Government, 2022).

The case of PreViaje shows how one single dynamic policy addresses the three strategies suggested by the UNWTO (2020) to deal with COVID-19: managing the crisis and mitigating the impact, providing stimulus and accelerating recovery, and preparing for tomorrow (Fig. 24.4).

First, we identified the emergence of PreViaje as a response to the COVID-19 pandemic. We highlighted the steps and conditionalities that this complex policy features. Second, we introduced the consolidation of the program to accelerate recovery. Here we noticed that not only did the program continue but it also expanded, with special benefits for retired people. Last, we showed the major changes and implementation challenges that the policy underwent in order to prepare for tomorrow. Changes included off-peak season travel windows, compulsory price caps for hotels and increases in maximum money returns. Meanwhile, challenges in implementation were related to communication to the private sector, communication to potential beneficiaries, development of web platform, validation of submissions, credit reception and use, and assistance and response to complaints.

PreViaje is a one-of-a-kind program that subsidizes the selling of tourism services to residents travelling within Argentina. The policy has shown adaptability over time, emerging as a response to the pandemic and undergoing significant changes. With its more tailored approach to improving domestic tourism development, PreViaje seems here to stay.

Table 24.1. Challenges during PreViaje's implementation.

Challenge	Details	Response
Communication to the private sector	• Making the private sector understand the program. • Engaging the private sector to participate. • Discouraging overprices as a result of the program.	• Providing training sessions to the private sector around the country on the program's details. • Setting maximum price caps to hotels per destination.
Communication to potential beneficiaries (consumers)	• Making potential beneficiaries understand the program. • Gaining beneficiaries' trust concerning credit granting.	• Generating an engaging but also clear promotional campaign for the program.
Development of web platform	• Developing a new web platform for the program. • Making the platform suitable for massive visitor figures.	• Establishing partnerships with a university (Universidad de Buenos Aires) for the development of a web platform suitable for massive visits.
Validation of submissions	• Verifying receipts to avoid fraud. • Checking details and dates in receipts to validate eligibility. • Checking beneficiaries' retirement status for special benefits.	• Establishing partnerships with state agencies in charge of businesses' registration and retirement registries. • Excluding cash payments as eligible.
Credit reception and use	• Creating accounts and cards for beneficiaries to receive credit. • Transferring credit amounts to beneficiaries' accounts. • Making it possible for beneficiaries to use their credit with eligible providers.	• Establishing partnerships with the national bank for the creation of bank accounts and cards on behalf of beneficiaries. • Adding the option for electronic wallets instead of physical cards. • Providing training sessions to the private sector around the country on specific needs for credit use (e.g. Wi-Fi access).
Assistance and response to complaints	• Responding to enquiries by both the private sector and potential beneficiaries. • Assisting potential beneficiaries with platform use, especially the elderly.	• Providing training sessions to the public sector around the country. • Increasing staff in assistance centre.

Source: Own elaboration based on Escuela de Economía y Negocios – UNSAM (2023) and La Agencia de Viajes Argentina (2023).

Fig. 24.4. UNWTO strategies, PreViaje's edition and stages. Source: Own elaboration based on UNWTO (2020).

References

Argentina Productive 2030 (2023) Misión 11. Potenciar la actividad turística para el desarrollo territorial a partir de la gran biodiversidad local. Plan para el Desarrollo Productivo, Industrial y Tecnológico – Ministerio de Economía de la Nación. Available at: https://tableros.yvera.tur.ar/recursos/biblioteca/mision_11.pdf (accessed 2 May 2023).

Argentinian Government (2022) PreViaje creó 54 mil empleos [press release] . www.argentina.gob.ar, 7 December. Available at: https://www.argentina.gob.ar/noticias/previaje-creo-54-mil-empleos (accessed 2 May 2023).

Argentinian Government (2023) Exitosa edición de PreViaje: Un millón de turistas disfrutarán del beneficio. www.argentina.gob.ar, 28 April. Available at: https://www.argentina.gob.ar/noticias/exitosa-edicion-de-previaje-un-millon-de-turistas-disfrutaran-del-beneficio (accessed 2 May 2023).

Argentinian Statistics and Census National Institute (2022) Cuenta satélite de turismo de la argentina Años 2004 y 2016–2019. *Cuentas Nacionales* 6(3). Available at: https://www.indec.gob.ar/uploads/informesdeprensa/cst_02_2252E11293B7.pdf (accessed 3 December 2022).

Argentinian Tourism and Sports Ministry (2020) Turismo interno 2019 – Total país. Dirección Nacional de Mercados y Estadística, Subsecretaría de Desarrollo Estratégico. Available at: https://www.yvera.tur.ar/sinta/informe/documentos/descarga/5f3d9c2f5572a140769562.pdf (accessed 2 May 2023).

Argentinian Tourism and Sports Ministry (2022) *Informe de cadena de valor del turismo*. Dirección Nacional de Mercados y Estadística, Subsecretaría de Desarrollo Estratégico. Working document Number 5, September 2022. Available at: https://tableros.yvera.tur.ar/recursos/biblioteca/cadena_de_valor.pdf (accessed 2 May 2023).

Argentinian Tourism and Sports Ministry (2023a) *Propuesta territorial de La Ruta Natural*. Subsecretaria de Desarrollo Estratégic. Available at: https://tableros.yvera.tur.ar/recursos/biblioteca/propuesta_territorial_lrn.pdf (accessed 25 May 2023).

Argentinian Tourism and Sports Ministry (2023b) Tablero Padrón Único Nacional de Alojamiento (PUNA) [online data visualization tool]. Available at: https://tableros.yvera.tur.ar/puna/ (accessed 25 May 2023).

Escuela de Economía y Negocios – UNSAM (2023) Ciclo Mano a Mano: Previaje desde adentro. 23 June. Universidad Nacional de San Martín, Argentina. Available at: https://www.youtube.com/watch?v=Fj3SbcHqXWI (accessed 31 July 2023).

Garcia, C.I., Porto, N. and Ciaschi, M.O. (2024, Forthcoming) Wicked contexts pushing for change in Argentina: Tourism, labour informality and PreViaje policy. In: Pforr, C. and et al, s (eds) *Tourism Policy-Making in the Context of Contested Wicked Problems*. Emerald.

Hosteltur (2021) Argentina exporta su programa PreViaje a Ecuador, Peru y Paraguay. Hosteltur. www.hosteltur.com, 1 November. Available at: https://www.hosteltur.com/lat/147646_argentina-exporta-su-programa-previaje-a-ecuador-peru-y-paraguay.html (accessed 2 May 2023).

La Agencia de Viajes Argentina (2023) PreViaje 4: La política de incentivos no termina de convencer al sector. https://argentina.ladevi.info/, 14 April. Available at: https://argentina.ladevi.info/previaje/previaje-4-la-politica-incentivo-no-termina-convencer-al-sector-n50888 (accessed 2 May 2023).

UNWTO (United Nations World Tourism Organization) (2020) Supporting jobs and economies through travel and tourism: A call for action to mitigate the socio-economic impact of COVID-19 and accelerate recovery. UNWTO E-Library. Available at: https://webunwto.s3.eu-west-1.amazonaws.com/s3fs-public/2020-04/COVID19_Recommendations_English_1.pdf (accessed 2 May 2023).

UNWTO (United Nations World Tourism Organization) (2023) UNWTO Barometer January 2023. UNWTO E-Library. Available at: DOI: 10.18111/wtobarometereng.2023.21.1.1 (accessed 31 July 2023).

25 Buenos Aires Youth Tourism Access Policy

Silvia A. Tapia[1], Bárbara Catalano[2] and Maria Xaus[3]
[1]National Scientific and Technical Research Council, Universidad de Buenos Aires, Buenos Aires, Argentina; [2]National Scientific and Technical Research Council, Universidad Provincial de Córdoba, Córdoba, Argentina; [3]Universidad de Buenos Aires, Buenos Aires, Argentina

Summary

In Argentina, in the south of Latin America, student tourism has been associated with trips for young people who finish secondary school level. Such trips have a mountain and snow destination during the winter season and are organized privately between travel agencies and student groups, without the intervention of schools or state aid. Although sustained for decades, given their high costs, access to this type of tourism has become increasingly associated with upper-middle socio-economic sectors. During the COVID-19 pandemic, in 2021, the government of Buenos Aires, one of the largest and most populous provinces of Argentina, established the programme 'End of Term Trip', a local public policy of student tourism aimed at secondary school students of that province. Through a free trip package, it links the aims of universality and equality with a tourism policy that promotes new destinations, favours the work of travel agencies, and boosts off-season tourism. Based on documented data and an interview with a programme official this case study presents a tourism policy that seeks to boost the tourism sector and, at the same time, encourages the participation of young people from different social sectors in recreational tourism practices and their access to leisure and tourism rights.

The Value and Interest of the Case Study

The case of the End of Term Trip Programme shows the potential and limitations in the implementation of a local tourism policy with a macro-contextual justification of the impact suffered by tourism activity during the COVID-19 pandemic and the consequent need to revitalize the sector in highly affected areas, as well as an objective associated with the extension of young people's rights to leisure, tourism, and travel.

Background

The possibilities and ways of travelling and doing tourism are a distinguishing axis among young people from different social sectors.

In Argentina, in the south of Latin America, because of socio-economic inequalities, and workers' demands developed since the mid-20th century, tourism was considered an activity and a public problem that required State intervention (Piglia, 2011). In the region, this country was a pioneer in rights acknowledgement associated with holidays and free time. First, the upper classes, the aristocracy, and then the working class, acquired access to travel as a practice with motives of leisure and rest. In this context, state interventions targeted domestic tourism, especially social and school tourism aimed at workers and young people, and international tourism, by promoting investments and agreements between employers, unions, and the state (Pastoriza and Pedetta, 2009; Pastoriza, 2011).

The province of Buenos Aires (PBA) functioned as a laboratory of social tourism, from which the national government generated a broader strategy of significant political impact (Piglia, 2012). One of its coastal cities, Mar del Plata, was presented as an iconic model where the first attempts to organize and create a tourist centre were tested. It was a democratizing process of access to leisure, accompanied by the reduction of the working day, paid vacations, the development of the railway and the social ascent of the working class (Pastoriza and Pedetta, 2009).

The PBA is in the centre of the country (Fig. 25.1). It represents the largest jurisdiction in Argentina. Equivalent in size to countries such as Italy, the PBA has a territory with multiple administrative units and is composed of 135 municipalities. The province presents important socio-demographic, environmental, and socio-economic diverse characteristics. This heterogeneity has impacted differentiating the social and productive development of the regions within the PBA (CEPAL, 2019).

Its current population is 17,569,053, 38% of the country's total. A total of 61.8% of these inhabitants concentrate in the 'Great Buenos Aires', representing the main population core of the country (Provincial Statistics Direction, 2023). This sector of the PBA refers to a group of 24 municipalities in a metropolitan area that holds the major industrial activity in the province. It has been associated with an image of a territory of popular cultural and social integration, and at the same time, with a negative and conflictive image connected to crime and insecurity, urban polarization, poverty, and inequalities (Kessler, 2015).

In the PBA district, 'Tourism represents a key economic sector that boosts local and regional development, making the province of Buenos Aires the leading tourist destination in the country' (GPBA, 2023). Among its main destinations, the beaches on the Atlantic coast are the most visited. However, in the last two decades, the tourist offer has diversified, giving rise to new attractions associated with the valorization of cultural and historical heritage (Pérez Winter, 2014).

Within the framework of ongoing research that seeks to identify the links between tourism, (im)mobilities, rights, and inequalities, based on the experiences of young secondary school students from popular neighbourhoods of the Metropolitan Area of Buenos Aires, regarding student tourism, as well as the policies and regulations linked to this type of trip, the case of the *End of Year Trip programme* is discussed. Launched in 2020, currently in its third edition, the programme was created to facilitate universal access to graduate trips for young students in their last year in secondary schools located in the PBA and, at the same time, promote the reactivation of tourism in that jurisdiction.

The analysis of this programme shows the economic and social scope of tourism policies that emerged from the measures imposed by the COVID-19 pandemic, regarding state interventions against unequal youth experiences in the exercise of rights associated with tourism. To present this case, we draw on contributions from the *mobility justice* approach (Sheller, 2018). We understand, along with Sheller, mobility justice conceptualizations as a political and ethical issue and an approach to think about different ways of transitioning towards more just mobilities.

In this regard, we find it important to observe uneven levels of *motility*. It refers to the capability for mobility and control over the mobility of Others that social groups have. It also implies different *accessibility* to various kinds of spaces, experiences, and social goods that result from racial, gendered, classed, sexed, and other social conditions, that are also products of those hierarchical systems of differentiation.

To gather information about the case, a qualitative approach was adopted that used primary and secondary sources: the programme website, associated regulations (decree, resolutions, and laws), articles from two of the most relevant national digital newspapers in Argentina (La Nación and Página/12) that, with positions for and against, covered its creation during 2021, 2022, and 2023. We

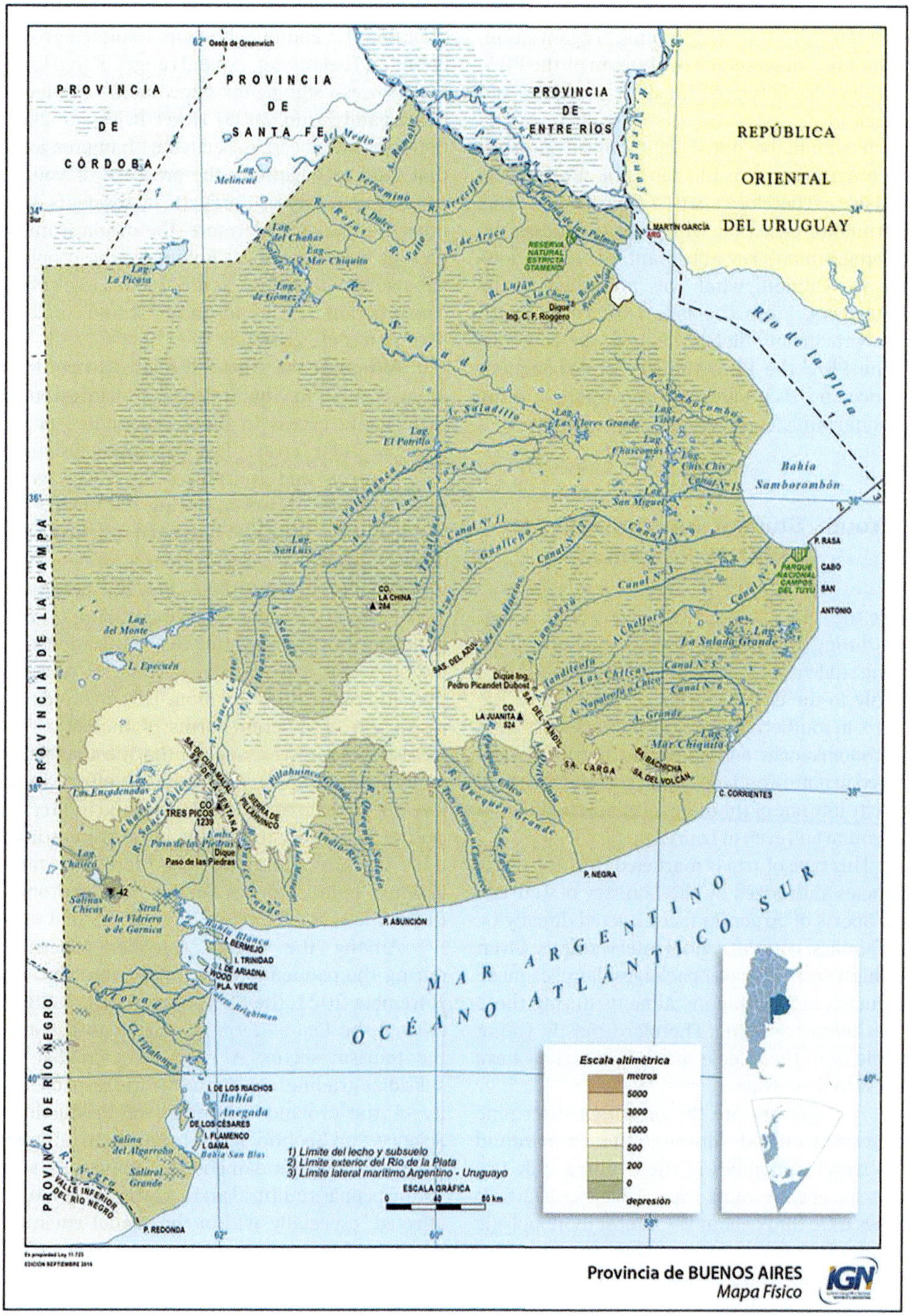

Fig. 25.1. Map of the Province of Buenos Aires, Argentina. Courtesy of the National Geographic Institute, Argentina.

also interviewed a programme official as a key informant of the programme organization. Within the Subsecretary of Tourism of the PBA, he intercedes between the state, students, and agencies.

It should be noted that official data on the programme's results are not yet publicly available. Therefore, we reconstructed this information based on what was mentioned by the programme's key informant, the regulations and, in addition, what was published in the newspapers. From the latter, interviews with other provincial officials, such as the governor of the PBA, the PBA Ministry of Production, Science and Technological Innovation, and the National Minister of Tourism, were retrieved.

Youth, Student Tourism, and the COVID-19 Pandemic

Since the 1980s, especially in the winter season and during the last year of secondary school, young students (17–19 years old) have made trips mainly to the city of Bariloche, province of Rio Negro, in southern Argentina. Bariloche has a big gastronomic offer and important tourist services related to nature and mountain sports. These turn this city into one of the main tourist destinations in Argentina and even in Latin America.

This type of trip is marketed through travel agencies authorized by the Ministry of Tourism and Sports of Argentina, contracted directly by the families, without school interventions. Given the high costs of travel packages, families make payments in monthly instalments during the 2 years before travelling. Therefore, mostly, young people from the middle and upper sectors have access to these trips.

The agencies are the ones that determine the services offered. Although there are around 200 travel companies in the country, only five agencies operate most trips (Vallejos, 2020). It allows us to think about the almost monopolistic nature of such companies. They organize trips of 7–10 days long, which include transportation, accommodation and an intense routine of day and night leisure activities for young people. One of the main activities is associated with night outings to nightclubs, which are the greatest attraction of the trip.

These itineraries are lived as experiences to celebrate the end of schooling, reinforce group bonds and belonging, expand ranges of freedom, and process significant educational moments (Rocha and Surdo, 2005). Nevertheless, graduation trips have been associated with impressions that tend to negativize the presence of young people in the city of Bariloche by residents and other tourists. Furthermore, the media usually present negative images linking young people's behaviour with risky practices related to alcohol consumption, night outs and/or sexual conduct (Eymann *et al.*, 2007).

Although the trips are related to government regulation, the State has not intervened to facilitate access to those experiences or to regulate their costs. The existing legislation only controls and guarantees the travel agencies' activities. However, during the COVID-19 pandemic, the state also regulated the permitted dates to travel.

It is important to note that, in Argentina, in March 2020, after the declaration of the pandemic by the World Health Organization, the national government determined social, preventive, and mandatory isolation. This involved the restriction of different forms of mobility. The tourism sector was one of the most affected sectors by such policy measures. It affected the levels of income of people directly and indirectly dedicated to tourism, mainly in tourist destinations. Local communities were forced to take action to redirect tourist flows to their territories (Izquierdo *et al.*, 2020).

Among the tourist activities restricted during the pandemic were graduation trips. In November 2021, the PBA's Government (GPBA) created the *End of Term programme* to activate the tourism sector. A policy was created to subsidize graduation trips under the responsibility of the province's Ministry of Production, Science and Technological Innovation. Among the programme fundamentals, it stands out how young people's cultural and social activities were affected, especially within the school environment. In this sense, the programme recognizes graduation trips as a 'ritual' that enables young people to access tourism as a social and cultural right.

Unlike the traditional graduation trips to Bariloche city, the programme redirects this type of trip to destinations situated only within the

PBA. It especially proposes that young people travel to cities that are not common tourist destinations in the province. In this regard, as our key informant stated, any city of the province with different historical patrimony, natural landscape, or tourist development may become a destination for the End of Term Trips. This is related to the aim of 'boosting tourism, extending the season, reactivating the economy and promoting inclusion, and re-engagement for 220,000 students' (GPBA, 2021). Below, we describe the programme's main features.

About the programme

This initiative offers free group travel packages. The GPBA finances all costs associated with the trips and acts as a mediator between travellers and travel agencies. The itineraries, of 4 days and three nights, include two excursions per day, all meals, transfers, two excursions and medical assistance. They have different tourist destinations of beaches, mountains, lakes, or rural environments within the PBA. The budget for each student is 50,000 Argentinian pesos (approximately US$200) for all the services included (Lucesole, 2022).

The beneficiaries are students in their final year of state-run and private secondary education in the province, including special education for students with disabilities and formative centres for youngsters and adults. According to the resolution of the programme, 'Within the universe of tourism activities, *student tourism* is understood as the set of activities carried out by young people and adolescents during their educational training period, with the aim of satisfying cultural and leisure motivations' (GPBA, 2023, p. 2).

To take part in the programme, a group must gather at least 10 applicants. In addition, it includes one accompanying parent bonus trip for every 20 students. Beneficiaries can express their destination and a travel date of preference. However, the selected destinations must be 600 km or less from the student's school. According to our key informant, this rule was implemented to prevent the students from spending excessive hours travelling, instead of enjoying their destination.

Participating in the programme consists of three moments: registration, approval, and execution of the trip. Registration is online through a form in which school and personal data are recorded. After registration closes, the destination and contact details of the agency that organizes the itinerary are confirmed by e-mail.

According to the official we interviewed, as well as the comments of other authorities retrieved from the newspapers, young people went through several difficulties during their last school year isolated by the pandemic. Because of that, the authorities understand these trips as a way of reconnecting the students not only socially and affectionally but also as experiences of learning, strengthening ties and closure of their secondary education. With these travel opportunities, they sought to guarantee youth rights to tourism and leisure.

This policy is related to the Support and Productive Reactivation of Tourism Activity Law which aims to boost domestic tourism through tax measures, fiscal stimulus, and financial support. Because of this, another of the central foundations of the programme is based on making trips to local destinations in periods of low season and managed in partnership with private agencies of the province that organize all the services offered. As the key informant mentioned, the programme asks the agencies to offer at least 10% of destinations different from the traditional ones.

End of Term Trip: Mobilizing Rights and Opportunities

Youth on the move

The programme is presented as a universal policy that implies the responsibility of the State to guarantee the access and exercise of a right, in this case, to leisure and tourism. For this purpose, it enables mechanisms to give equal opportunities for all young students from secondary schools, both state-run and private, within the province (GPBA, 2021). It should be noted that, at the beginning of the programme, in 2021, of the total number of schools at this level (3981), 64% were state-run and 36% private. The latter includes more than

70% that gain a state subsidy (National Ministry of Education, 2023).

When comparing the number of young people who have participated in the programme to the total number of potential beneficiaries (approximately 220,000 students), two situations can be distinguished: those who only enrolled and those who finally travelled. The PBA government highlights the success of the two previous editions of the programme in terms of the number of young people, mostly from public schools, who managed to visit different parts of the province: about 20,000 of 30,000 enrolled, travelled in the first edition (Lucesole, 2022) and 110,000 of 130,000 enrolled travelled in the second one (GPBA, 2023).

Differences between the first and second editions can be linked to the moment of the year when the trips were made. In the first edition, for the group of students who finished secondary school in 2020, the dates of the itineraries were set for 2021: 'Since tourism activity had not yet recovered after the pandemic' (programme official). These dates made it difficult to travel because many of the students were engaged in other activities: 'As reported by the authorities, many of them gave up. As indicated, the main reason was the beginning of university' (Lucesole, 2022).

Due to the difficulties of the first edition, changes were made to facilitate participation in the programme. Dates were modified so that the trips took place during the school year, considering the needs and capacities of the tourism sector. At the same time, the registration process was improved: (a) dates for inscription were extended to reach more students, and (b) the registration was digitalized, since filling out a form by hand and uploading it to the web involved using tools that not all had access to.

Nevertheless, according to our key informant, the problem with registration remains unsolved. Even with a relevant justification, those who have not accurately filled out the form or have not registered on time cannot travel. The reason is that travel insurance has already been contracted and, without it, students cannot travel. This precludes exceptions, even if they are duly substantiated.

The programme sought to address two limitations of young people's mobilities after the pandemic: the negative psychosocial consequences of the pandemic and the socio-economic difficulties experienced by many of the families of PBA students that prevent them from paying for a graduation trip.

In this regard, various public officials justify the programme's creation to alleviate such situations. Thus, for example, the minister of the agency responsible for the programme mentioned: 'It means good news for a group of teenagers who had to finish secondary school in a pandemic. Many of them could not go on a graduate trip. Because they did not have the resources to do so, because it meant a limitation for the family' (Página/12, 2022a). Similarly, in response to criticism of this policy from opposition politicians, the governor of the province said: 'They pretend that only those who are able to pay can visit other places' (Página/12, 2021).

The programme not only aimed to favour access to travel for young people to different destinations. The key informant affirmed that the programme also sought to ensure the quality of services offered, such as three-star or more hotels, travel insurance and recreational activities. Although, as he pointed out, there is still a need to improve the offer of night outings that are of great interest to young people.

Tourism sector on the move

From the beginning of the programme and during its implementation, the current government was questioned by the opposition politicians. They criticized the use of public funds for financing trips for young people. Instead, they proposed allocating these funds to other purposes considered more appropriate, such as the construction of schools or the purchase of computers for students. In response to these criticisms, the governor and the programme officials highlighted how travel not only beneficiated young people but also revitalized the tourism sector. The PBA minister of Production affirmed 'The End-of-Year Trips programme guarantees rights, generates work and income and allows the tourist season to be extended' (Página/12, 2022b).

The launch of the programme addressed a constant difficulty in the field of tourism: seasonality. It is characterized by the low demand of visitors in off-season periods of certain destinations. This situation has repercussions on business and individual income and produces unstable labour conditions in the sector, which ends up being informal and precarious. The programme seeks to alleviate seasonality by redirecting groups of students towards the low-season months. This process contributes to the revitalization of the sector and mitigates the idle capacity generated in the months when cities do not receive many visitors.

In turn, the programme seeks to deconcentrate tourism, in geographical terms, to attenuate the unequal tourism development among the province's municipalities. Given that purpose, the flows of students are redirected to the less visited or popular destinations that have natural and cultural attractions. As the key informant points out: 'What you want is to be equitable, right? The programme's purpose is that the population reaches many destinations, many businesses and that much of the sector receives tourist movement out of season'. However, despite the search for diversification, according to the results of the programme's first edition released in newspapers, the destinations most chosen by young people were coastal cities, the already most visited tourist destinations in the province, such as Villa Gesell, Pinamar, Miramar, Mar del Plata, San Bernardo, Santa Teresita or San Clemente, and only a few cities that present others characteristics, such as Sierra de la Ventana and Tandil, with a hilly landscape (Lucesole, 2022).

Another aim of the programme's implementation is to reactivate the economic-commercial and operational activities of tourism agencies. They are significant actors in the intermediation of tourism services such as hotels, transport, accommodation, and recreational activities. According to the official interviewed, after suffering the devastating effects of the pandemic, this policy helped the local agencies to recover from the economic crisis. In many cases, they needed to expand their fleet, infrastructure, and human resources, to respond to the programme demand. To participate in the programme, they must respond to the national student tourism legislation (Law No 25599, Argentina, 2002) that determines obligations for the trips such as an authorized group coordinator or young travellers' insurance (Lucesole, 2021).

This policy reactivates the tourism sector through monetary injection in the network of actors involved in the productive chain of tourism activity (hotels, gastronomy, travel agencies, recreational activities, transport, etc.). GPBA officials, to justify the benefits of the programme, emphasized how the money was not aimed only at young people but also at the different elements of the touristic chain, including money that the groups of young people invested in the destinations during their visits (Fernández, 2023).

Conclusions

The case of the End of Term Trips programme emerges as a response to problems arising in the context of the COVID-19 pandemic that affected social groups in different ways. The singularity of this programme was to present itself as a tourism policy able to identify, at the same time, two actors as beneficiaries: young people and agents of the tourism sector.

Consequently, the programme has managed to *mobilize* young people, agents, destinations, and temporalities, questioning what or who moves, from and to where, in what way and under what conditions (Sheller, 2018). Given a diagnosis of a situation before the pandemic, two scenarios emerged: many of the province's young people did not have access to graduation trips, especially for economic reasons, and several destinations with interesting attractions were not visited or only received tourists on limited dates during the year. To change this, the programme proposed equal opportunities for both actors. By mobilizing economic resources, young people can travel, and destinations can offer their places as possible destinations, articulating with local agencies which also benefited from this policy. Because of the success of the high demand for trips by students of the province and its positive effects on the tourism sector, the local government decided to release the third edition of the programme for the current year, 2023.

It is worth noting how the possibility of having the right to move, to know new places and to share cultural and recreational experiences with other young people introduced by the programme enables new images of students as subjects of rights instead of the negative discourses that present them as problematic and at risk. Given the gratuity and universality of the proposal, the programme put on the agenda the discussion about young people's rights to leisure, travel, and tourism. However, the need to constantly justify the dual function of the policy, that is, benefiting young people and the tourism sector, limits the programme officials' argument against criticisms that do not consider the validity of youth's rights to enjoy or to travel.

It is important to note that, in December 2023, a new national government with a neoliberal approach took over in Argentina. This entails the shrinking of the state through the reduction of public investment, the limitation of the exercise of rights for middle and popular sectors, the privatization of public assets, and the strengthening of large business sectors, among other things. In this context, Governor Axel Kicillof was re-elected in the PBA, continuing his administration. However, due to the anticipated decrease in the public budget for the provinces, as the governor stated, it is still not possible to determine which policies, including those related to tourism, can be sustained in the future.

References

Argentina (2002) Law No 25599. Tourist Travel Agencies.

Comisión Económica para América Latina y el Caribe (CEPAL) (2019) Territorio y desarrollo en la Argentina: las brechas estructurales de desarrollo en la provincia de Buenos Aires. Documentos de Proyectos (LC/TS.2019/30), Santiago 114 p.

Eymann, A., Busaniche, J., Mulli, V., Paz, M. *et al.* (2007) Estudiantes secundarios y su viaje de egresados: prevalencia de hábitos, conductas de riesgo y enfermedades en los años 1997 y 2004. *Archivos Argentinos de Pediatria* 105(1), 17–22.

Fernández, L. (2023) Kicillof anunció la tercera etapa del programa Viajes de Fin de Curso. *Página/12*, 5 January. Available at: https://www.pagina12.com.ar/513300-kicillof-anuncio-la-tercera-etapa-del-programa-viajes-de-fin (accessed 5 April 2023).

Gobierno de la Provincia de Buenos Aires (2021) Decree No 973. Buenos Aires Student Tourism Programme creation.

Gobierno de la Provincia de Buenos Aires (2023) Resolution no 83. buenos aires student tourism programme.

Izquierdo, S.I., Diez Alberdi, A.M. and Camio, M.I. (2020) Resiliencia en destinos turísticos. Un análisis de las ciudades de Tandil y Mar del Plata, Argentina frente a la COVID-19. *PID Magazine* 8(3), 14–40.

Kessler, G. (2015) *Historia de la provincia de Buenos Aires: el Gran Buenos Aires*. Edhasa, Buenos Aires.

Lucesole, M.J. (2021) Axel Kicillof abre el registro de viajes egresados gratis cuando faltan 14 días para las elecciones. *La Nación*, 1 November. Available at: https://www.lanacion.com.ar/politica/axel-kicillof-abre-el-registro-de-viajes-egresados-gratis-cuando-faltan-14-dias-para-las-elecciones-nid01112021/ (accessed 5 April 2023).

Lucesole, M.J. (2022) La Provincia destinaría unos $6500 millones para viajes de egresados. La Nación. 1 September. Available at: https://www.lanacion.com.ar/sociedad/la-provincia-destinaria-unos-6500-millones-para-viajes-de-egresados-nid01092022/#:~:text=As%C3%AD%2C%20en%20caso%20de%20que,de%2070.000%20millones%20de%20pesos (accessed 5 April 2023).

National Ministry of Education (2023) Integrated System for Consultation of Educational Data and Indicators (SICDIE). Available at: https://data.educacion.gob.ar/nivel/secundario-comun (accessed 12 May 2023).

Página/12 (2021) Axel Kicillof volvió a defender los viajes de egresados gratuitos y se cruzó con Diego Santilli. 11 October. Available at: https://www.pagina12.com.ar/373955-axel-kicillof-volvio-a-defender-los-viajes-de-egresados-grat (accessed 5 April 2023).

Página/12 (2022a) Arrancaron los primeros viajes de egresados gratis en Buenos Aires. 29 March. Available at: https://www.pagina12.com.ar/471427-arrancaron-los-primeros-viajes-de-egresados-gratis-en-buenos (accessed 5 April 2023).

Página/12 (2022b) Viajes de egresados gratis en Provincia de Buenos Aires: el gobierno bonaerense lanzó una nueva edición del programa. 29 June. Available at: https://www.pagina12.com.ar/431892-viajes-de-egresados-gratis-en-provincia-de-buenos-aires-el-g (accessed 5 April 2023).

Pastoriza, E. (2011) *La conquista de las vacaciones. Breve historia del turismo en la Argentina*. Edhasa, Buenos Aires.

Pastoriza, E. and Pedetta, M. (2009) Lo que el pueblo necesita. Turismo social y Peronismo. Argentina, 1930-1943. *Études Caribéennes* 13–14.

Pérez Winter, C. (2014) La provincia de Buenos Aires (Argentina): de un espacio a un territorio turístico. *Turismo – Visão e Ação* 16(3), 544–572.

Piglia, M. (2011) The awakening of tourism: The origins of tourism policy in Argentina, 1930–1943. *Journal of Tourism History* 3, 57–74.

Piglia, M. (2012) En torno a los Parques Nacionales: primeras experiencias de una política turística nacional centralizada en la Argentina (1934–1950). *Pasos. Revista de Turismo y Patrimonio Cultural* 10(1), 61–73.

Provincial Statistics Direction, Province of Buenos Aires Government (2023) *Análisis de los Resultados Provisionales del Censo Nacional De Población,. Hogares y Viviendas 2022 En La Provincia de Buenos Aires*.

Rocha, N. and Surdo, R. (2005) Turismo estudiantil masivo en Bariloche-Argentina. Estrategias de recreación en las discotecas. *Estudios y perspectivas en turismo* 14(1), 5–19.

Sheller, M. (2018) Theorising mobility justice. *Tempo Social* 30, 17–34.

Vallejos, S. (2020) Coronavirus en la Argentina: ¿viajes de egresados en septiembre? una pequeña ilusión para los estudiantes secundarios. *La Nación*. 27 May. Available at: https://www.lanacion.com.ar/sociedad/coronavirus-argentina-viajes-egresados-septiembre-pequena-ilusion-nid2370410/ (accessed 5 April 2023).

26 Pedal Power to the People? Cycle Tourism and Local Economic Development in Cuba

Erin Sharpe
Department of Recreation and Leisure Studies, Brock University, St. Catharines, Ontario, Canada

Summary

Cycle tourism is a rapidly growing niche sector of the tourism industry and has been highlighted for its potential as a tool for socio-economic development, particularly in rural communities. Although tourism is well established in Cuba, it is primarily beach-based and massively controlled by state entities. However, Cuba has recently made a series of legislative changes intended to grow private sector small business opportunities. These changes open up the possibility for Cuban people to play a greater economic role in tourism than previously, if opportunities can be identified and challenges overcome. This case study considers the potential of cycle tourism as a form of people-powering tourism in Cuba. Using interview, field-based, and textual data, this case study maps the current state of cycle tourism, its actors and their interactions, and the opportunities, current challenges, and requirements to grow people-powering cycling tourism in Cuba.

The Value and Interest of the Case Study

By focusing on the niche sector of cycle tourism, readers of this case study will gain insight into the complex political, legal, and economic factors that control and distribute the benefits produced through tourism in Cuba. Attending to the political and economic context of tourism in Cuba encourages readers to consider the ways that tourist behavior produces economic benefits and for whom, and how tourism policy and practice can be (re) oriented to produce more empowering outcomes for the Cuban people.

Cycle Tourism and Local Economic Development

Akama and Kieti (2007) argue that local socio-economic development through tourism requires local empowerment and participation in tourism, including participation in the tourism market in meaningful ways. One of their critiques of tourism in Mombassa, Kenya, for example, was that local people were rarely involved in the provision of core and profitable tourism and hospitality services (e.g. transportation, accommodation, food services, management of tourism facilities). Resorts and other forms of 'enclave tourism' (Goodwin, 1998, p. 4) have been criticized as a form of internal colonialism where tourism resources in a destination mostly benefit outsiders while the majority of local people derive only marginal benefits. For tourism to offer local socio-economic benefits, local communities must be involved in core tourism and hospitality services so that revenue has the opportunity to leak into local communities where its impact can be compounded. According to Goodwin (1998), tourism features that promote local socio-economic development include: (1) local ownership and management of tourism resources; (2) enhancement of linkages and partnerships to the local economy; and (3) local sources of supply to tourism establishments. Community control over-representation in marketing and promotion is also a facet of local empowerment (Akama and Kieti, 2007).

One of the forms of tourism that has been highlighted for its potential for socio-economic development is cycle tourism. Cycle tourism is tourism in which bicycles are used as the primary means of transportation, although not necessarily exclusively (Petino *et al.*, 2021), with the cycle tourer traveling to a different destination by cycling each day (Gazzola *et al.*, 2018). Currently a rapidly growing segment of the tourism market, the emergence of e-bicycling modalities has led some to suggest that 'although niche, it is a sector with exponential potential' (Petino *et al.*, 2021, p. 4).

One reason that cycle tourism has been linked with local socio-economic development is because cycle tourists seek areas with calm traffic combined with landscapes immersed in nature. As such, cycling moves tourists out of tourist enclaves and into remote and marginal areas with a lesser tourism presence (Petino *et al.*, 2021). The slower speed of travel by bicycle opens up greater opportunities for encounters with local landscapes, communities, and culture (Gazzola *et al.*, 2018). Tourers spend money on direct tourism activities (e.g. overnight stays, culinary experiences, visiting local sites) (Gazzola *et al.*, 2018). Cycle tourists' daily spending is comparable to that of other tourists, and cycle tourism can contribute in particular to rural economies that have not previously enjoyed mainstream tourism development (Piket *et al.*, 2013). For these reasons, tourism planners have started to look at cycle tourism strategically as a form of sustainable rural tourism. Consequently, they have, for example, added cycling infrastructure and created cycling itineraries. Researchers have documented the potential of cycling tourism for sustainable rural and remote development in regions of Sicily (Petino *et al.*, 2021), northern Italy (Gazzola *et al.*, 2018), and Croatia (Brščić *et al.*, 2021). Although initial findings are encouraging, they noted that cooperation and coordination among local stakeholders are required for cycle tourism to flourish (Gazzola *et al.*, 2018).

This case study considers the potential of cycle tourism as a tool for local socio-economic development in Cuba. Although tourism is well established in Cuba, it is primarily beach and city-based and massively controlled by state entities. However, Cuba holds promise as a setting for cycle tourism, which could spread tourism benefits into a wider array of areas in the country. Further, Cuba has recently made a series of legislative changes intended to grow small business opportunities for Cubans. These changes open up the possibility for a wider array of Cuban people to play a more meaningful role in the tourism economy than previously, if opportunities can be identified and challenges overcome.

Using interview, field-based, and textual data collected in Cuba in spring 2023, this case study maps the current state of cycle tourism, its actors and their interactions, and the opportunities, challenges, and possibilities for local socio-economic development through cycle tourism in Cuba. By focusing on the niche sector of cycle tourism, readers will gain insight into the complex political, legal, and economic factors that control and distribute the benefits produced through tourism in Cuba. Attending to the political and economic context of tourism in Cuba encourages readers to consider the ways that tourist behavior produces economic benefits and for whom, and how tourism policy and practice can be (re)oriented to produce more empowering outcomes for the Cuban people.

Tourism in Cuba – Background

The evolution of tourism in Cuba corresponds to the country's economic and socio-political history. In the first half of the 19th century, tourism flourished. Cuba was the Caribbean's top tourist destination, receiving just over one-quarter of all visitors to the region. Eighty-five percent of tourists to Cuba were North Americans, who were drawn to the island for its beaches, gaming, and prostitution (Salinas *et al.*, 2018). Tourism changed dramatically with the 1959 Castro-led Cuban Revolution, as the revolutionary government viewed tourism as a tool of colonialism and imperialism. The Castro regime expropriated hotels, reopened beaches to Cuban nationals, and seized the property of US companies (Salinas *et al.*, 2018). For the next 30 years, Cuba remained an island that rejected international tourism, associating it with negative social impacts (e.g. hedonism, social, and economic inequality).

Cuba began to rethink its position on tourism as the economic conditions in the country began to decline. The collapse of the Soviet Union in the late 1980s/early 1990s had an immediate and detrimental impact on the Cuban economy given that the USSR was involved in 80% of the country's exports and imports and so the withdrawal of trade and subsidies from the Soviet Union resulted in a 35% decline in GDP (Salinas *et al.*, 2018). Soon after, Cuba began to develop tourism intensively with the objective of obtaining foreign currency to regenerate the economy (Espino, 2000). The change of policy direction led to widespread construction of hotels and hotel services, and tourism infrastructure.

In 1994, the Ministry of Tourism (MINTUR) was created for the purposes of planning, regulating, and controlling the tourism sector, which included managing the tourism development process through the establishment of State-owned tourism companies (Espino, 2000). However, realizing that it lacked the know-how to promote and develop tourism, the state drew up new legislation (Act 77) that permitted foreign investment in tourism as long as the tourism property remained majority Cuban-owned. The legislation resulted in the creation of mixed companies and joint use of facilities with companies, mainly from Spain (e.g. Melia, Iberostar, H10) but also Canada (Blue Diamond) (Figueras, 2020). In the 1990s, the average annual growth rate of tourism in Cuba was 19% in terms of visitors and 26% in terms of revenue (Salinas *et al.*, 2018).

However, attitudes toward tourism in the revolutionary state were that it was a 'temporary solution', implemented to weather the difficulties of the 'special period' in order to 'save the homeland, the revolution, and socialism' (Castro, 1991, p. 9). Tourism development was confined to specific regions of the country (Camaguey, Holguin, Santago de Cuba, Jardines del Rey, and Varadero – See Fig. 26.1 for prioritized tourism zones) so as not to negatively influence everyday Cuban socialist life. The type of tourism developed in Cuba through this period was predominately enclosure-style, all-inclusive 'package' experiences that involved little diffusion of tourism benefits to the local population.

Continued economic struggles in Cuba led to additional transformations in the tourism sector. One was a state effort to diversify tourism products beyond the all-inclusive 'sun, sea, and sand' tourism products that dominated the Cuban tourism landscape. In 2014, the state passed Resolution 50, which defined and codified the state's interest in establishing three new forms of tourism: nature/eco, adventure, and rural/agro tourism. Resolution 50 created a pathway for communities to submit proposals for tourism development and established a new category of tour guide: the nature, adventure, and rural tourism guide. Nature, adventure, and rural tourism proposals were assessed based on environmental and social sustainability. This included the extent to which the tourism products created high-quality local jobs and whether the economic and social benefits were fairly distributed through the host community (2014). Established concurrently was ECOTUR, a national travel agency specializing in outdoor and adventure tourism including diving and fishing, birdwatching, hiking, horseback riding, jeep safari, agrotourism, and cycle tourism.

Another major transformation to the tourism sector has been the expansion of opportunities for private enterprise in a sector that has been dominated by state ownership and operations. Two early private business opportunities that were authorized were home-based restaurants called *paladars* (1995) and home-based accommodations called *casas particulares* (1997). Both paladars and casas particulares have become popular with tourists as an alternative to state-run hotels, resorts, and

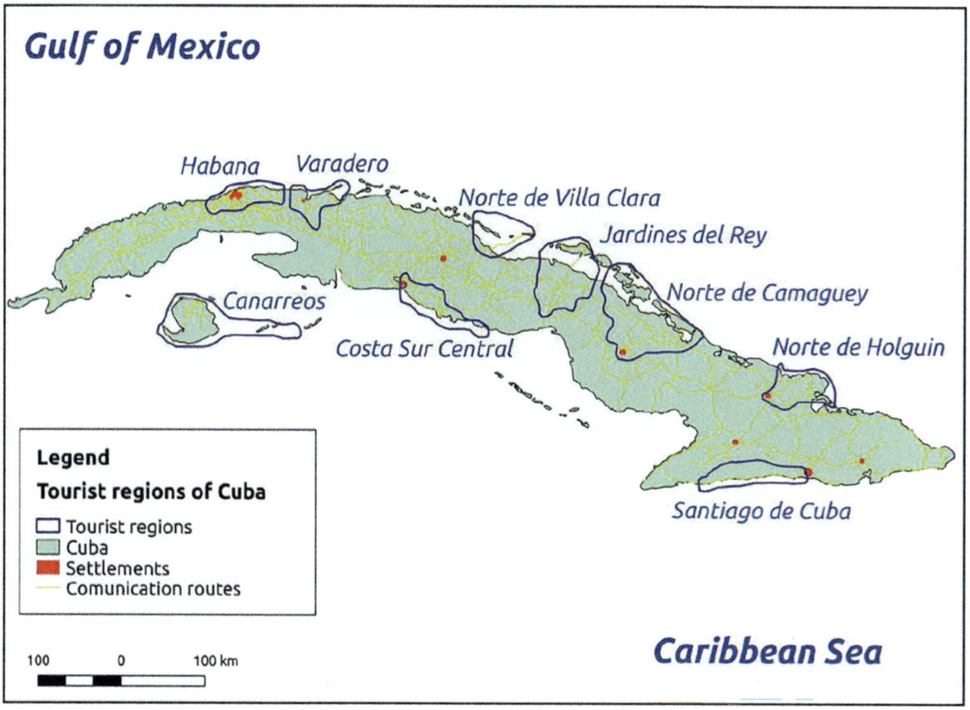

Fig. 26.1. Prioritized tourism zones in Cuba, 2016. Source: Salinas *et al.* (2018).

restaurants. Casas particulares now account for one-quarter of all international overnight stays in Cuba (Pirone, 2021). Both casas particulares and paladars offer locals much more opportunity to participate in and directly benefit from the tourist economy. Further, these businesses offer secondary economic benefits via the services that support their operation (e.g. food procurement, cleaning, transportation).

The Cuban government has continued to expand opportunities for private enterprise, most notably in 2021 when the state (through Resolution 132) dramatically expanded the list of economic activities permitted in the private sector. By 2021, the state had authorized not only private accommodation and restaurants but also transportation and bicycle rental and repair businesses. However, kept on the 'banned list' were the essential tourism activities of tour guiding, tour operator services, and travel agency services (Gaceta Oficial 46). Currently, what this means for cycle tourism is that although tourists are able to rent bicycles from a local enterprise, these enterprises are not able to obtain licenses to offer tourism products such as organizing cycling excursions or urban cycling tours as these products can only be offered by licensed tour guides that work for state-owned operations. Clearly, state law (intentionally) limits the opportunities for cycle tourism within locally owned, private enterprises.

Cycle Tourism in Cuba – Current Landscape

Conditions favorable for the development of cycling tourism in Cuba include: exceptional geography (a variety of terrain, good quality roads, and bike paths; roads without a lot of cars or noise); enviable climate (warm, minimal rain); good transportation connectivity (close to airports, bus, trains); and basic services available (housing, health, food) (Bakogiannis *et al.*, 2020; Mundet *et al.*, 2022). Cuba is a large Caribbean nation with high road connectivity and little vehicular traffic (car ownership is

Fig. 26.2. Cuba cycling landscape. Source: Author.

111/1000 people (Warren and Enoch, 2010)) (Fig. 26.2). It boasts nine UNESCO World Heritage sites, ranging from unique landscape features (Vinales) to urban settings with stunning pre-colonial (Habana Vieja) and colonial architecture (Trinidad). Other attractive features include its hospitable and educated population, extensive road and airport infrastructure, its electrical, communications, and drinking water networks across the country, and its rapidly expanding accommodation and restaurant sector (Salinas *et al.*, 2018).

However, even with these favorable conditions, the cycle tourism industry in Cuba remains small as the country has had only limited success in expanding its global perception as a destination for adventure, nature-based, and rural tourism. The cycle tourism industry is also diverse, with not only Cuban state and private entities but also international actors involved in the provision of cycle tourism opportunities. For example, a cycle tourist can book a cycle holiday directly with the state, through the state-owned travel agency ECOTUR, which then organizes food, accommodation, transportation, and excursions through other state-owned entities (restaurant, hotel, bus, state-run tourist facilities). Alternatively, a cycle tourist is able to plan a trip on their own, for example, by renting bicycles from a local bicycle rental shop and sourcing their own food, accommodation, and transportation. Cycle tourists can also book an excursion through an international tour operator who plans the on-the-ground logistics of the trip on their behalf. Table 26.1 provides an overview of the three main types of cycle tourism providers operating in Cuba.

It is important to recognize the distinctions between these three types of cycle tourism providers because each type has different implications for local economic development. For example, a cycle tour booked through a state-owned tour company, which then contracts other state-owned entities to provide services, results in nearly 100% absorption of tourist spending by the state vs local economies. Similarly, booking a tour through an international tour operator that organizes cycling excursions in Cuba offers little local economic benefit. In general, most international tour operators establish contracts with the state to provide on-the-ground cycling services for its clients. The result is that the

Table 26.1. Types of cycle tourism operators in Cuba and their contributions to local economic development.

State operator	International operator partnering with state	Private operator in Cuba
State-owned and operated. Partners with other state-owned enterprises for ancillary services (accommodation, food services, transportation). Employs licensed tour guides. Tourist spending nearly fully absorbed by state. Contribution to local economic development is very low.	Cycle tours sold by private international operators who contracts state operators to deliver on-the-ground services (tour guiding, accommodation, food, transportation). Employs licensed tour guides. Tourist spending is split between international private operator and state operator. Contribution to local economic development is very low.	Private Cuban-owned businesses (almost always with international support). May informally help tourists with booking accommodation, food services, and transportation. Cannot employ licensed tour guides, though guiding may happen under different licenses (e.g. mechanic). Tourist spending leaks into local community through spending on bicycle rental, accommodation, food, and transportation, although tourist discernment is required. Contribution to local economic development is potentially high.

tourist receives a similar kind of cycling experience as had they booked the excursion directly through the state operator ECOTUR, although they are likely to receive a different pre- and post-travel experience due to the client-oriented service of the international operator. Depending on the details of how a trip is organized, any revenue is likely to be shared primarily between the foreign operators that promote and sell the excursion, and the state that provides transportation, food, accommodation, tour guides, and cycling support services.

In contrast to operators that partner with the state, cycle tourism organized through local operators that are more embedded in the local community offer greater potential for local economic development. Legally, locally owned bicycle rental businesses are permitted to offer outfitting services to tourists who then take on different degrees of self-organization. However, in typical Cuban style, some companies find ways to work around these restrictions, for example, by offering tourism services through licenses they are permitted to obtain, such as 'mechanic' or 'event planner'. This opens up a slightly wider range of services that can be offered to the cycle tourist through local enterprises. For tourists looking for higher levels of support, private operators can organize private transportation (taxi), accommodation (casas particular), and cycling support (mechanic). However, large group tours are not feasible as only state-operated tour companies are permitted to book bus transportation and hotel accommodation. Because these local businesses operate solely in the private sector, they encounter restrictions that limit their ability to start and scale up their enterprises. Individuals interested in opening a bicycle-related business also face extreme difficulties in securing the capital required to do so as Cuban banks are reticent to extend loans to new businesses. Thus, the overwhelming majority of small businesses in Cuba – in cycling and beyond – have some form of international participation (extended family, friend, spouse living internationally) to help finance, operate, and supply the business with essential goods that are difficult to obtain in Cuba (e.g. bicycles, tires, helmets).

Cycle tourists who self-organize tours are also well-positioned to support the local economy with their spending. Whether this potential is realized depends on the tourist's ability to discern between state and private providers offering the same services (e.g. accommodations, food, taxi transportation). Thus, tourists need to arrive with some knowledge on the local economic context and basic Spanish fluency.

Factors Influencing Local Economic Development Through Cycle Tourism

The previous section highlighted the diversity within the cycle tourism industry, and the relationships between the type of cycle tourism provider and the potential for local economic development. This section highlights some of the factors in the current context that have influenced cycle tourism and its potential to foster local economic development. Some of the main factors discussed are tourism zones, prioritization of state-owned tourism, human resource instability, growing international investment in private properties, and increasing interconnectivity.

Tourism zones

In Cuba, one of the main ways that the state regulates tourism is through its delineation of zones across the country in which tourism is being actively developed and tourist activity is encouraged. In cycle tourism, tourists travel through and between areas zoned for tourism. Whereas tourism services (e.g. accommodation) inside tourism zones are plentiful, outside of tourism zones they are scarce. For cycle tourism, the zone approach to tourism limits the options available to travel without motorized assistance as travel distances between zones are at times too far to reach by bicycle. This is one of the reasons that cycle tourism operators organize some, if not the majority, of in-country travel by bus vs bicycle. The lack of services outside of zones designated for tourism activity reduces the potential of cycle tourism to be a catalyst for rural economic development.

Prioritization of state-owned tourism

Tourism is a crucial part of the Cuban economy and the current state strategy is to capture as much tourism spending as possible. For example, all hotels, tour bus companies, tourist attractions, and most restaurants are state-owned. Although the state is increasingly legislating opportunities for Cuban-owned private businesses, it continues to set limits on non-state economic activity, for example, by granting (or not) business licenses and controlling access to credit. State procurement companies also prioritize state tourism companies to receive vital goods, such as food, gasoline, and other material items. For small tourism businesses, the instability and unpredictability of access to vital goods limits planning and development, unless the business is able to establish its own procurement channels, for example, through contacts abroad.

Human resource instability

The current economic conditions of Cuba are very unstable. As with most tourist-related destinations, the COVID-19 pandemic hit Cuba hard. Tourism dropped by 95% from 2019 to 2020, while it has been rebounding it has yet to reach pre-pandemic levels. Cuba is also experiencing massive inflation [401% in 2022 (World Bank, 2023)] and shortages of food, gas, and electricity. Leaving to find a better life, Cuban out-migration is at record levels (Augustin and Robles, 2022). Out-migration is highest among those who are young, healthy, mobile, speak multiple languages, and are entrepreneurial – in other words, all of the qualities ideal for cycle tourism-related enterprises. As a result, there is a lot of human resource fluctuation in the industry. Staff hiring and training is a constant process, making it more difficult for businesses to grow and develop.

International investment in private tourism properties

One of the most impactful ways that cycle tourists support local economic development is through their use of locally-based private services, specifically for their food (paladars), accommodation (casas particulares), and transportation needs (private taxi). Services such as these are a good way for visitors to reach beyond the 'tourism bubble' and gain access to a more authentic Cuba, as well as provide a substantial source of revenue to the Cubans that operate them (Simoni, 2018). The income earned on these services far exceeds state salaries even with the current 20% tax rate, and offers local economic spinoff benefits as private business owners spend in local economies on services in support of their businesses (e.g. cleaning or cooking services, food procurement). Further, new regulations have opened up opportunities for private businesses to procure food from agricultural cooperatives directly rather than through state procurement channels – another path that offers greater benefit to the local agricultural economies (Deere and Royce, 2019). At the same time, Pirone (2021) has noted an increased international presence in the Cuban private sector as the lucrative possibilities of Cuban tourism become recognized globally. Pirone (2021) noted that in the past 5 years, Italian nationals have created over 20 commercial businesses in Havana alone. Many are casas particulares, which are purchased through various arrangements with Cuban nationals (marriage, purchasing in their name) or through contracts with the state. These various ownership arrangements reflect the difficult reality of Cuban nationals to enter the private sector on their own. However, these arrangements also reduce the local economic benefit of these popular forms of small business.

Increasing interconnectivity

In the past 5 years, Cuba has undergone a massive rise in internet connectivity, from 29% of the population in 2014 to 71% in 2020 (ITU, 2023). Increased access to the internet, as well as to communication platforms such as WhatsApp, Telegram, and Facebook Messenger, have been extremely beneficial to Cubans, facilitating cooperation in an already highly cooperative society. WhatsApp groups, for example, have made it much easier for goods to flow directly among Cubans, rather than through state channels. For example, the introduction of Airbnb to Cuba in 2015 made it possible for Cuban casa owners to communicate directly with potential tourists traveling from abroad. The impact of access to this platform was immediate and meteoric (Schwartz and Hernandez-Tapia, 2022). The increased direct access to privately owned accommodation (and the networks of the private owners) has made it easier for cycle tourists to self-organize tours, outside of state and international tour operators.

Cycle Tourism in Cuba – Pedal Power to the People?

Overall, cycle tourism in Cuba is a niche sector that has the potential to foster local economic development, if it is undertaken with care and a discerning eye. Compared to state- and foreign-dominated beach resorts and hotel tourism, cycle tourism has much greater potential to engage Cuban citizens in the core economic activities of tourism, including bicycle rental and service, food and accommodation, and on-ground transportation. However, local economic development requires some skills and attributes on the part of the tourist. One is a basic fluency with the Spanish language, given that outside of top tourism zones (Fig. 26.1), second language proficiency is limited in Cuba. Another requirement is the ability of tourists to discern between state, foreign, and locally owned services and properties for activities in which all parties are involved (e.g. transportation, food). Ultimately, because cycle tourism is a niche tourism activity and because the state continues to restrict private business growth, the potential of cycle tourism to foster economic development, while perhaps greater than other forms of tourism, remains modest.

References

Akama, J.S. and Kieti, D. (2007) Tourism and socio-economic development in developing countries: A case study of Mombasa Resort in Kenya. *Journal of Sustainable Tourism* 15(6), 735–748.

Augustin, E. and Robles, F. (2022) Cuba is depopulating': Largest exodus yet threatens country's future. *The New York Times,* 10 December, 2022. Available at: https://www.nytimes.com/2022/12/10/world/americas/cuba-us-migration.html (accessed 3 May 2023).

Bakogiannis, E., Vlastos, T., Athanasopoulos, K., Christodoulopoulou, G., Karolemeas, C. *et al.* (2020) Development of a cycle-tourism strategy in greece based on the preferences of potential cycle-tourists. *Sustainability* 12(6), 2415.

Brščić, K., Lovrečić, K. and Šugar, T. (2021) Impact of cycle tourism development on tourist destination – attitudes of local stakeholders. *Tourism in Southern and Eastern Europe* 6, 137–154.

Castro, F. (1991) Discurso pronunciado en el IV Congreso del PCC, el 10 de octubre de 1991 en Sntiago de Cuba. Granma, La Habana, pp. 7–11.

Deere, C.D. and Royce, F.S. (2019) Provisioning Cuba's private restaurants: Linking tourism and agriculture. Tourism Planning & Development 16(6), 675–695.

Espino, M.D. (2000) Cuban tourism during the special period. Cuba in Transition 10, 360–373.

Figueras, M.A. (2020) Foreign participation in the development of tourism in Cuba. *Colombia Law School Cuba Capacity Building Project*. Available at: https://horizontecubano.law.columbia.edu/news/foreign-participation-development-tourism-cuba (accessed 7 May 2023).

Gazzola, P., Pavione, E., Grechi, D. and Ossola, P. (2018) Cycle tourism as a driver for the sustainable development of little-known or remote territories: The experience of the apennine regions of Northern Italy. *Sustainability* 10(6), 1863.

Goodwin, H. (1998) Sustainable tourism and poverty elimination. In: *DFID/DETR Workshop on Sustainable Tourism and Poverty*, Vol. 13. Department for International Development, London.

International Telecommunication Union (ITU) (2023) Individuals using the internet. Available at: https://datahub.itu.int/data/ (accessed 7 June 2023).

Mundet, L., Marin, J. and Figueroa, A. (2022) How to develop a road cycling tourism destination: Girona as a case study. *Journal of Outdoor Recreation and Tourism* 39, 100566.

Petino, G., Reina, G. and Privitera, D. (2021) Cycling tourism and revitalization in the sicilian hinterland: A case study in the taormina–etna district. *Sustainability* 13(18), 10022.

Piket, P., Eijgelaar, E. and Peeters, P. (2013) European cycle tourism: A tool for sustainable regional rural development. *Applied Studies in Agribusiness and Commerce, AGRIMBA* 7(2–3), 1–5.

Pirone, T. (2021) Airbnb lands in Havana: The arrival, adaptation, and early impact of a gringo platform in Cuba. *South Atlantic Quarterly* 120(4), 853–869.

Salinas, E., Mundet, L. and Salinas, E. (2018) Historical evolution and spatial development of tourism in Cuba, 1919–2017: What is next? *Tourism Planning & Development* 15(3), 216–238.

Schwartz, L. and Hernandez-Tapia, L. (2022) How Airbnb reshaped Cuba's tourism economy in its own image. *Rest of World*. Available at: https://restofworld.org/2022/airbnb-cuba-tourism-industry/ (accessed 28 May 2023).

Simoni, V. (2018) Business, hospitality, and change in Cuba's private tourism sector: A view from casas particulares in Viñales. *Tourism Planning and Development* 15(3), 293–312.

Warren, J.P. and Enoch, M.P. (2010) Island transport, car ownership and use: A focus on practices in Cuba, Malta, Mauritius and Singapore. *Island Studies Journal* 5(2), 193–216.

World Bank (2023) Inflation, GDP Deflator (Annual %) – Cuba. Available at: https://data.worldbank.org/indicator/NY.GDP.DEFL.KD.ZG?locations=CU (accessed 7 June 2023).

Index

Printed and bound by CPI Group (UK) Ltd, Croydon, CR0 4YY
11/01/2026
14804882-0004